AF716756

Fusarium

Genomics, Molecular and Cellular Biology

Edited by

Daren W. Brown and Robert H. Proctor

Bacterial Foodborne Pathogens and Mycology Research
USDA-ARS-NCAUR
USA

Caister Academic Press

Copyright © 2013

Caister Academic Press
Norfolk, UK

www.caister.com

British Library Cataloguing-in-Publication Data
A catalogue record for this book is available from the British Library

ISBN: 978-1-908230-25-6

Description or mention of instrumentation, software, or other products in this book does not imply endorsement by the author or publisher. The author and publisher do not assume responsibility for the validity of any products or procedures mentioned or described in this book or for the consequences of their use.

All rights reserved. No part of this publication may be reproduced, stored in a retrieval system, or transmitted, in any form or by any means, electronic, mechanical, photocopying, recording or otherwise, without the prior permission of the publisher. No claim to original U.S. Government works.

Cover design adapted from Figure 7.1

Printed and bound in Great Britain

Contents

Current books of interest

Omics in Soil Science	2014
Mollicutes: Molecular Biology and Pathogenesis	2014
Genome Analysis: Current Procedures and Applications	2014
Bacterial Toxins: Genetics, Cellular Biology and Practical Applications	2013
Cold-Adapted Microorganisms	2013
Prions: Current Progress in Advanced Research	2013
RNA Editing: Current Research and Future Trends	2013
Real-Time PCR: Advanced Technologies and Applications	2013
Microbial Efflux Pumps: Current Research	2013
Cytomegaloviruses: From Molecular Pathogenesis to Intervention	2013
Oral Microbial Ecology: Current Research and New Perspectives	2013
Bionanotechnology: Biological Self-assembly and its Applications	2013
Real-Time PCR in Food Science: Current Technology and Applications	2012
Bacterial Gene Regulation and Transcriptional Networks	2012
Bioremediation of Mercury: Current Research and Industrial Applications	2012
Neurospora: Genomics and Molecular Biology	2012
Rhabdoviruses	2012
Horizontal Gene Transfer in Microorganisms	2012
Microbial Ecological Theory: Current Perspectives	2012
Two-Component Systems in Bacteria	2012
Foodborne and Waterborne Bacterial Pathogens	2012
Yersinia: Systems Biology and Control	2012
Stress Response in Microbiology	2012
Bacterial Regulatory Networks	2012
Systems Microbiology: Current Topics and Applications	2012
Quantitative Real-time PCR in Applied Microbiology	2012
Bacterial Spores: Current Research and Applications	2012
Small DNA Tumour Viruses	2012
Extremophiles: Microbiology and Biotechnology	2012
Bacillus: Cellular and Molecular Biology (Second edition)	2012
Microbial Biofilms: Current Research and Applications	2012
Bacterial Glycomics: Current Research, Technology and Applications	2012
Non-coding RNAs and Epigenetic Regulation of Gene Expression	2012
Brucella: Molecular Microbiology and Genomics	2012
Molecular Virology and Control of Flaviviruses	2012
Bacterial Pathogenesis: Molecular and Cellular Mechanisms	2012
Bunyaviridae: Molecular and Cellular Biology	2011
Emerging Trends in Antibacterial Discovery: Answering the Call to Arms	2011

Full details at www.caister.com

Contributors

Daren W. Brown
Bacterial Foodborne Pathogens and Mycology Research
USDA-ARS-NCAUR
Peoria, IL
USA

daren.brown@ars.usda.gov

Lanelle R. Connolly
Department of Biochemistry and Biophysics, Program for Molecular and Cellular Biology, and Center for Genome Research and Biocomputing (CGRB)
Oregon State University
Corvallis, OR
USA

connolll@science.oregonstate.edu

Kim Hammond-Kosack
Wheat Pathogenomics
Department of Plant Biology and Crop Science
Rothamsted Research
Harpenden
UK

kim.hammond-kosack@rothamsted.ac.uk

Linda J. Harris
Eastern Cereal & Oilseed Research Centre
Agriculture & Agri-Food Canada
Ottawa, ON
Canada

linda.harris@agr.gc.ca

H. Corby Kistler
USDA ARS Cereal Disease Laboratory
University of Minnesota
St. Paul, MN
USA

hckist@umn.edu

John F. Leslie
Department of Plant Pathology
Kansas State University
Manhattan KS
USA

jfl@ksu.edu

Li-Jun Ma
Department of Biochemistry and Molecular Biology
University of Massachusetts Amherst
Amherst, MA
USA

lijun@biochem.umass.edu

Michael Freitag
Department of Biochemistry and Biophysics and Center for Genome Research and Biocomputing (CGRB)
Oregon State University
Corvallis, OR

michael.freitag@cgrb.oregonstate.edu

Takumi Nishiuchi
Advanced Science Research Center
Kanazawa University
Kanazawa
Japan

tnish9@staff.kanazawa-u.ac.jp

Thérèse Ouellet
Eastern Cereal & Oilseed Research Centre
Agriculture & Agri-Food Canada
Ottawa, ON
Canada

therese.ouellet@agr.gc.ca

Kyle R. Pomraning
Department of Biochemistry and Biophysics and Program for Molecular and Cellular Biology, and Center for Genome Research and Biocomputing (CGRB)
Oregon State University
Corvallis, OR
USA

pomranosaurus@gmail.com

Robert H. Proctor
Bacterial Foodborne Pathogens and Mycology Research
USDA-ARS-NCAUR
Peoria, IL
USA

robert.proctor@ars.usda.gov

Martijn Rep
Swammerdam Institute for Life Sciences
University of Amsterdam
Amsterdam
The Netherlands

m.rep@uva.nl

Kristina M. Smith
Department of Biochemistry and Biophysics
Oregon State University
Corvallis, OR
USA

smitkris@science.oregonstate.edu

Rajagopal Subramaniam
Eastern Cereal & Oilseed Research Centre
Agriculture & Agri-Food Canada
Ottawa, ON
Canada

rajagopal.subramaniam@agr.gc.ac

Brett A. Summerell
Royal Botanic Gardens
Sydney
Australia

brett.summerell@rbgsyd.nsw.gov.au

Frances Trail
Department of Plant Biology
Michigan State University
East Lansing, MI
USA

trail@msu.edu

Bettina Tudzynski
Institute for Biology and Biotechnology of Plants
Molecular Biology and Biotechnology of Fungi
Westfälische Wilhelms-University
Münster
Germany

bettina.tudzynsb@uni-muenster.de

Martin Urban
Wheat Pathogenomics
Department of Plant Biology and Crop Science
Rothamsted Research
Harpenden
UK

martin.urban@rothamsted.ac.uk

Joseph P. Whalen
Department of Biochemistry and Biophysics and Program for Molecular and Cellular Biology, and Center for Genome Research and Biocomputing (CGRB)
Oregon State University
Corvallis, OR
USA

josephwhalen@gmail.com

Philipp Wiemann
Institute for Biology and Biotechnology of Plants
Molecular Biology and Biotechnology of Fungi
Westfälische Wilhelms-University
Münster
Germany;
Department of Medical Microbiology and Immunology
University of Wisconsin–Madison
Madison, WI
USA

pwiemann@wisc.edu

Preface

The fungus *Fusarium* has tremendous detrimental impacts on crop production, through its ability to cause plant disease, and on the health of humans and livestock, through its ability to produce toxins. Plant pathologists, mycologist and chemists have long recognized this ubiquitous pathogen for its wide host range, ability to cause a variety of plant diseases, and production of secondary metabolites that vary markedly in structure and biological activity. The toxins fumonisins and trichothecenes are a particular concern because of their repeated associations with toxicoses. Basic research on *Fusarium* biology has resulted in tremendous advances in understanding genetic and biochemical processes that enable the fungus to infect plants and cause disease and to synthesize secondary metabolites, including toxins.

In this book, an international group of *Fusarium* researchers review current research in critical areas of *Fusarium* biology. The opening chapter is an introduction to the organism. Subsequent chapters deal with molecular and genetic components involved in sexual development, plant pathogenesis and secondary metabolism. At every stage of research, in the field or at the bench, or while applying genomic, proteomic or metabolomic approaches, the goal has been, and continues to be, identification of strategies to reduce crop diseases and improve human health.

Daren W. Brown and Robert H. Proctor

An Overview of *Fusarium*

1

John F. Leslie and Brett A. Summerell

Abstract

The fungal genus *Fusarium* is notorious as a pathogen of plants, animals and humans and as a producer of secondary metabolites causing toxicoses resulting from consumption of contaminated food by humans and other animals. This chapter outlines the taxonomic status of the genus and provides an overview of the different species concepts that are used either singularly or in combination to define a species within *Fusarium*. As a result of the economic and social impacts of *Fusarium* with respect to disease and mycotoxins, there is a vast amount of knowledge of the biology of many *Fusarium* species such that the genus is now extensively used as model species to explore many fundamental aspects of fungal biology and ecology.

What is *Fusarium*?

Fusarium is a form genus of ascomycete fungi first described by Link in 1809 as *Fusisporium*. Members of the genus are numerous and can be recovered from plants and soil worldwide as pathogens, endophytes and saprophytes. Members of the genus are notorious for their capabilities as plant pathogens, although work with native plants and soil in undisturbed areas suggests that the number of species not associated with known diseases may far outnumber those that cause disease. Most members of the genus produce an array of secondary metabolites, which vary widely in chemical form (Desjardins, 2006). A number of the secondary metabolites are important as mycotoxins that are toxic and/or carcinogenic to humans and domesticated animals (Marasas *et al.*, 1984; Desjardins, 2006), may have a role in plant disease (Proctor *et al.*, 1995; Harris *et al.*, 1999; Glenn *et al.*, 2008), and may be regulated in commercial and international trade (Verstraete, 2008).

History of the genus

Members of the genus *Fusarium* have been difficult to identify almost since the genus was first described. From the original description in 1809 until the publication of *Die Fusarien* by Wollenweber and Reinking (1935) nearly 1000 taxa were described, many based solely on a host association. Wollenweber and Reinking consolidated all of these taxa into 16 sections containing 65 species, 55 varieties and 22 forms and generated the framework in which all of the taxonomical approaches to the genus since then have been based. For the next approximately 50 years taxonomic arguments were dominated by debates over the relative importance of existing morphological characters, usually spore morphology, and newly identified morphological characters, e.g. conidiogenous cells (Booth, 1971). The concept of single-spore purification of cultures (Hansen and Smith, 1932) and the recognition that both the media on which a strain was grown and the conditions under which it was incubated (Snyder and Hansen, 1947) became major themes that changed the way in which plant pathologists approached their work with these fungi. Various authorities presented taxonomic systems recognizing very different numbers of species as valid. For example, Snyder and Hansen (1940, 1941, 1945, 1954) eventually reduced the number of *Fusarium* species to nine, whereas Gerlach and

Nirenberg (1982) recognized approximately 90 species and varieties and Nelson *et al.* (1983) recognized 30 species with another 16 species classed as insufficiently documented. Leslie and Summerell (2006), while not a monograph of the genus, provided descriptions of 70 different species with a recognition that there were a number of species that were not included in that volume. Since then a number of new species of *Fusarium* have been discovered and described, e.g. *F. gaditjirrii* (Phan *et al.*, 2004), *F. lyarnte* and *F. werrikimbe* (Walsh *et al.*, 2010) and *F. ananatum* (Jacobs *et al.*, 2010).

In the 1970s, the first cracks in the morphology only definition of *Fusarium* species began to appear with the description of mating populations (= biological species) within both *Fusarium solani* (Matuo and Snyder, 1973) and *Fusarium moniliforme* (Hsieh *et al.*, 1977). These ideas were carried forward by others and led to the first formal recognition of a cryptic species, *Fusarium thapsinum*, based on differences in mating capabilities in 1997 (Klittich *et al.*, 1997). Biological species recognition was followed the next year by the first use of a phylogenetic species concept to assist in the description of a series of species in the *Gibberella fujikuroi* species complex (Nirenberg and O'Donnell, 1998) in a manner consistent with the Taylor *et al.* (2000) proposal for recognition of phylogenetic species. Application of both biological and phylogenetic species concepts have helped resolve old problems in the genus, e.g. the need to redefine the entities found within *Fusarium moniliforme* (Seifert *et al.*, 2003), but new problems have been created in their wake. There also remain nomenclatural disputes regarding some groups that all parties agree are the same entity, but for which questions of priority and the adequacy of the description remain unsettled. These nomenclatural disputes can be quite confusing to those not heavily immersed in the taxonomy of these organisms.

Taxonomy of *Fusarium*

The taxonomy of *Fusarium* focuses primarily on providing a framework to identify existing species, understanding the relationships between species within the genus and on the identification and characterization of new species of the genus. This process may, or may not, result in a rapid means of assigning a strain to a species particularly as the rigor of the taxonomic assessment varies significantly. Morphological, biological and phylogenetic species concepts have all been used in species descriptions, and species that can be identified with at least two of these approaches are generally much more robust than those that rely on only a single set of characters. Much of current practice is to look for a single gene whose sequence can be used as a diagnostic tool, preferably for every species in the genus, with the gene encoding transcription elongation factor 1α (*tef-1*) currently the most broadly used gene for this purpose (Geiser *et al.*, 2004). In effect this reduces the entire species concept for the genus to the sequence of a single gene. While such a reduction certainly eases diagnostics, it almost certainly will lead to difficulties when anything other than known sequences tied to well-established species are encountered. After all, the name associated with a species in a database is only as good as the person who did the initial identification (which was probably not based on a sequence). For this reason diagnostics based on a single gene such as *tef-1* should only be considered as a guide and, even though potentially very useful for routine diagnostics care must be taken in the interpretation of the results obtained. Even worse, there will be, if we sequence enough, cases in which we identify strains from different species with identical or near-identical sequences. The resolution to such problems requires going beyond the single gene that is valued so highly by diagnosticians, and often requires expert judgement rather than the application of a formal rule.

Species concepts

A species concept is a set of characters that is used both to circumscribe and to describe a species and to differentiate one species from another. The species concept used provides the framework about which questions regarding the species can be asked and answered. Any given species may be described by one or by more than one species concepts. Initially, species concepts grew out of the Biblical concept of a 'kind', which could vary over time, but which were fundamentally unique entities. Assuming the species concepts are using independent characters, the more species concepts that can be applied and give the same answer the more robust the species is and the more likely it is to maintain it is integrity.

Morphological species concepts

Morphological species concepts are based on differences in observable features of the fungus. In *Fusarium* the primary morphological features are the shape and size of the macroconidia, microconidia and chlamydospores, and the type and presence of conidiogenous cells (usually for the microconidia) (Leslie and Summerell, 2006). Other characters that are secondary to these characters include the nature of the hyphae in the culture (e.g. circinate morphology), pigments produced and secreted (usually on potato dextrose agar medium), detectable odours, growth rate and secondary metabolite production. Traditionally, most diagnosticians have relied heavily on these morphological characters when making their identifications.

Biological species concepts

Biological species were introduced by Mayr (1940, 1963) to help explain problems observed in field populations of animals, and are based on the concept that individuals who can mate and mix their genetic information to form progeny should be in the same species. The initial terminology used for biological species was mating populations, a terminology that persists today for members of *Fusarium solani*. The mating populations within *F. moniliforme* have been subject to more efforts in naming, e.g. Leslie (1991), with groups sometimes being defined as mating populations (Klittich and Leslie, 1992; Britz *et al.*, 1999) before being described as species (Klittich *et al.*, 1997). This group of species is now generally known as the *Gibberella fujikuroi* species complex, with most species now having two names (Samuels *et al.*, 2001) – a *Fusarium* name for the asexual (anamorph) stage and a *Gibberella* name for the sexual (teleomorph) – although the mating terminology is still used on some occasions.

As the biological species concept is based upon the production of the sexual fruiting structures it is important to be aware of the anamorph – teleomorph connections in this group of fungi. Most species have a sexual stage that belongs within the definition of *Gibberella* although there are a number that fall into genera such as *Albonectria, Cyanonectria, Haematonectria, Cosmospora* and *Neocosmospora* (Rossman *et al.*, 1999). It is now generally accepted that all of these species form what is considered a monophyletic '*Fusarium*' clade and that it is preferential to use the *Fusarium* name as it contains the most information for the user of the taxonomy. Recent changes in the International Code of Botanical nomenclature as it relates to anamorphic fungi will result in the discontinuation of this awkward and unusual dual naming system. Newly described species that have a sexual stage will include a description of the sexual stage as a part of the *Fusarium* species description, a convention that some authors have already have begun to follow, e.g. Amata *et al.* (2010).

Biological species require sexual crosses as evidence that two strains belong to the same species. Ready implementation of this requirement means that highly female-fertile tester strains must be available along with clear protocols. For many species within the *G. fujikuroi* species complex these strains are available for minimal or no charge through the Fungal Genetics Stock Center (FGSC) at the University of Missouri-Kansas City (www.fgsc.net). The functional test is to use the standard tester strain as the female parent in the cross and the unknown field strain as the male parent. Standard crossing protocols rely on carrot agar and the presence of black light (Klittich and Leslie, 1988) and require patience and incubator space for the several weeks from the making of the initial cross until ascospores are observed. Unfortunately, biological species concepts cannot be used with many species in *Fusarium* as no sexual stage has been observed under laboratory conditions. The ability to identify mating type alleles via PCR amplification (Kerényi *et al.*, 2004) has made the search for the sexual stage much easier than it used to be. When available, biological species concepts provide the greatest confidence in a species definition as the organism is telling us where the limits of a species are rather than these limits being inferred from other characters that may not be clearly delimited.

Phylogenetic species concepts

Phylogenetic species concepts as currently implemented rely almost exclusively on DNA sequences of one or a relatively few genes. Usually these genes encode proteins and have one to several introns in the genomic sequence that is amplified by PCR for analysis. The exon portions of the sequence are usually widely conserved, which enables a common primer pair to be used to amplify the sequence of interest from a number of different species. Sequences from various strains are aligned and analysed with a computer program to create a tree. Variants are presumed to be selectively neutral, and are used to trace lineages. The lineages coalesce within the tree as you move from the present (with the existing set of sequences) back through evolutionary history. At each node where lineages coalesce, the amount of variation present is reduced. Branching patterns are used to infer speciation events, and in some cases when in time these speciation events occurred.

Many recent *Fusarium* species descriptions have relied substantially on phylogenetic species concepts, including most of the newly described species within the *G. fujikuroi* and *F. solani* species complexes. In most species where enough strains have been examined, multiple sequences have been identified for genes that are sufficiently variable to be used for identification purposes. Potential problems at this point include (i) that the genes evaluated may not reflect species differences and (ii) that the number of positions at which differences occur may not be the most important indicator of species differences, as the critical point may be whether a particular change, or set of changes has occurred. These problems make it difficult to write a species description that goes beyond saying the sequences for the species being described cluster with one another, but do not cluster with sequences derived from other species, and can sap confidence in the species being described.

With intraspecific sequence variation possible, where to draw the line to separate two different species often is not clear. In such cases it helps tremendously to have an external character to validate the hypothesis generated by the phylogenetic analysis. Sometimes, once groups are resolved, a morphological character that is consistent with the split can be identified or a sexual stage can be identified that enables a more robust species definition. In general, biological and phylogenetic species definitions result in the same groups being included in a species. When these species concepts conflict, care must be taken in describing species, since it may mean that there is greater or lesser gene flow possible than would have been anticipated based just on the name(s). Thus, there is no one rule that can always be followed and be guaranteed to determine whether two strains are in the same or different species – human evaluation of the data remains essential.

Polyphasic species descriptions

All of the species of *Fusarium* described in the past decade have been defined on the basis of a combination of morphological, biological and phylogenetic characters through an approach defined as a polyphasic species description. The relative contribution of each dataset to the description varies by species as it is depends on character availability. For example, biological species data are not available for many of the recently described species, so the species definitions were based on morphological and phylogenetic data. The contribution of the phylogenetic data has expanded considerably as the technology has become more accessible and multiple gene regions are sequenced and evaluated. Robust species descriptions may now include data from four or more gene regions. We think that robust species descriptions should include as many of these datasets as possible and that each dataset should be based on at least 20 non-clonal isolates.

Currently accepted species and taxonomic discrepancies

Numerous species of *Fusarium* have been described, although the number that cause serious diseases or produce mycotoxins are but a small fraction of those that have been validly described. The listing of Leslie and Summerell (2006) contains the major entities that are widely accepted as species. This list has some caveats, however. First, some important species had not been described in time to be included, e.g. *F. langsethiae* (Torp and Nirenberg, 2004). Second, two of the species on this list, *F. brevicatenulatum* and *F. pseudoanthophilum* have recently been synonymized as they have a common sexual stage (Amata *et al.*, 2010). Third, the various phylogenetic species within *F. graminearum* are not listed as separate species because all of those tested are significantly cross-fertile with the female fertile tester strains of Lee *et al.* (2003). Fourth, some species, e.g. those with teleomorphs in *Cosmospora* and *Neocosmospora* (Rossman *et al.*, 1999), were deliberately excluded since it was unclear at the time of publication whether these fungi were going to be removed from *Fusarium* and put into a new genus of their own. Finally, there are controversies over naming priority and the adequacy of the description of some species that lead to a group having different names resulting from a dispute among taxonomists, even though all concerned agree that they are talking about the same entity. The two most prominent problems of this sort are *F. crookwellense*–*F. cerealis* and *F. semitectum*–*F. incarnatum*–*F. pallidoroseum*.

Impact of *Fusarium*

Plant pathogens

Fusarium species cause an array of diseases that affect agriculture and horticulture in all parts of the world. Several of the diseases, such as head blight of wheat (Windels, 2000) and panama disease of bananas (Ploetz, 1990), have had nearly devastating economic and sociological impacts on the farmers and communities that rely on these crops for their livelihoods. The plant diseases caused by *Fusarium* species are not restricted to any particular region or cropping scenario and can be as problematic in temperate, commercial agriculture as they are in subsistence tropical agriculture. The one factor that all of these disease scenarios share, regardless of the *Fusarium* species or the crop involved, is that the options for control usually are limited and difficult to implement.

Fusarium can cause various types of diseases, including vascular wilts, head and seed blights, stem rots, root and crown rots and canker diseases, with some species capable of simultaneously causing multiple or overlapping disease syndromes depending on the host and the environment. Vascular wilt diseases are typically caused by members of the *Fusarium oxysporum* species complex known as *formae speciales*. Plants usually are infected via the root system with the fungus obstructing the vascular system and reducing or preventing the flow of water from the roots to the upper plant, which leads to plant wilting. *Formae speciales* are not taxonomic entities, but rather are groups of isolates recognized for their ability to cause disease in a specific set of host plant species. *Formae speciales* have evolved through convergent evolution the ability to cause diseases and need not be, and often are not, members of a monophyletic clade (Baayen *et al.*, 2000). Other species closely related to *F. oxysporum*, e.g. *F. xylarioides*, also can cause vascular wilt diseases in some plant species.

Head, grain and seed blights are diseases predominantly restricted to cereal crops. These diseases are caused by diverse species but the most important are *F. graminearum* on wheat, *F. verticillioides* on maize, and *F. thapsinum* on sorghum. These diseases are typified by yield loss and can be further complicated by the *in situ* production of mycotoxins that contaminate the grain produced. Stem rot diseases also occur on a broad range of hosts, but again are the most important on cereals such as sorghum, maize, wheat and barley. These diseases result in yield losses in two ways. First, in rotted or degraded stems, nutrient and water flow to the grain are reduced along with the yield. Second, by weakening the stem tissue the plants are more susceptible to lodging, i.e. in which the plant falls over, and is more difficult to harvest. Typical diseases include stalk rots of maize caused by *F. verticillioides* and stalk rots of sorghum caused by *F. thapsinum*.

Root and crown rots are probably the most widespread type of disease caused by *Fusarium* species. In many cases the causal agents of these diseases may be difficult to determine, as they may occur in plants affected by other factors or that are simultaneously infected by weak pathogens or saprophytes. Species such as *F. pseudograminearum*, *F. culmorum* and *F. solani* cause diseases in which the roots and crown of the host plant are rotted resulting in an insufficient or ineffective root system and increased susceptibility to lodging and collapse.

Canker diseases involve the formation of

lesions on stems and branches and are most prominent on woody perennial species in horticulture and forestry. Pathogens such as *F. circinatum*, the cause of pitch canker in pine, and *F. decemcellulare*, which causes cankers on tropical fruit trees, are good examples of the pathogens in this category.

Most of the control practices recommended for plant diseases caused by *Fusarium* are dependent on either modifications to the agronomic practices, e.g. ploughing or reduction of host residue, rotation to a non-host(s) of the specific pathogen, or through the development of host resistance. Chemical control measures have been spectacularly unsuccessful for *Fusarium* diseases and there are very few fungicides currently available that could be described as economically effective control strategies for these diseases.

Human and animal pathogens

Fusarium spp. also are known as direct pathogens of both humans and some domesticated animals. Members of either the *F. oxysporum* or *F. solani* species complexes are the most common disease causing agents, although a number of other species have been implicated on at least an occasional basis. *F. solani* is currently best known for its ability to cause keratitis and its association with a contact lens solution (Bullock and Khamis, 2010). Both *F. oxysporum* and *F. solani* are more commonly associated with dermatophytic infections (Leslie and Summerell, 2006) and infections of nails and hooves, but even these infections are usually rare enough that the detection of one may merit a case description style publication. Most strains of *Fusarium* grow poorly, if at all, at 37°C. However, those that do and that become dispersed infections in humans are usually fatal as there are no generally effective antifungals available for the treatment of such well-distributed infections (Leslie and Summerell, 2006).

Diseases in other animals usually parallel those seen in humans, with *F. solani* being the only *Fusarium* species that was pathogenic to immunocompetent mice (Mayayo *et al.*, 1999). They can be particularly serious in animals with an exoskeleton, e.g. shrimp (Hose *et al.*, 1984) or turtles, whose external shell is subject to degradation by *Fusarium* (Cabañes *et al.*, 1997). Most of the common syndromes of animals associated with *Fusarium* are not due to a direct disease in the animal being caused by the fungus, but rather are correlated with the consumption of a food product that has been colonized by the fungus and contaminated with one or mycotoxins.

Mycotoxins

Strains of *Fusarium* can synthesize literally hundreds of different secondary metabolites, most of whose function is completely unknown (see Chapters 7–9). A few of these secondary metabolites have been associated with toxicoses in domesticated animals or with nutritional diseases in humans. The association of nutritional diseases in humans with *Fusarium* toxicoses dates back to ancient Greece (Schoental, 1994) and carries through to the present time, e.g. neural tube defects in newborn children in the Rio Grande valley associated with fumonisins (Missmer *et al.*, 2006).

The best known *Fusarium* mycotoxins are the trichothecenes, the fumonisins, the zearalenone and the gibberellic acids. The trichothecene class of mycotoxins includes the better-known deoxynivalenol (DON), nivalenol (NIV), diacetoxyscirpenol (DAS) and T-2 toxin. The last two are on the US government's list of select agents that might be/have been weaponized for biological warfare (Mirocha *et al.*, 1983; Rosen and Rosen, 1982). The biosynthetic pathways for the main products and many of the secondary products of all of these pathways are known (Desjardins, 2006). The genes that encode the biosynthesis of the compounds also have been identified and usually at least some of the regulatory loci that operate on the pathways are known as well. Although there is general knowledge of the effects of various environmental parameters on mycotoxin biosynthesis, the means by which these environmental signals are received by the producing cells and processed internally to alter mycotoxin production is at best poorly understood.

Fusarium as a model research organism

Since its first description over 200 years ago the greatest interest in *Fusarium* has been for the diseases it can cause, the toxins it can produce and the differentiation and correct identification of the appropriate causal agents. In the last 20–30 years, several *Fusarium* species have become amenable to routine laboratory manipulations and have been the subject of detailed genetic and physiological studies. The classical genetic maps of *F. verticillioides* (Xu and Leslie, 1996; Jurgenson *et al.*, 2002b) and *F. graminearum* (Jurgenson *et al.*, 2002a; Gale *et al.*, 2005; Lee *et al.*, 2008), the available genome sequences at the Broad Institute (www.broadinstitute.org/annotation/genome/*Fusarium*_group/MultiHome.html), and the ability to make controlled crosses under laboratory conditions (Klittich and Leslie, 1988; Bowden and Leslie, 1999; Lee *et al.*, 2003) has opened these two species up to virtually all genetic studies except for the study of ordered tetrads, which are not produced in *Fusarium* due to a spindle overlap during meiosis. There are a number of transformation methods, selectable markers and vectors available that can lead to gene replacement, disruption or ectopic integration of the vector, and high quality genomic DNA is relatively easy to recover by using any of several protocols. These fungi produce a plethora of secondary metabolites (Frisvad *et al.*, 2008; Desjardins, 2006). Some of these secondary metabolites have a huge economic impact, but also result from unusual biosynthetic pathways whose enzymes, intermediates and regulation are still not understood in great detail. Primary metabolism also has some differences in its regulation (Leslie, 1986, 1987) that also warrants further study. Finally the host–pathogen relationship that occurs between *Fusarium* and its hosts is different than that seen for other model plant pathogens such as *Magnaporthe oryzae* (Kankanala *et al.*, 2007; Valent and Khang, 2010). In many cases, *Fusarium* infections occur and cause no obvious plant reaction (Oren *et al.*, 2003). It is only when the plant is stressed that the fungus switches from being an endophyte to being a pathogen capable of doing significant damage and contaminating seed with toxins. The regulation of this mode of fungal establishment and disease initiation is important and will need to be understood to identify critical points in the process that could be disrupted or perhaps prolonged to obtain effective disease control and limit toxin contamination.

Another area in which *Fusarium* is poised to serve as an important model system is in the area of evolutionary biology. The protracted taxonomic discussions and studies of the past 200 years have resulted in *Fusarium* being one of the most carefully evaluated of the fungal genera. The recent increase in species due to the addition of biological and phylogenetic species concepts for defining species has put an independent layer of confirmation under this work, changing some but also leaving much intact. As a result, the genus contains a number of species with known host preferences and in some cases with relatively reliable estimates of geographic distribution. Closely related species, e.g. *F. fujikuroi* and *F. proliferatum* (Leslie *et al.*, 2004) or *F. subglutinans* and *F. circinatum* (de Vos *et al.*, 2007), can be crossed and progeny evaluated for genetic regions that are associated with speciation. In some cases, e.g. *F. graminearum*, incipient species may be identifiable even if the entities are not completely resolved. Some species are sufficiently widely distributed that they may be able to offer insights into on-going allopatric speciation processes in response to various selective environments. Finally, there is the possibility of collecting many new additional species from natural habitats and subsistence farming scenarios that could provide significant connections between existing groups and yield new insights into the selection processes and pressures that led to the development of the species currently of economic importance in many commercial farming systems.

References

Amata, R.L., Burgess, L.W., Summerell, B.A., Bullock, S., Liew, E.C.Y., and Smith-White, J.L. (2010). An emended description of *Fusarium brevicatenulatum* and *F. pseudoanthophilum* based on isolates recovered from millet in Kenya. Fungal Divers. *43*, 11–25.

Baayen, R.P., O'Donnell, K., Bonants, P.J., Cigelnik, E., Kroon, L.P.N.M., Roebroeck, E.J.A., and Waalwijk, C. (2000). Gene genealogies and AFLP analyses in the *Fusarium oxysporum* complex identify monophyletic and non-monophyletic *formae speciales* causing wilt and rot disease. Phytopathology *90*, 891–900.

Booth, C. (1971). The Genus *Fusarium* (Kew, Surrey, UK: Commonwealth Mycological Institute).

Bowden, R.L., and Leslie, J.F. (1999). Sexual recombination in *Gibberella zeae* (*Fusarium graminearum*). Phytopathology *89*, 182–188.

Britz, H., Coutinho, T.A., Wingfield, M.J., Marasas, W.F.O., Gordon, T.R., and Leslie, J.F. (1999). *Fusarium subglutinans* f. sp. *pini* represents a distinct mating population in the *Gibberella fujikuroi* species complex. Appl. Environ. Microbiol. *65*, 1198–1201.

Bullock, J.D., and Khamis, H.J. (2010). A retrospective statistical analysis of the *Fusarium* keratitis epidemic of 2004–2006. Ophthalmic Epidemiol. *17*, 179–184.

Cabañes, F.J., Alonso, J.M., Castellá, G., Alegre, F., Domingo, M., and Pont, S. (1997). Cutaneous hyalohyphomycosis caused by *Fusarium solani* in a loggerhead sea turtle (*Caretta caretta* L.). J. Clin. Microbiol. *35*, 3343–3345.

de Vos, L., Myburg, A.A., Wingfield, M.J., Desjardins, A.E., Gordon, T.R., and Wingfield, B.D. (2007). Complete genetic linkage maps from an interspecific cross between *Fusarium circinatum* and *Fusarium subglutinans*. Fungal Genet. Biol. *44*, 701–714.

Desjardins, A.E. (2006). *Fusarium* Mycotoxins: Chemistry, Genetics and Biology (APS Press, St. Paul, Minnesota).

Frisvad, J.C., Andersen, B., and Thrane, U. (2008). The use of secondary metabolite profiling in chemotaxonomy of filamentous fungi. Mycol. Res. *112*, 2231–2240.

Gale, L.R., Bryant, J.D., Calvo, S., Giese, H., Katan, T., O'Donnell, K., Suga, H., Taga, M., Usgaard, T.R., Ward, T.J., and Kistler, H.C. (2005). Chromosome complement of the fungal plant pathogen *Fusarium graminearum* based on genetic and physical mapping and cytological observations. Genetics *171*, 985–1001.

Geiser, D.M., Jimenez-Gasco, M., Kang, S., Makalowska, I., Veeraraghavan, N., Ward, T.J., Zhang, N., Kuldau, G.A., and O'Donnell, K. (2004). *FUSARIUM*-ID v. 1.0, A DNA sequence database for identifying *Fusarium*. Eur. J. Plant Pathol. *110*, 473–479.

Gerlach, W., and Nirenberg, H. (1982). The genus *Fusarium* – A pictorial atlas. Mitteilungen aus der Biologischen Bundesanstalt Für Land- und Forstwirtschaft (Dahlem, Berlin) *209*, 1–405.

Glenn, A.E., Zitomer, N.C., Zimeri, A.M., Williams, L.D., Riley, R.T., and Proctor, R.H. (2008). Transformation-mediated complementation of a *FUM* gene cluster deletion in *Fusarium verticillioides* restores both fumonisin production and pathogenicity on maize seedlings. Mol. Plant Microbe Interact. *21*, 87–97.

Hansen, H.N., and Smith, R.E. (1932). The mechanisms of variation in imperfect fungi: *Botrytis cinerea*. Phytopathology *22*, 953–964.

Harris, L.J., Desjardins, A.E., Plattner, R.D., Nicholson, P., Butler, G., Young, J.C., Weston, G., Proctor, R.H., and Hohn, T.M. (1999). Possible role of trichothecene mycotoxins in virulence of *Fusarium graminearum* on maize. Plant Dis. *83*, 954–960.

Hose, J.E., Lightner, D.V., Redman, R.M., and Danald, D.A. (1984). Observations on the pathogenesis of the imperfect fungus *Fusarium solani* in the California Brown shrimp *Penaeus californienis*. J. Invertebr. Pathol. *44*, 292–303.

Hsieh, W.H., Smith, S.N., and Snyder, W.C. (1977). Mating groups in *Fusarium moniliforme*. Phytopathology *67*, 1041–1043.

Jacobs, A., van Wyk, P.S., Marasas, W.F.O., Wingfield, B.D., Wingfield, M.J., and Coutinho, T.A. (2010). *Fusarium ananatum* sp. nov. in the *Gibberella fujikuroi* species complex from pineapples with fruit rot in South Africa. Fungal Biol. *114*, 515–527.

Jurgenson, J.E., Bowden, R.L., Zeller, K.A., Leslie, J.F., Alexander, N.J., and Plattner, R.D. (2002a). A genetic map of *Gibberella zeae* (*Fusarium graminearum*). Genetics *160*, 1452–1460.

Jurgenson, J.E., Zeller, K.A., and Leslie, J.F. (2002b). An expanded genetic map of *Gibberella moniliformis* (*Fusarium verticillioides*). Appl. Environ. Microbiol. *68*, 1972–1979.

Kankanala, P., Czymmek, K., and Valent, B. (2007). Roles for rice membrane dynamics and plasmodesmata during biotrophic invasion by the blast fungus. Plant Cell. *19*, 706–724.

Kerényi, Z., Moretti, A., Waalwijk, C., Oláh, B., and Hornok, L. (2004). Mating type sequences in asexually reproducing *Fusarium* species. Appl. Environ. Microbiol. *70*, 4419–4423.

Klittich, C.J.R., and Leslie, J.F. (1988). Nitrate reduction mutants of *Fusarium moniliforme* (*Gibberella fujikuroi*). Genetics *118*, 417–423.

Klittich, C.J.R., and Leslie, J.F. (1992). Identification of a second mating population within the *Fusarium moniliforme* anamorph of *Gibberella fujikuroi*. Mycologia *84*, 541–547.

Klittich, C.J.R., Leslie, J.F., Nelson, P.E., and Marasas, W.F.O. (1997). *Fusarium thapsinum* (*Gibberella thapsina*): A new species in section *Liseola* from sorghum. Mycologia *89*, 643–652.

Lee, J., Lee, T., Lee, Y.-W., Yun, S.-H., and Turgeon, B.G. (2003). Shifting fungal reproductive mode by manipulation of *Mating type* genes: Obligate heterothallism of *Gibberella zeae*. Mol. Microbiol. *50*, 145–152.

Lee, J., Jurgenson, J.E., Leslie, J.F., and Bowden, R.L. (2008). Alignment of genetic and physical maps of *Gibberella zeae*. Appl. Environ. Microbiol. *74*, 2349–2359.

Leslie, J.F. (1986). Utilization of nitrogen sources by *Gibberella zeae*. Mycologia *78*, 568–576.

Leslie, J.F. (1987). Nitrate nonutilizing mutant of *Gibberella zeae*. J. Gen. Microbiol. *133*, 1279–1287.

Leslie, J.F. (1991). Mating populations in *Gibberella fujikuroi* (*Fusarium* section *Liseola*). Phytopathology *81*, 1058–1060.

Leslie, J.F., and Summerell, B.A. (2006). The *Fusarium* Laboratory Manual (Blackwell Professional, Ames, Iowa).

Leslie, J.F., Zeller, K.A., Wohler, and Summerell, M., B.A. (2004). Interfertility of two mating populations in the *Gibberella fujikuroi* species complex. Eur. J. Plant Pathol. *110*, 611–618.

Link, H.F. (1809). Mag. Ges. Naturf. Freunde, Berlin 3, 3.

Marasas, W.F.O., Nelson, P.E., and Toussoun, T.A. (1984). Toxigenic *Fusarium* Species: Identity and

Mycotoxicology (The Pennsylvania State University Press, University Park, Pennsylvania).
Matuo, T., and Snyder, W.C. (1973). Use of morphology and mating populations in the identification of special forms in *Fusarium solani*. Phytopathology *63*, 562–565.
Mayayo, E., Pujol, I., and Guarro, J. (1999). Experimental pathogenicity of four opportunist *Fusarium* species in a murine model. J. Med. Microbiol. *48*, 363–366.
Mayr, E. (1940). Speciation phenomena in birds. Am. Nat. *74*, 249–278.
Mayr, E. (1963). Animal Species and Evolution (Harvard University Press, Cambridge, MA).
Mirocha, C.J., Pawlosky, R.A., Chatterjee, K., Watson, S., and Hayes, W. (1983). Analysis of *Fusarium* toxins in various samples implicated in biological warfare in Southeast Asia. J.A.O.A.C. *66*, 1485–1499.
Missmer, S.A., Suarez, L., Felkner, M., Wang, E., Merrill, A.H., Rothman, K.J., and Hendricks, K.A. (2006). Exposure to fumonisins and the occurrence of neural tube defects along the Texas-Mexico border. Environ. Health. Perspect. *114*, 237–241.
Nelson, P.E., Toussoun, T.A., and Marasas, W.F.O. (1983). *Fusarium* Species: An Illustrated Manual for Identification (Pennsylvania State University Press, University Park, PA).
Nirenberg, H.I., and O'Donnell, K. (1998). New *Fusarium* species and combinations within the Gibberella fujikuroi species complex. Mycologia *90*, 434–458.
Oren, L., Ezrati, S., Cohen, D., and Sharon, A. (2003). Early events in the *Fusarium verticillioides*–maize interaction characterized by using a green fluorescent protein-expressing transgenic isolate. Appl. Environ. Microbiol. *69*, 1695–1701.
Phan, H.T., Burgess, L.W., Summerell, B.A., Bullock, S., Liew, E.C.Y., Smith-White, J., and Clarkeson, J. (2004). *Gibberella gaditjirri* (*Fusarium gaditjirri*) sp. nov., a new species from tropical grasses in Australia. Stud. Mycol. *50*, 261–272.
Ploetz, R.C., ed. (1990). *Fusarium* Wilt of Banana (APS Press, St. Paul, MN).
Proctor, R.H., Hohn, T.M., and McCormick, S.P. (1995). Reduced virulence of *Gibberella zeae* caused by disruption of a trichothecene toxin biosynthetic gene. Mol. Plant Microbe Interact. *8*, 593–601.
Rosen, R.T., and Rosen, J.D. (1982). Presence of four *Fusarium* mycotoxins and synthetic material in 'Yellow Rain': Evidence for the use of chemical weapons in Laos. Biomed. Mass Spectrom. *9*, 443–450.
Rossman, A.Y., Samuels, G.J., Rogerson, C.T., and Lowen, R. (1999). Genera of Bionectriaceae, Hypocreaceae and Nectriaceae (Hypocreales, Ascomycetes). Stud. Mycol. *42*, 1–248.
Samuels, G.J., Nirenberg, H.I., and Seifert, K.A. (2001). Perithecial species of *Fusarium*. In B.A. Summerell, J.F. Leslie, D. Backhouse, W.L. Bryden, and L.W. Burgess, eds., *Fusarium*: Paul E. Nelson Memorial Symposium (APS Press, Saint Paul, MN), pp. 1–14.
Schoental, R. (1994). Mycotoxins in food and the plague of Athens. J. Nutr. Med.*4*, 83–85.
Seifert, K.A., Aoki, T., Baayen, R.P., Brayford, D., Burgess, L.W., Chulze, S.N., Gams, W., Geiser, D.M., de Gruyter, J., Leslie, J.F., *et al.* (2003). The name *Fusarium moniliforme* should no longer be used. Mycol. Res. *107*, 643–644.
Snyder, W.C., and Hansen, H.N. (1940). The species concept in *Fusarium*. Am. J. Bot. *27*, 64–67.
Snyder, W.C., and Hansen, H.N. (1941). The species concept in *Fusarium* with reference to section *Martiella*. Am. J. Bot. *28*, 738–742.
Snyder, W.C., and Hansen, H.N. (1945). The species concept in *Fusarium* with reference to *Discolor* and other sections. Am. J. Bot. *32*, 657–666.
Snyder, W.C., and Hansen, H.N. (1947). Advantages of natural media and environments in the culture of fungi. Phytopathology *37*, 420–421.
Snyder, W.C., and Hansen, H.N. (1954). Variation and speciation in the genus *Fusarium*. Ann. N. Y. Acad. Sci. *60*, 16–23.
Taylor, J.W., Jacobson, D.J., Kroken, S., Kasuga, T., Geiser, D.M., Hibbett, D.S., and Fisher, M.C. (2000). Phylogenetic species recognition and species concepts in fungi. Fungal Genet. Biol. *31*, 21–32.
Torp, M., and Nirenberg, H.I. (2004). *Fusarium langsethiae* sp. nov. on cereals in Europe. Int. J. Food Microbiol. *95*, 247–256.
Valent, B., and Khang, C.H. (2010). Recent advances in rice blast effector research. Curr. Opin. Plant Biol. *13*, 434–441.
Verstraete, F. (2008). EU legislation on mycotoxins in food and feed: overview of the decision-making process and recent and future developments. In Mycotoxins, Leslie, J.F., Bandyopadhyay, R., and Visconti, A., eds. (CABI, Wallingford, Oxfordshire, UK), pp. 77–99.
Walsh, J.L., Laurence, M.H., Liew, E.C.Y., Sangalang, A.E., Burgess, L.W., Summerell, B.A., and Petrovic, T. (2010). *Fusarium*: two endophytic novel species from tropical grasses from northern Australia. Fungal Divers. *44*, 149–159.
Windels, C.E. (2000). Economic and social impacts of Fusarium head blight: changing farms and rural communities in the Northern Great Plains. Phytopathology *90*, 17–21.
Wollenweber, H.W., and Reinking, O.A. (1935). Die Fusarien, ihre Beschreibung, Schadwirkung und Bekampfung (Verlag Paul Parey, Berlin, Germany).
Xu, J.-R., and Leslie, J.F. (1996). A genetic map of *Fusarium moniliforme* (*Gibberella fujikuroi* mating population A). Genetics *143*, 175–189.

Sex and Fruiting in *Fusarium*

2

Frances Trail

Abstract

Fusarium spp. represent an array of sexual lifestyles: asexual, homothallic and heterothallic. The recent availability of genomic resources for several *Fusarium* species has inspired intense research on these organisms, including a better understanding of sporulation. Although studies have clarified the arrangement of the *MAT* idiomorphs among these species, little is known about the role of *MAT* genes in sex and fruiting body development. For most *Fusarium* species, the sexual cycle does not predominate in the field. However, *F. graminearum*, a homothallic species, relies on sexual development for spore dissemination to host plants. Recent functional studies have revealed genes involved in many aspects of perithecium development in this species. This chapter will focus on morphogenic aspects of sexual development, summarize the function of genes that have been shown to affect development, and speculate about the ecological and evolutionary implications of sexual lifestyles.

Introduction

Fusarium species provide researchers with a closely related group of organisms that represent diverse sexual lifestyles (asexual, homothallic and heterothallic) and diverse approaches to interactions with their hosts. In the last decade, the genus *Fusarium* has risen to prominence among the fungi that are both critical to agriculture and have genomic resources. Currently, the genomes of *Fusarium graminearum*, *F. oxysporum*, *F. solani* and *F. verticillioides* have been sequenced and are publicly available (Cuomo *et al.*, 2007; Coleman *et al.*, 2009; Ma *et al.*, 2010). *F. fujikuroi* has strong genetic resources, and is likely to be sequenced soon. Of the four species of *Fusarium* with sequenced genomes, two have *Gibberella* teleomorphs (*G. zeae*: anamorph *F. graminearum*; and *G. moniliforme*: anamorph *F. verticillioides*), one has a *Nectria* teleomorph (*N. haematococca*: anamorph *F. solani* f. sp. *pisi*) and one has no known sexual stage (*F. oxysporum* f. sp. *lycopersici*). Sequence availability of the *Fusarium* spp. has stimulated research into evolution and diversity. Together with the ability to easily generate random and targeted genetic mutants, these *Fusarium* species provide ideal systems to study host–pathogen interactions, asexual and sexual development and morphogenesis. Despite the worldwide importance of these agricultural pathogens, and the importance of the sexual cycle to disease development, both for recombination and dispersal, our understanding of the sexual process remains limited.

The role of the sexual cycle in *Fusarium* spp. is varied. On one end, *F. graminearum* relies on ascospores as a critical part of the primary inoculum leading to head blight disease of wheat and barley. Sexual development was shown to be essential for disease production. Wheat plants exposed to mating type gene (MAT) mutants in the field (Desjardins *et al.*, 2006) had significantly reduced disease levels compared with inoculation with the wild-type. Ascospores are also important, but not proven essential, as propagules for dispersal in *F. solani* f. sp. *pisi* (VanEtten, 1978). For the majority of *Fusarium* species that produce fruiting bodies in culture, sexual development is uncommonly observed in the field. Of the 12 species with *Gibberella* teleomorphs reviewed by Desjardins (2003), fruiting is common only in *G. zeae* while occasionally observed for *G. fujikuroi* in the field. The remaining species (*G. avenaceae, G. baccata, G. circinata, G. coronicola, G. intermedia, G. moniliformis, G. nygamai, G. pulicaris, G. subglutinans, G. thapsina*) all have been rarely or never observed to produce fruiting bodies in nature. Nevertheless, most of these species do produce perithecia in the lab on natural or artificial substrates (Leslie *et al.*, 2006). Whether or not species that appear to have little sexual development in the field still use unobserved sexual development for recombination remains to be investigated.

Of all of the *Fusarium* species, *F. graminearum* has the best-studied sexual cycle, and thus will feature prominently in this review. Because of its economic impact worldwide as a pathogen of small grains, *F. graminearum* was the third filamentous fungus to be sequenced with public funds in the United States. The availability of the genome and other 'omic resources, as well as the ease of genetic manipulation of the fungus has made it an ideal organism in which to study sexual development in relation to pathogenicity. During sexual development, *Fusarium* undergoes many of the morphogenic processes of more complex eukaryotes, providing a simple system in which to study membrane function, secretory pathways, cytoskeleton dynamics, cell adhesion, cell polarity, signalling, cell cycle, apoptosis, differentiation, and stem cells (Fig. 2.1). Exploration of these aspects of sexual development in fungi in general has just begun.

Recently, much progress has been made in the genetic characterization of sexual development in *Fusarium* species, however, other systems have a long and rich history of studies in ascocarp development. In particular, studies of *Sordaria* spp. and *Neurospora crassa* have shed light on perithecium development. Therefore, we will use these models as comparisons for the *Fusarium* species. The reader is referred to recent review articles for a more comprehensive review of development in these species (Pöggeler *et al.*, 2006; Lord and Read, 2011).

Morphology and phylogenetics

Fungi have historically been categorized morphologically by their fruiting bodies. While hyphae have very few defining characteristics, fungal fruiting bodies are generally easily distinguishable between species. However, for *Fusarium* species, reliance on fruiting body morphology to distinguish species has resulted in large species complexes, which phylogenetic tools have begun to resolve into multiple species (O'Donnell *et al.*, 2004, 1998). The fungi belonging to the *Fusarium* anamorphs are now considered to be part of the subphylum Pezizomycotina, within the class of perithecium-producing ascomycetes Sordariomycetes (Spatafora *et al.*, 2006). They are in the order Hypocreales, a group characterized by 'occasional fleshy stromata (sing. stroma; a hyphal matrix on which the perithecia form), spherical, fleshy, brightly coloured, walled perithecia having ostioles lined with periphyses, rarely ornamented by hairs, apical paraphyses, thin-walled asci, with simple openings, colourless or pale-brown ascospores, which are usually multicellular' (Kirk *et al.*, 2001; refer to Fig. 2.2). *F. graminearum* forms purplish-black perithecia due to a combination of two pigments (Gaffoor *et al.*, 2005). *F. verticillioides* has blue-black perithecia (See Chapter 8). The perithecia of both species form multicelled, unbranched apical paraphyses, meaning that paraphyses develop from tissue in the top of the centrum and grow downwards. Their structure, in comparison with *Neurospora crassa*, is summarized in Fig. 2.2.

Bistis *et al.* (2003) delineated all of the cell types in *N. crassa*. For the sexual phase, 15 cell types were described including: ascogonium, enveloping hyphae (vegetative hyphae that surround the ascogonium and develop the protoperithecial wall), trichogyne, ascogenous hyphae, terminal crozier cell, basal crozier cell, ascus mother cell, ascus, ascospore, paraphyses, centrum pseudoparenchymal cells (become the inner wall of the perithecium), outer wall cells, hair cells, periphyses, wall cells of the perithecium neck. The trichogyne is the female element and the point of cellular fusion and recognition of the male element. For *Fusarium* teleomorphs, hair cells are absent or rare, there are no neck (beak) cells, and the ostiole is hard to distinguish on the outer surface of the spherical mature perithecium (Fig. 2.2). Trichogynes occur in few species. In *F. graminearum* and *F. verticillioides*, apical paraphyses grow down from the top of the centrum, but attach at the base of the asci, so in squash mounts, they appear as swollen hyphae radiating outward between the asci (Trail and Common, 2000; Sikhakolli *et al.*, 2012).

In culture, development in *F. graminearum* takes approximately 144 h from induction of the sexual stage to development of mature perithecia with actively discharging spores. Development

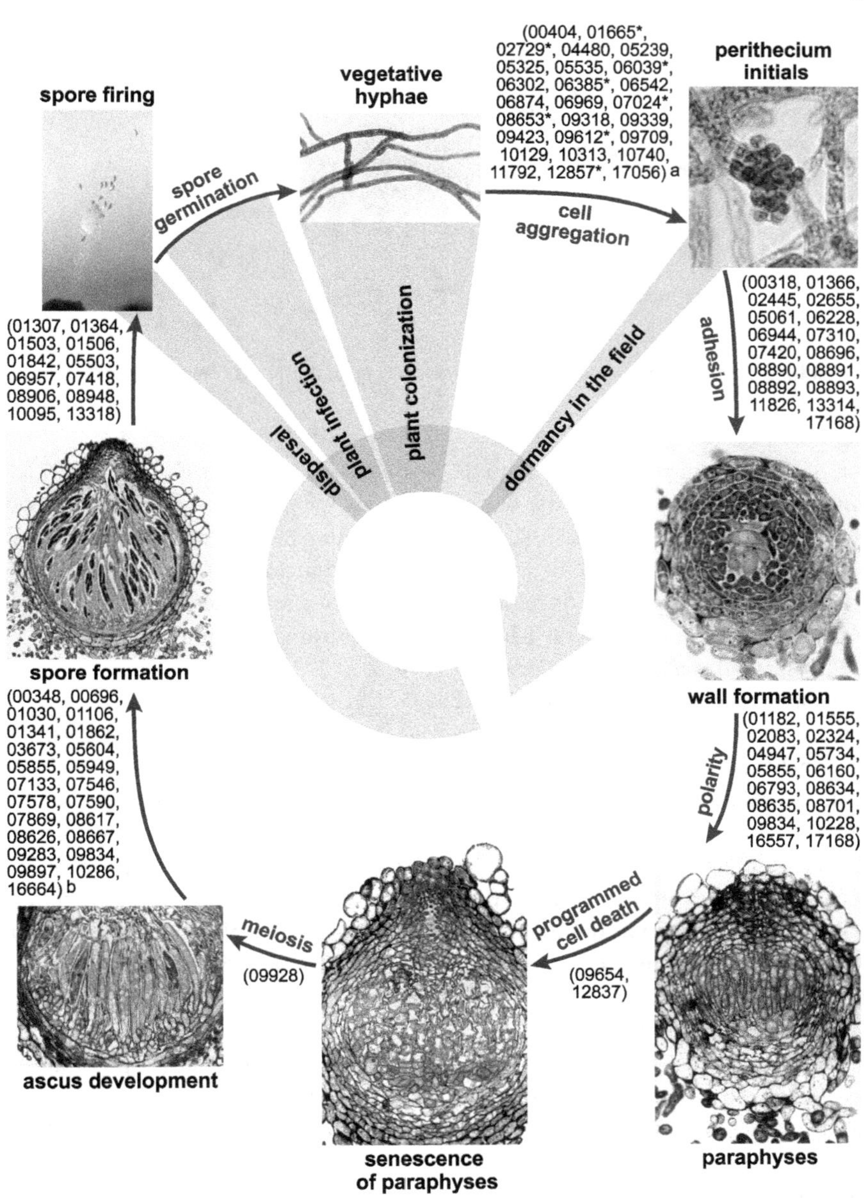

Figure 2.1 Stages of perithecium development in *Fusarium graminearum*, showing developmental process, morphogenic processes, and location of stage on host. Numbers correspond to genes affecting development at that stage (gene designations from http://mips.helmholtz-muenchen.de/genre/proj/FGSG_DB/). Additional genes encoding protein kinases whose mutants do not development perithecia and are not shown above are: (a) FGSG_00362, FGSG_00408, FGSG_02795, FGSG_04053, FGSG_04484, FGSG_05484, FGSG_05547, FGSG_06326, FGSG_06385, FGSG_06970, FGSG_07295, FGSG_07329, FGSG_08691, FGSG_09612, FGSG_09903, FGSG_10037, FGSG_10066, FGSG_10313, FGSG_10381, FGSG_01188; and (b) whose mutants are defective in ascospore morphology: FGSG_00337, FGSG_01058, FGSG_01347, FGSG_01641, FGSG_04382, FGSG_05418, FGSG_06878, FGSG_07251, FGSG_07344, FGSG_08468, FGSG_09274, FGSG_09897, FGSG_11812 (Wang *et al.*, 2011). Mutants of the following genes encoding transcription factors affected perithecium numbers and/or delayed maturation: FGSG_00729, FGSG_00930, FGSG_01176, FGSG_01576, FGSG_02608, FGSG_02719, FGSG_05475, FGSG_06382, FGSG_06427, FGSG_07052, FGSG_07546, FGSG_07928, FGSG_09019, FGSG_12970 (Son *et al.*, 2011). Two acetyl CoA reductase genes (FGSG_00330, and FGSG_01743) yielded mutant strains with reduced numbers of perithecia (Lee *et al.*, 2011). Mutants of the following transcription factors had increased the numbers of normally functioning perithecia: FGSG_02814,FGSG_ 09308, FGSG_06356 (Son *et al.*, 2011).

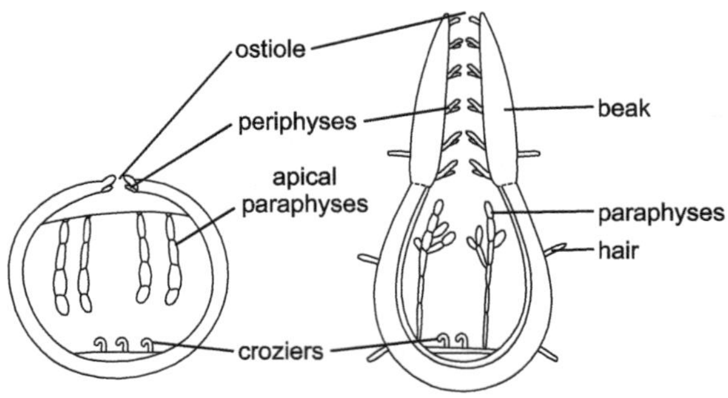

Figure 2.2 Structural comparison of perithecia of *Fusarium* spp. (left) and *Neurospora* spp. (right).

occurs most reliably on carrot agar for both *F. graminearum* and *F. verticillioides* with the sexual phase being induced by addition of 2.5% Tween 60 solution, applied to the surface of a well-colonized Petri dish, followed by incubation under light (Bowden and Leslie, 1999; Klittich and Leslie 1988; Cavinder *et al.*, 2012). For some strains of *F. graminearum*, including PH-1 (Trail and Common, 2000; the strain for the reference genome), a synchronously developing lawn of perithecia will cover the surface of the agar. Sexual development is accompanied by suppression of asexual sporulation, thus providing ideal material for studies of sexual development. In *F. verticillioides*, conditions for suppression of conidiation are not known, and conidiation occurs simultaneously with perithecium development (Sikhakolli *et al.*, 2012). The developmental progression of *F. graminearum* (Trail and Common, 2000), *F. verticillioides* (Sikhakolli *et al.*, 2012) and *F. solani* (Hanlin, 1971) are similar. Perithecia undergo morphogenic processes similar to development of more complex eukaryotes, sequentially developing four 'tissue types': perithecium wall, ascogenous hyphae, paraphyses and periphyses, and asci with ascospores (Figs. 2.1 and 2.2). Ascogonia are produced first and are fertilized, in the heterothallic species, or develop without fertilization in homothallic *F. graminearum*. Enveloping hyphae surround the ascogonium, and form the wall (peridium), ascogenous hyphae remaining inside. Then paraphyses and periphyses develop, filling the central cavity (centrum). As the asci develop, the paraphyses undergo spontaneous senescence, but remain between the asci, possibly assisting in forcible discharge (Trail and Common 2000; Sikhakolli *et al.*, 2012; Trail 2007). The ascogenous hyphae expand as the paraphyses recede, becoming the predominant tissue in the centrum and producing asci and ascospores (Fig. 2.1). In all species, ascospores are forcibly discharged into the air from the mature perithecia.

Life cycle in the field and niche associations

Sexual propagules in fungi are usually associated with the onset of unfavourable conditions, such as winter, drought or nutritional stress. Sexual propagules can result from genetic recombination, and thus also play a fundamental role of increasing genetic diversity in the life cycle. Gilbert *et al.* (2008) show that, in the laboratory, ascospores remain viable following exposure to extremes of temperature and humidity, however, their ability weather adverse conditions in the field and disperse is questionable. Examination of *F. graminearum* and some *Nectria* species, including *N. galligena*, has found that they overwinter as mycelia (Guenther and Trail, 2005; Jones and Aldwinkle, 1990). In these species, the small, fleshy perithecia are short-lived and do not appear to have a role in overwintering. *F. graminearum* overwinters as a perithecium initial embedded in the substomatal cavity in wheat. These specialized hyphae have been shown to store significant quantities of lipids, which are later used to fuel perithecium formation (Guenther *et al.*, 2009). Following the formation of ascogonia in the substomatal cavity, the fungus may overwinter, a period of dormancy that does not develop in culture (Guenther and Trail, 2005). Lipid stores may also be important in survival over the winter months. Survival of species with rare sexual stages is likely to be as mycelia or chlamydospores and the possible role of stored lipids in the ability of these species to survive has not been explored.

In fungi, asexual propagules are produced throughout the life cycle, with generally less investment for each propagule than for sexual spores, and dispersal is their sole function. For the Fusaria and most fungi, the asexual stage and hyphae are produced continuously as long as a food source is available. Notable for *Fusarium* species is the consistent and prolific production of conidia, with some species producing two types, both macro- and microconidia. *Fusarium* species produce conidia in sporodochia, which are clusters of conidia-producing cells in a slimy mass. The characteristic fusoid shape of the conidia gave the genus its name and, together with the mucoid coating, makes the spores ideal for splash dispersal (Deacon, 2006).

It is unclear what role the anamorph and teleomorph play in niche adaptation of these fungal species, although one can speculate. *F. graminearum*, a homothallic species, appears to rely on sexual development as an important dispersal agent (at least in the Midwestern US; Shaner, 2003; Desjardins *et al.*, 2006). Ascospores have also been shown to be an important inoculum source for stem rot of pepper caused by *Fusarium solani* f. sp. *piperis* (teleomorph *Nectria haematococca* f. sp. *piperis*; Ikeda, 2010; Hamada *et al.*, 1988), which also produces abundant perithecia in the field. On the other hand, *F. verticillioides* has infrequent perithecium production, and produces both microconidia and macroconidia. *F. verticillioides* operates as an endophyte and, being seed borne, has less need for becoming airborne for dispersal. Since the overwhelming majority of species that have *Fusarium* anamorphs have rare fruiting bodies in nature, it appears that the asexual phase, with slimy sporodochia and a focus on splash dispersal, has been successful for dissemination. However, even rare sexual recombination can contribute greatly to genetic diversity, so these species may maintain the ability to recombine for rare encounters, thus the function of the sexual phase may be solely for genetic recombination in the primarily asexual species.

Mating and sexual compatibility

The first unique identifiable sexual structure of filamentous ascomycetes is the ascogonium, a coiled structure that comprises the female element. In many ascomycetes, the ascogonium produces a hyphal extension, the trichogyne, which is the location of plasmogamy with the male element. The male element can be a micro- or macroconidium or a hyphal fragment, which fuses to the female element. Each element donates a nucleus. Following plasmogamy, nuclei remain distinct, and each cellular division maintains the nuclear pairing, thus initiating the dikaryon. In *Fusarium* teleomorphs, trichogynes have been described for *F. solani* and *F. verticillioides*. Modified narrow hyphae protruding from the ascogonium were proposed to serve as trichogynes in *F. solani* and in *F. verticillioides*, although no fertilization event was demonstrated in either case (Dyer *et al.*, 1993; Sikhakolli *et al.*, 2012), but in *N. haematococca* var. *cucurbitae*, trichogynes were shown to grow chemotropically towards conidia (Bistis, 1979). It is possible that differentiation into a typical trichogyne does not occur in most *Fusarium* life cycles.

The trichogyne is receptive to the male pheromone, prompting fusion with the male gamete. Pheromones have been best studied in *Saccharomyces cerevisiae*, which produces an œ-factor pheromone and an a-factor pheromone. Pheromones are generated by pheromone precursor genes, and bind to specific receptors. Despite its homothallic disposition, *F. graminearum* can outcross and has pheromone precursor genes, designated *ppg1* (homologous to the *S. cerevisiae* œ-factor pheromone precursor gene) and *ppg2* (homologous to the *S. cerevisiae* a-factor) and the corresponding receptor genes are *pre2* and *pre1*, respectively. The process of fertilization during outcrossing has not been observed. However, in the absence of trichogynes in *Fusarium* teleomorphs, it is presumed that pheromones would enhance fusion of the male element to the ascogonium. Lee *et al.* (2008) reported that *ppg1* is expressed in *F. graminearum* germinating conidia and mature ascospores, but *ppg2* (now designated FGSG_16834) expression was not detected in any cells examined. They found that *pre2* was expressed in all tissues tested with the exception of vegetative hyphae, whereas *pre1* was not expressed in any cell types tested. Examination of the recent transcriptional profiles of the stages of development shown in Fig. 2.1 (Sikhakolli *et al.*, 2012) showed that both *pre1* and *pre2* are expressed at all of these stages, peaking during ascus and ascospore formation, and opening up the possibility of a role in later stages of development. In this dataset, *ppg1* increased in expression across the developmental phases, peaking during the final phase of spore maturation. Expression of *ppg2* was not determined owing to the absence of the gene in the dataset, probably because of its short length (66 nt; Lee *et al.*, 2008).

Mutations in *ppg1* or *pre2* resulted in significantly reduced fertility in self-fertilization experiments. Mutants still formed mature perithecia, but fewer were formed than in the wild type and numerous small empty walled structures were present (Lee *et al.*, 2008). Mutants of *ppg2* or *pre1* had no observable phenotype. The study also incorporated examination of *MAT1-1* and *MAT1-2* mutants in combination with *ppg1* and *pre2* mutants. Results indicated that the pheromones and receptors are only functional during the fertilization step. In another study, Kim *et al.* (2008) reported expression of *ppg1* during the initial stages of perithecium formation and *ppg2* was expressed during a very short window, two days after induction of sexual development. Furthermore, deletion of the pheromone genes or their receptors did not affect fertility of the deletion mutants. However, the deletion of both *ppg1* and *pre2* (or *pre1*) resulted in numerous sterile perithecia that were intermixed with fertile perithecia. There are slight differences between the mutant phenotypes described by Kim *et al.* (2008) and Lee *et al.* (2008). Presumably these differences may be explained by the effect of the mutations within the context of the different parent strains (genotypes) used.

Mating processes in fungi are controlled by the MAT locus. Organization and content of genes in and around the MAT locus have been characterized in a number of Ascomycota (reviewed in Heitman *et al.*, 2007). In *F. graminearum*, the locus consists of four genes: *MAT1-1-1, MAT1-1-2, MAT1-1-3* and *MAT1-2-1* (Table 2.1; Yun *et al.*, 2000). The first three genes correspond to the œ-idiomorph in *S. cerevisiae*, the latter corresponds to the a-idiomorph. *MAT1-1-1, MAT1-1-3*, and *MAT1-2-1* encode DNA binding proteins. *MAT1-1-2* encodes the MATœ-1 homologue and has no recognizable functional domain. Kim *et al.* (2008) showed that deletion of *MAT1-1* and *MAT1--2* resulted in loss of expression of *ppg1*, and that transcription of *ppg2* increased in the *MAT1-1* deletion strain during perithecium development, but was lacking in the *MAT1-2* mutant. In the *MGV1* (encoding a MAP kinase) mutant expression of *ppg1* was high-throughout sexual development. The pheromone gene transcripts were not detected in either of the *MAT* mutants nor the *MGV* mutant. Thus both *MGV1* and the *MAT* genes appear to regulate expression of the pheromone genes and their receptors. Furthermore, the *MGV1*-based control of transcription of pheromone and pheromone receptor genes may explain the phenotype of *MGV1* mutant (Table 2.1).

Homothallism in *F. graminearum* has been determined to be due to the presence of the *MAT1-1* and *MAT1-2* idiomorphs in selfing strains (Yun *et al.*, 2000). It appears that both sets of *MAT* genes are essential for self-compatibility. Disruption of the entire *MAT* locus results in infertility (Desjardins *et al.*, 2004). However, 51% of the deletion transformants produced the perithecium wall, but lacked development of sexual structures inside, suggesting that the *MAT* locus is not involved in wall formation, and that wall formation can be triggered by optimal conditions for sexual development. This is in keeping with the notion that the wall is formed from enveloping hyphae that do not originate from the ascogonium (Moore-Landecker, 1996). Disruption of either *MAT1-1* or *MAT1-2* results in self-incompatibility, with the strain functionally heterothallic (Lee *et al.*, 2003). *MAT1-1; mat1-2* strains can cross with *mat1-1; MAT1-2* strains and result in fertile, wild-type perithecia. In addition, either mutant can cross with wild-type selfing strains and result in formation of fertile perithecia, and normal asci with segregation of the wild-type and mutant locus.

In *F. graminearum, MAT1-1-3* and *MAT1-2* are expressed mainly during sexual development,

Table 2.1 Genes involved in sexual development and function. Genes are listed based on their most prominent mutant phenotype

Gene name/FGSG number	Species under study	Knockout phenotype related to sexual development	Gene function	Reference
Mating				
PRE2/FGSG_02655	*F. graminearum*	Reduced female fertility. Fewer mature perithecia when selfed	Pheromone	Lee *et al*. (2008), Kim *et al*. (2008)
PPG1/FGSG_05061	*F. graminearum*	Reduced male fertility. Fewer mature perithecia; many perithecia did not reach maturity	Pheromone receptor	Lee *et al*. (2008)
MAT1 locus: FGSG_08890 to 08893	*F. graminearum*	Self sterile; produced small sterile 'perithecia' on carrot agar	MAT1-1/1-2 mutant	Desjardins *et al*. (2004)
MAT genes: *MAT1-1* and *MAT 1-2*/ FGSG_08890 to 08893	*F. graminearum*	Self-sterile strains; can outcross	Mating-type idiomorphs	Lee *et al*. (2003)
MAT1-2-1/ FGSG_08893	*F. verticillioides*	Inability to outcross as male or female	Mating type gene	Keszthelyi *et al*. (2007)
Ascogonia				
FSR1/FGSG_01665	*F. graminearum* and *F. verticillioides*	No perithecia+^	Probably signal transduction scaffold protein	Shim *et al*. (2006)
CHS7/FGSG_02729	*F. graminearum*	No perithecia+^	Chitin synthase	Kim *et al*. (2009)
FGSG_04480	*F. graminearum*	No protoperithecia	Conserved hypothetical proteins with N-terminal transcriptional regulatory motif	Lee *et al*. (2010)
FGSG_05239	*F. graminearum*	No protoperithecia	Related to G protein coupled receptor like protein: rhodopsin-like	Lee *et al*. (2010)
FGSG_05325	*F. graminearum*	Reduced protoperithecia; few small, sterile perithecia	Cons. Hypothetical protein	Lee *et al*. (2010)
GPA1/FGSG_05535	*F. graminearum*	Female sterility	G-protein alpha subunit 1	Yu *et al*. (2008)
ACL2/FGSG_06039 *ACL1*/FGSG_12857	*F. graminearum*	No perithecia+^	ATP citrate lyase	Lee *et al*. (2010)
FIG1/FGSG_06302	*F. graminearum*	No perithecium initials	Low-Affinity Calcium uptake	Cavinder and Trail (2012)
MAP1-FMK1-GPMK1/ FGSG_06385	*F. graminearum*	No perithecia are formed+^	Pathogenicity MAP kinase	Urban *et al*. (2003), Jenczmionka *et al*. (2003)
TOP1/FGSG_06874	*F. graminearum*	No perithecia or perithecium initials	Topoisomerase I	Baldwin *et al*. (2010)

Table 2.1 (Continued) Genes involved in sexual development and function. Genes are listed based on their most prominent mutant phenotype

Gene name/FGSG number	Species under study	Knockout phenotype related to sexual development	Gene function	Reference
FBP1/FGSG_06969	*F. graminearum*	Female sterile; as males, incomplete octads of spores	F-box protein: involved in protein degradation	Han *et al*. (2007)
RAS2/FGSG_07024	*F. graminearum*	No perithecia⁺^	Ras GTPase	Bluhm *et al*. (2007)
CHS5/FSGS_08653	*F. graminearum*	No perithecia⁺^	Chitin synthase	Kim *et al*. (2009)
BAR1/FGSG_09423	*F. graminearum*	No perithecia are formed⁺^	Ceramide synthase	Rittenour *et al*., 2011)
HOG1/FGSG_09612	*F. graminearum*	No perithecia formed⁺	MAPK for osmotic stress	Oide *et al*., 2010)
STUAp/FGSG_10129	*F. graminearum*	No perithecia⁺^	APSES domain-containing transcription factor	Lysøe *et al*. (2011)
MGV1/FGSG_10313	*F. graminearum*	No perithecia are formed; reduced hyphal growth^	MAP kinase for penetration and sporulation	Hou *et al*. (2002)
ATG8/FGSG_10740	*F. graminearum*	No perithecia are formed⁺^	Autophagy-related protein	Josefsen *et al*. (2012)
FGSG_17056 (08320 in ref.)	*F. graminearum*	No protoperithecia	Cytochrome P450	Lee *et al*. (2010)
Perithecium development				
FGSG_00348	*F. graminearum*	Long delay in maturation of perithecia	Argonaute-like post-transcriptional gene silencing protein QDE-2	Lee *et al*. (2010)
FGSG_01862	*F. graminearum*	Delayed perithecium maturation	Cons. Hypothetical protein	Lee *et al*. (2010)
PKS12/FGSG_02324	*F. graminearum*,	*F. graminearum*: perithecia are flat black instead of purple-black, and hyphae in medium are black	Polyketide synthase	Gaffoor *et al*. (2005)
JGI_33672	*F. solani*	Loss of red pigment of perithecia	Polyketide synthase	Graziani *et al*. (2004)
FGSG_03673	*F. graminearum*	Reduced numbers of perithecia; perithecia of mixed sizes	Related to carboxypeptidase	Lee *et al*. (2010)
FGSG 07578	*F. graminearum*	Reduced numbers of perithecia; delayed maturation	Conserved hypothetical protein	Lee *et al*. (2010)
FGSG_07590	*F. graminearum*	Reduced numbers of perithecia; delayed maturation	Taurine dioxygenase: expressed during sulfate starvation	Lee *et al*. (2010)
FGSG_07869	*F. graminearum*	Reduced numbers of perithecia	Conserved hypothetical protein; dehydrogenase	Lee *et al*. (2010)

Gene name/FGSG number	Species under study	Knockout phenotype related to sexual development	Gene function	Reference
FGSG_16664 (06484 in ref.)	*F. graminearum*	Reduced numbers of perithecia	Hypothetical model	Lee *et al.* (2010)
PGL1/PKS3/ FGSG_17168	*F. graminearum; F. fujikuroi; F. solani*	Perithecia are yellow-pink	Polyketide synthase	Gaffoor *et al.* (2005), Chaisrisook and Leslie, (1990), Proctor *et al.* (2007)
Spore development/ascus structure				
ZIF1/FGSG_01555	*F. graminearum*	No asci or ascospores as female parent	b-ZIP transcription factor, ascomycete specific	Wang *et al.* (2011)
MYT2 FGSG_07546	*F. graminearum*	Larger perithecia	Transcription factor	Lin *et al.* (2012)
ROA/FGSG_08667	*F. graminearum*	Abnormal spores	Ketopantoate reductase	Kim *et al.* (2010)
RGSC/FGSG_09283	*F. graminearum*	Abnormal spores	Regulator of G-proteins	Park *et al.* (2012)
PDC1/FGSC_09834	*F. graminearum*	No asci or ascospores	Pyruvate decarboxylase	Son *et al.* (2012
SNF1/FGSG_09897	*F. graminearum*	Fewer perithecia; matured slowly; asci with 1–8 abnormal spores	Sucrose non-fermenting protein kinase from yeast/ regulation in glucose starvation	Lee *et al.* (2009)
SYN2/FGSG_09928	*F. graminearum*	Loss of female fertility. Ascus shape abnormal; ascospore delimitation delayed	Syntaxin 2: protein involved in fusion of vesicles with plasma membrane	Hong *et al.* (2010)
RUM1/FGSG_16557	*F. graminearum*	No asci or ascospores	Related to regulator protein rum1	Kim *et al.* (2011)
Spore discharge				
CCH1/FGSG_01364	*F. graminearum*	Reduced ascospore discharge; abnormal ascospores	L-type Ca ion channel	Hallen and Trail (2008)
RGSB/FGSG_01503	*F. graminearum*	Reduced ascospore discharge	Regulator of G-proteins	Park *et al.* (2012)
MID1/FGSG_07418	*F. graminearum*	Reduced ascospore discharge; Abnormal ascospores	Stretch activated Ca ion channel	Cavinder *et al.* (2011)
SSK1/FGSG_08948	*F. graminearum*	Reduced ascospore discharge; tetrad formation; fewer perithecia	Histidine kinase response regulator	Oide *et al.* (2010)

*Protein kinases and transcription factors from two major functional studies (Son *et al.*, 2011; Wang *et al.*, 2011) are not included in this table owing to space constraints, but are included in Fig. 2.1.
+Presence of perithecium initials not examined.
^Pleiotropic effects that may result in inability to generate sex cells.
#ppg2 was not found to have a mutant phenotype.

and the transcripts are at very low levels during other conditions (Hallen *et al.*, 2007). In contrast, these studies indicate that *MAT1-1-1* and *MAT1-1-2* transcripts are at very low levels during all stages of sexual development and during several conditions of vegetative culture. These results are at odds with the phenotypic effects seen in the *MAT1-1* mutants. It is possible that very low transcript levels are needed for gene activity. Genes regulated by MAT1-2 were analysed by Lee *et al.* (2006), who used suppressive subtractive hybridization between a *mat1-1* mutant and wild type to find 291 genes down-regulated in the mutant. Many of these genes were found to be expressed under several conditions, including vegetative growth, in GeneChip studies (Hallen *et al.*, 2007).

In the *F. verticillioides* genome, the genes of the *MAT1-1* locus are present (FVEG_02491, FVEG_02490, FVEG_02489; http://www.broadinstitute.org/annotation/genome/*Fusarium*_group) and the *MAT1-2* gene has been previously described (Martin *et al.*, 2010). Comparative expression studies during perithecium development in *F. verticillioides* and *F. graminearum* revealed that *MAT1-2* is expressed only during early development in *F. verticillioides* and in *F. graminearum MAT1-2* expression peaks late in development (Sikhakolli *et al.*, 2012). These differences may reflect differences in the roles of nuclei in heterothallic and homothallic species. In both species, the *MAT1-1* genes peak in activity during later stages of maturation, indicating a role for these genes in development.

The *F. solani* genome contains all three *MAT1-1* genes, *MAT1-1-1, MAT1-1-2, MAT1-1-3* (neha_08499, neha_08498, neha_08497, respectively; www.jgi.doe.gov/nectria). As only one strain was sequenced, the *MAT1-2* idiomorph is not present in the reference genome. In *F. fujikuroi*, the *MAT* idiomorphs are present in different strains (Yun *et al.*, 2000). The HMG moiety of *MAT1-2* loci of *F. fujikuroi* and other species within the *Gibberella fujikuroi* species complex (*F. verticillioides, F. subglutinans, F. proliferatum, F. thapsinum, F. nygamai*) were cloned using primers specific to the species complex (Kerényi *et al.*, 1999). All species appeared to be heterothallic based on the presence of this fragment in half the strains examined and the cosegregation of the fragment with the *MAT1-2* phenotype. Perhaps not surprisingly, the genome of *F. oxysporum*, with no known sexual phase, also contains a *MAT* locus, and the *MAT1-1-1, MAT1-1-2* and *MAT1-1-3* genes are all present (http://www.broadinstitute.org/annotation/genome/*Fusarium*_group). *MAT1-2, MAT1-1-1* and *MAT 1-1-2* were previously identified in several *forma specialis* of *F. oxysporum* (Yun *et al.*, 2000; Arie *et al.*, 2000; Lievens *et al.*, 2009). These studies indicate the potential for a sexual cycle in this fungus. Pairing of genetically identified compatible strains of long-thought asexual species *Aspergillus parasiticus* and *A. flavus* resulted in successful sexual crosses (Horn *et al.*, 2009a: 2009b), indicating the possibility of a similar strategy for *F. oxysporum*.

In *F. graminearum*, examination of the asci from successful crosses of *mat1-1* or *mat1-2* to wild type resulted in all asci being recombinant(Lee *et al.*, 2003). The absence of wild-type asci was interpreted to suggest that the two different nuclei recognize each other and that there is a preference for karyogamy between distinct nuclei rather than selfing of the wild type (Debuchy and Turgeon, 2006; Lee *et al.*, 2003). However, the experimental design does not support this conclusion. Fertilization of 7-day-old cultures of the *mat1-1* or *mat1-2* was performed with conidial suspensions of the wild type. The mutant mycelia are well established, thus providing an environment in which the wild-type nuclei have to serve as males. Because of the nutritional resources needed for sexual development, already established cultures do not provide wild-type conidia with the resources to develop their own mycelia. A better test for preferential recognition would be to fertilize established wild-type cultures with a gradient of mutant spores from very few to an overabundance. A corresponding increase in recombinant perithecia with addition of increasing numbers of spores would indicate a predilection for outcrossing.

Development of a perithecium

Sexual development in *Fusarium* is a continuous process, and unlike in some other ascomycetes, once the perithecium begins to form, the process is continuous until maturation is completed. Fig.

2.1 illustrates the stages of development in *F. graminearum*. The term 'protoperithecium' is often used to describe the initial female structure that develops into the perithecium. However, the term was originally defined as 'a young but walled perithecium before ascus formation' by Ellis (1960). Ellis describes the initial coils that lead to the protoperithecium as the 'protoperithecium initials'. Others have shortened this to 'perithecium initial' (i.e. Jensen, 1982). In the following discussion, we will use the term 'protoperithecium' to refer to the spherical young perithecium, and the term 'ascogonium' to refer to the early, unwalled coiled structure (Fig. 2.2). The protoperithecium is the spherical walled structure that develops from the ascogonium.

Inside the protoperithecium, dikaryotic ascogenous hyphae form and begin the transition to mature asci (Beckett, 1981; Read and Beckett, 1996). Nuclear pairing has been demonstrated in formation of the ascogenous hyphae just after plasmogamy in *Podospora* (Turgeon and Debuchy, 2007). It is likely that the same occurs in the *Fusarium* teleomorphs. In homothallic species, paired nuclei have been observed during crozier formation (Trail and Common, 2000), but where the initial pairing occurs is uncertain. It has been suggested that nuclei in homothallic species may take on the role of opposite mating types in dikaryotic stages (Turgeon and Debuchy, 2007). In *F. graminearum*, following induction of the sexual cycle, wide hyphae filled with lipid bodies are generated, which give rise to the ascogonia. In these hyphae, binucleate cells have been observed at a much higher frequency than in vegetative hyphae (Trail and Common, 2000). Whether these represent true dikaryons or not is unknown, but it is apparent that these wide hyphae are the first stage in development of the sexual cycle and that the stored lipid is used for perithecium development, and probably overwintering (Guenther *et al.*, 2009).

Genetics of development

Homeobox genes are the master regulators of development in eukaryotes (Meyerowitz, 2002). In fungi, homeobox proteins play a crucial role in mating, with both ascomycetes and basidiomycetes containing homeobox proteins in their mating type loci and genes related to development (Liu *et al.*, 2010; Arnaise *et al.*, 2001; Hull *et al.*, 2005; Bakkeren *et al.*, 2008). MAT1-1-1 is a MADS-box type DNA-binding transcription factor; MAT1-1-3 and MAT1-2-1 are both HMG (high mobility domain) proteins that also bind to DNA and facilitate unwinding and bending (Martin *et al.*, 2010). As with flowers, fungal sexual development is patterned radially and there does not appear to be a parallel to dorsal-ventral/adaxial/abaxial development characteristic of animals.

In the vast majority of fungi, extensive morphogenesis occurs during fruiting body formation. During this time, developmental processes occur that are absent during much of the rest of the life cycle. Many of these processes are common in development of higher eukaryotes, including polarity, cell aggregation, adhesion, differentiation, programmed cell death, and compartmentalization resulting in a fairly complex multicellular structure. In ascomycetes, following formation of the ascogonium, hyphae arising from the parent hyphae, from the ascogonium, or from surrounding areas enclose the ascogonium and form a wall (Read, 1983). The wall of the perithecium, called the peridium, is multilayered, and the hyphae forming these lose their hyphal appearance and become 'conglutinate' cells (Read, 1994; Lord and Read, 2011). An inner wall and outer wall can be distinguished in the *Fusarium* perithecia (Trail and Common, 2000, Sikhakolli *et al.*, 2012; Dyer *et al.*, 1993). The cells of the inner wall are thinner walled than those of the outer wall. The outer wall cells senesce following expansion of the perithecium. Adhesion between cells and wall layers occurs, but the basis of this adhesion and the extracellular matrix is not well understood. However, it is known that the outer layer of the hyphal wall is composed of glycoproteins, with β-glucans and chitin forming the inner wall (DeGroot *et al.*, 2005). Representatives of one group of these proteins, called GPI (glycosylphosphatidylinositol)-modified cell wall proteins have been identified in *Fusarium oxysporum* (Schoffelmeer *et al.*, 1999). The *Fusarium* extracellular matrix protein 1 (FoFEM1) apparently links 1,6,-β-glucans and 1,3-β-glucans.

These proteins are thought to play important roles in attachment with the hosts in pathogenic *Candida* species (DeGroot *et al.*, 2005; Hoyer *et al.*, 2001). Yeast has sexual adhesion proteins such as the sexual agglutinins (Lipke and Kurjan, 1992), which are GPI-modified cell wall proteins. Whether or not homologues of these genes are involved in the conglutinate walls is not yet known.

Following wall formation, the young perithecium develops polarity. In *Fusarium*, the orientation appears to be related to the substrate, not light or gravity. In *Neurospora*, the beak is oriented to light, but *Fusarium* has no beak, and there is no evidence of light orientation for the ostiole. Development of the paraphyses occurs from the cells at the top of the centrum (central cavity) in *Fusarium* teleomorphs. These are termed 'apical paraphyses' or 'pseudoparaphyses' because of their distinct origin, and are characteristic of the Order Hypocreales. In *Fusarium graminearum*, and most likely the other *Fusarium* teleomorphs, the apical paraphyses undergo programmed cell death, to allow room for the asci to form (Sikhakolli *et al.*, 2012). However, the membranes remain intact and become engorged during squash mounts, indicating the presence of osmolytes within the cells. The apical paraphyses attach to the base of the developing asci and the cells near the top of the centrum break in *F. graminearum* and *F. verticillioides* (Trail and Common, 2000; Sikhakolli *et al.*, 2012). In *Sordaria*, it has been noticed that paraphyses fuse with each other (Lord and Read 2011); a similar attachment mechanism may be at work in *Fusarium*. It is likely that these structures serve to enhance the pressure inside the perithecia, directing the asci upwards. Similarly, the periphyses lining the ostiole become senesced during the final stages of perithecium development, resulting in the opening of the ostiole.

The availability of the genome sequence of *F. graminearum* (Cuomo *et al.*, 2007) and the *Fusarium* GeneChip (Gueldener *et al.*, 2006) has made it possible to determine the expression pattern during formation of each of the perithecium 'tissues' (ascogenous cells, wall, pseudoparaphyses, asci and ascospores). Development of each tissue is separated by approximately 24 h in culture comprising a lawn of near-synchronously developing perithecia, making it possible to identify uniquely expressed genes for each tissue type (Hallen *et al.*, 2007). During ascus development, the largest number of unique genes is expressed compared to any other stage, most likely because meiosis takes place during this time. In comparison with many other expression studies and conditions performed by us and others, over 2000 genes (approximately one-eighth of the genome) were identified that were unique to sexual development (Hallen *et al.*, 2007). It will be interesting to determine the presence of these genes in asexual *Fusarium* species and their consistency in other fungal species.

F. graminearum has become a model fungus for the study of sexual development. The ability to easily disrupt specific genes by homologous recombination and the homothallic lifestyle enable easy phenotypic analysis of genes important to sexual development. Numerous genes have now been identified that have effects on sexual development (Table 2.1 and Fig. 2.1). These genes interrupt or change perithecium maturation from the formation of perithecium initials (not always assessed) through ascospore shape, to the ability to forcibly discharge spores (Fig. 2.1). These genes are related to mating type and pheromone synthesis, involved in signalling, encode transcriptional regulators or affect regulatory processes, are related to metabolism, are structural or are of unknown function. Two very large studies have included the functional analysis of 96 protein kinase genes (Wang *et al.*, 2011) and 657 transcription factors (Son *et al.*, 2011) in *F. graminearum*. However, this group of genes is insignificant compared with the 2000 identified found to be unique to sexual development. For instance, our understanding of the adhesion process involved in formation of the peridium, paraphyses and their senescence, and delineation of spores is minimal. A recent study of transcriptional profiles in *F. graminearum* and *F. verticillioides* showed that the majority of genes expressed during stages of sexual development are ORFan genes and conserved genes of unknown functions (Sikhakolli *et al.*, 2012). Fig. 2.1 shows the distribution of genes identified as critical to the perithecium developmental stages. Note that relatively few genes have been identified as essential to ascus development, the stage where the most genes unique to perithecium development have been shown to be expressed (Hallen *et al.*, 2007), highlighting the need for increased study of this developmental process.

As mentioned above, one of the roles of the sexual cycle is dispersal. Ascospores of *Fusarium* teleomorphs are forcibly discharged into the air and carried by wind. In *F. graminearum*, this phenomenon has been well studied, and ascospores are propagules of long-range dispersal (Shaner, 2003). The mechanism of forcible ascospore discharge is not well understood in any ascomycete (Trail, 2007), but *F. graminearum* has been a model system for study of the physiology and genetics of ascus function (Trail *et al.*, 2005). Light enhances the number of spores discharged, but is not essential for discharge (Trail *et al.*, 2002). Three genes have been identified which clearly affect ascus function, but do not cause phenotypic effects on spore formation. Two of these are calcium ion channels, Mid1 and Cch1. Disruption of these genes results is almost complete loss of discharge (Hallen and Trail, 2008; Cavinder *et al.*, 2011). Disruption of *Cch1* in *F. graminearum* resulted in mutants with significantly reduced forcible ascospore discharge, slower growing hyphae and a low frequency of abnormal ascospore segmentation (Hallen and Trail, 2008). A similar phenotype was observed in *Mid1* mutants. Supplementation of cultures of these mutants with exogenous calcium rescued forcible discharge, suggesting that the machinery for discharge was present, but the signals were not functional in the absence of calcium messenger. Recently, the *ssk1* gene, encoding a response regulator for the HOG1 pathway involved in high osmolarity sensing and response, was disrupted in *F. graminearum* and the progeny reported to be deficient in ascus function, although sexual development was also profoundly suppressed (Oide *et al.*, 2010). Finally, a round spore mutant, resulting from the disruption of *Roa*, encoding a ketopantoate reductase (Table 2.1) was found to discharge spores longer distances (Min *et al.*, 2010). The effect was concluded to be due to a change in the size of the ascus and the spores, which resulted in a change in the biomechanics of the discharge process. Although these mutations give some small insight into ascus function, much more work needs to be done to completely understand how the structure supports the function.

Relationships between fruiting and secondary metabolism

The relationship between fungal development and secondary metabolism has been best elucidated in *Aspergillus nidulans*, where the global regulatory genes *laeA* and *velA* have been shown to inhibit both sexual development and secondary metabolism (reviewed in Amaike and Keller, 2011). Homologues for these genes have been identified and functionally characterized in several *Fusarium* species. In *F. verticillioides*, disruption of *Lae1* reduced biosynthesis of mycotoxins fumonisin, fusaric acid and fusarin C and was shown to reduce production of the reddish pigmentation bikaverin (Butchko *et al.*, 2012), which gives perithecia a purplish cast (Brown *et al.*, 2012). In *F. fujikuroi*, functional analysis of *Vel1* and *Lae1* demonstrated that Vel1 is a negative regulator of bikaverin, and a positive regulator of fumonisins and fusarin C (Wiemann *et al.*, 2010). Interestingly, the *vel1* mutant produced significantly more perithecia than wild type, indicating it may function as a negative regulator of perithecium formation as well. Although the *Vel1* gene has been functionally disrupted in *F. graminearum*, the strain used in one study did not produce perithecia (Merhej *et al.*, 2012) and in a second study, this attribute was not examined (Jiang *et al.*, 2011), therefore the effect of this gene on perithecium development is not known.

Several of the polyketides made by *Fusarium* spp. have been implicated to be involved in sexual development. There had been some investigation of the polyketide mycotoxin, zearalenone functioning as a mating pheromone in *F. graminearum* (Wolf and Mirocha, 1973), but disruption of the zearalenone biosynthetic genes had no detectable effect on sexual reproduction (Gaffoor and Trail, 2006; Kim *et al.*, 2005). Other evidence suggests that zearalenone may affect sexual development in other fungi (Nelson, 1971). The black pigment of the perithecia has been shown to be produced by a polyketide synthase in *F. graminearum* (Gaffoor *et al.*, 2005), and *F. verticillioides* (Brown *et al.*, 2012) and pigments called fusarubins are responsible for the dark perithecial colour of *F. fujikuroi* (Chaisrisook and Leslie, 1990; Studt *et al.*, 2012). The perithecia of *F. graminearum* have a purple cast that is the results of accumulation of aurofusarin (Gaffoor *et al.*, 2005, Kim *et al.*, 2005; Maltz *et al.*, 2005). Another pigment gives the perithecia of *F. solani* their reddish colour (Graziani *et al.*, 2004).

The ecological role of the sexual cycle

Research on self-compatibility/incompatibility systems has been done primarily in plants. Schemske and Lande (1985) demonstrated a bimodal distribution of selfing and outcrossing species, where intermediate rates of outcrossing are limited. This distribution is maintained by differential selective forces acting on each mating system, with population bottlenecks promoting inbreeding, and selection for outcrossing in large populations with inbreeding depression. In plants, self-compatibility is also known to lead to evolutionary dead-ends, where self-incompatible species yield a higher species abundance (Goldberg *et al.*, 2010). Furthermore, self-incompatibility is commonly linked to clonality in plants. Thus self-incompatible plants often have clonal means of reproducing to assure reproduction in the absence of a viable partner (Vallejo-Marin and O'Brien, 2006). Finally, self-compatible species are more often found in unstable environments and environmental factors can promote occasional outcrossing in these species (Schemske and Lande, 1985). Thus, outcrossing may provide essential genetic variation for populations to adapt to environmental changes.

Self-compatibility is relatively rare in fungi. Homothallism has been observed mainly in the Zygomycetes and the Ascomycota and is rarer in the Basidiomycota. Interestingly, many of the dung fungi are homothallic species, supporting the observation that selfing is associated with unstable environments. Furthermore, few Basidiomycota are typical 'weedy' species. Among the *Fusarium* species, homothallism is by far the minority. The only homothallic *Gibberella* species is *Gibberella zeae*. *G. zeae* (*F. graminearum*) is also the only species that produces abundant perithecia in nature. It is possible that the abundance of fruiting bodies, and homothallism are related. For heterothallic species, it may be difficult to find a mate, so if fruiting body production is essential for the life cycle, self-compatible mating systems may be a selective advantage. The discharge of ascospores plays a critical role in the dispersal of inoculum for *F. graminearum*. In contrast, the self-incompatible species *F. verticillioides*, for example, is seed borne and is essentially an endophyte. Thus, it may not have the need to reproduce sexually for dispersal. Furthermore, *F. pseudograminearum* rarely produces fruiting bodies and is mainly a stem-base pathogen. Splash dispersal of conidia from the soil surface is likely to be a very effective source of inoculum for this pathogen. Although most of the *Fusarium* species appear to rarely produce perithecia in nature, it is possible that fruiting bodies are formed in locations (e.g. during growth on alternative hosts or in soil) where they are not observed. However, it has been suggested that heterothallism results in limited sexual interactions in fungi (Summerell *et al.*, 1998).

Sexual recombination can facilitate the assembly of multiple genes that enhance virulence. Although *F. graminearum* is homothallic, it is known to outcross in culture (Bowden and Leslie, 1999). When either *MAT1-1* or *MAT1-2* is deleted in *F. graminearum*, heterothallic mating results in only about 10% of the fertile perithecia of the wild-type selfing in culture. This implies that crossing is inefficient under the best of conditions (Lee *et al.*, 2003). A field study using *nit* tagged strains of *F. graminearum* that were carbendazim resistant showed that the outcrossing rate was only 13.9% (Chen and Zhou, 2009). Several studies have found distinct populations in the USA, Canada and Korea, suggesting limited outcrossing (Gale *et al.*, 2007; Ward *et al.*, 2008; Lee *et al.*, 2009) among strains in the field. On the other hand, work by Zeller *et al.* (2003, 2004) on Kansas populations suggests frequent outcrossing. Zeller *et al.* (2003) found an average of 1.8 strains infecting each head, indicating possibly a limited opportunity for partners. Weather conditions during dispersal and infection could greatly affect the availability of strains for outcrossing.

In *F. graminearum*, 2000 genes identified that are solely expressed during sexual development represents a sizeable portion of the genome and a significant genetic burden to maintain. For *F. graminearum*, which appears to depend on sexual reproduction to initiate disease, this burden is a critical one. However, other fungi reproduce sexually only rarely in nature, as far as we can determine, based on observations of fruiting bodies (admittedly a poor way to judge). If these genes are maintained in predominantly asexual populations, the question remains as to their importance for the populations in which they occur. The existence of cryptic sexual cycles in some fungi, such as *Aspergillus flavus* (Horn *et al.*, 2009a) and *A. parasiticus* (Horn *et al.*, 2009b) reveals our very poor understanding of the life cycles of the fungi that have them.

Conclusion

Fusarium species provide a great diversity of lifestyles for studying genetics of development and the impact of ecological niche on genetics and lifestyle. Fully understanding both aspects will provide us the best information for controlling these economically important organisms. As is illustrated here, our understanding of sexual development and its importance to the ecology and life cycles of the *Fusarium* species is in its infancy. However, this group of fungi has more genomic resources than most others. This should allow rapid progress in this area. Understanding of all aspects of these diverse life cycles is essential to developing means of controlling them.

Acknowledgements

The author thanks the many *Fusarium* workers, including the graduate students, undergraduates and postdocs in her lab who continue to inspire through their hard work. Marlene Camerson assisted with Fig. 2.1. Ralph Common assisted with preparing material for the photos in Fig. 2.1.

This work was supported by a grant from the National Science Foundation (0923794) and the Michigan AgBioResearch. In addition, the author would like to acknowledge continuing support by the USDA Wheat and Barley Scab Initiative.

References

Amaike, S., and Keller, N.P. (2012). *Aspergillus flavus*. Ann. Rev. Phytopath. *49*, 107–133.

Arie, T., Kaneko, I., Yoshida, T., Noguchi, M., Nomura, Y., and Yamaguchi, I. (2000). Mating-type genes from asexual phytopathogenic ascomycetes *Fusarium oxysporum* and *Alternaria alternata*. Mol. Plant Microb. Interact. *13*, 1330–1339.

Arnaise, S., Zickler, D., Poisier, C., and Debuchy, R. (2001). *Pah1*, A homeobox gene involved in hyphal morphology and microconidiogenesis in the filamentous ascomycete *Podospora anserina*. Mol. Microbiol. *39*, 54–64.

Bakkeren, G., Kamper, J., and Schirawski, J. (2008). Sex in smut fungi: structure, function and evolution of mating-type complexes. Fungal Genet. Biol. *45*, S15-S21.

Baldwin, T.K., Urban, M., Brown, N., and Hammond-Kosack, K.E. (2010). A role for topoisomerase I in *Fusarium graminearum* and *F. culmorum* pathogenesis and sporulation. Mol. Plant Microb. Interact. *23*, 566–577.

Beckett, A., (1981). The ascus with an apical pore: development, composition and function. In: Ascomycete systematics. The Luttrellian concept, Reynolds, D.R., ed. (Sprinter-Verlag, New York), pp. 7–26.

Bistis, G.N., and Georgopoulos, S.G. (1979). Some aspects of sexual reproduction in *Nectria haematococca* var. *cucurbitae*. Mycologia *71*, 127–143.

Bistis, G.N., Perkins, D.D., and Read, N.D. (2003). Different cell types in *Neurospora crassa*. Fungal Genet. Newsl. *50*, 17–19.

Bowden, R.L., and Leslie, J.F. (1999). Sexual recombination in *Gibberella zeae*. Phytopathology *89*, 182–188.

Brown, D.W., Butchko, R.A.E., Busman, M., and Proctor, R.H. (2012). Identification of gene clusters associated with fusaric acid, fusarin, and perithecial pigment production in *Fusarium verticillioides*. Fungal Genet. Biol. *49*, 521–532.

Butchko, R.A.E., Brown, D.W., Busman, M., Tudzynski, B., and Wiemann, P. (2012). Lae1 regulates expression of multiple secondary metabolite gene clusters in *Fusarium verticillioides*. Fungal Genet. Biol. *49*, 602–612.

Cavinder, B., Hamam, A., Lew, R.R., and Trail, F. (2011). Mid*1*, a mechanosensitive calcium ion channel, affects growth, development, and ascospore discharge in the filamentous fungus *Gibberella zeae*. Eukaryot. Cell *10*, 832–841.

Cavinder, B., Sikhakolli, U., Fellows, K.M., and Trail, F. (2012). Sexual development and ascospore discharge in *Fusarium* graminearum. J. Vis. Exp. *61*, e3895.

Cavinder, B., and Trail, F. (2012). Role of Fig*1*, a component of the Low-Affinitiy Calcium Uptake System, in growth and sexual development of filamentous fungi. Eukaryot. Cell *11*, 978–988.

Chaisrisook, C., and Leslie, J.F. (1990). A maternally expressed nuclear gene controlling perithecial pigmentation in *Gibberella fujikuroi* (*Fusarium moniliforme*). J. Heredity *81*, 189–192.

Chen, Y., and Zhou, M.-G. (2009). Sexual recombination of carbendazim resistance in *Fusarium graminearum* under field conditions. Pest Manag. Sci. *65*, 398–403.

Coleman, J.J., Rounsley, S.D., Rodriguez-Carres, M., Kuo, A., Wasmann, C.C., Grimwood, J., Schmutz, J., Taga, M., White, G.J., Zhou, S., *et al.* (2009). The genome of *Nectria haematococca:* Contribution of supernumerary chromosomes to gene expression. PloS Genet. *5*, e1000618.

Cuomo, C., Güldener, Xu, U., J.R., Trail, Turgeon, F., B.G., Di Pietro, A., J.D Walton, J.D., Ma, L.-J., Baker, S., Rep, M., *et al.* (2007). The *Fusarium* graminearum genome reveals a link between localized polymorphism and pathogen specialization. Science *317*, 1400–1402.

Deacon, J.W. (2006). Fungal Biology, 4th edn. (Blackwell Publishing, Oxford, UK), p .371.

Debuchy, R., and Turgeon, B.G. (2006). Mating-type structure, evolution, and function in Euascomycetes. In: The Mycota Growth, Differentiation and Sexuality. Kues, I., and Fischer, R. (eds.). (Springer Verlag, Berlin, Heidelberg), pp. 293–323.

DeGroot, P.W.J., Ram, A.F., and Klis, F.M. (2005). Features and functions of covalently linked proteins in fungal cell walls. Fungal Genet. Biol. *42*, 657–675.

Desjardins, A.E. (2003). *Gibberella* from (*Avenaceae*) to (*Zeae*). Annu. Rev. Phytopathol. *41*, 177–198.

Desjardins, A.E., Brown, D.W., Yun, S.-H., Proctor, R.H., Lee, T., Plattner, R.D., Lu, S.-W., and Turgeon, B.G. (2004). Deletion and complementation of the mating type (*MAT*) locus of the wheat head blight pathogen *Gibberella zeae*. Appl. Environ. Microbiol. *70*, 2437–2444.

Desjardins, A.E., Plattner, R.D., Shaner, G., Brown, D.W., Buechley, G., Proctor, R.H., and Turgeon, B.G. (2006). Field release of *Gibberella zeae* genetically modified to lack ascospores. (2006). In: Proceedings of the National Fusarium Head Blight Forum, Canty, S.M., Clark, A., and Van Sanford, D., eds. (University of Kentucky, North Carolina, Lexington), pp 39–44.

Dyer, P.S., Ingram, D.S., and Johnstone, K. (1993). Evidence for the involvement of linoleic acid and other endogenous lipid factors in perithecial development of *Nectria haematococca* mating population VI. Mycol. Res. *4*, 485–496.

Ellis, J.J. (1960). Plasmogamy and ascocarp development in *Gelasinospora calospora*. Mycologia 52, 557–573.

Gaffoor, I., Brown, D.W., Plattner, R., Proctor, R.H., Qi, W., and Trail, F. (2005). Functional analysis of the polyketide synthase genes in the filamentous fungus *Gibberella zeae* (anamorph *Fusarium graminearum*). Eukaryot. Cell. *4*,1926–1933.

Gaffoor, I., and Trail, F. (2006). Characterization of two polyketide synthase genes involved in zearalenone biosynthesis in *Gibberella zeae.* Appl. Environ. Microbiol. 72, 1793–1799.

Gale, L.R., Ward, T.J., Balmas, V., and Kistler, H.C. (2007). Population subdivision of *Fusarium graminearum* sensu stricto in the upper Midwestern United States. Phytopathology 97, 1434–1439.

Gilbert, J., Woods, S.M., and Kromer, U. (2008). Termination of ascospores of *Gibberella zea* after exposure to various levels of relative humidity and temperature. Phytopathology *98*, 504–508.

Goldberg, E.E., Kohn, J.R., Lande, R., Roberson, K.A., Smith, S.A., and Igic, B. (2010). Species selection maintains self-incompatibility. Science *330*, 439–495.

Graziani, S., Vasnier, C., and Daboussi, M.-J. (2004). Novel polyketide synthase from *Nectria haematococca.* Appl. Environ. Microbiol. *70*, 2984–2988.

Gueldener, U., K.-Seong, Y., Boddu, J., Cho, S., Trail, F., J.-Xu, R., Adam, G., H.-Mewes, W., Muehlbauer, G.J., and Kistler, H.C. (2006). Development of a *Fusarium graminearum* Affymetrix GeneChip For profiling fungal gene expression *in vitro* and *in planta.* Fungal Genet. Biol. *43*, 316–325.

Guenther, J., and Trail, F. (2005). The development and differentiation of *Gibberella zeae* (anamorph: *Fusarium graminearum*) during colonization of wheat. Mycologia *97*, 232–240.

Guenther, J.C., Hallen-Adams, H.E., Bucking, H., Shachar-Hill, Y and Trail, F. (2009). Triacylglyceride metabolism by *Fusarium graminearum* during colonization and sexual development on wheat. Mol. Plant Microb. Interact. *22*,1492–1503.

Hallen, H., Huebner, M., Shiu, S.-H., Guldener, U., and Trail, F. (2007). Gene expression shifts during perithecium development in *Gibberella zeae* (anamorph *Fusarium graminearum*), with particular emphasis on ion transport proteins. Fungal Genet. Biol. *44*, 1146–1156.

Hallen, H., and Trail, F.(2008). The L-type calcium ion channel, Cch*1*, affects ascospore discharge and mycelial growth in the filamentous fungus *Gibberella zeae* (anamorph *Fusarium graminearum*). Eukaryot. Cell 7, 415–424.

Hamada, M., Uchida, T., and Tsuda, M. (1988). Ascospore dispersion of the causal agent of Nectria blight of *Piper nigrum*. Ann. Phytopath. Soc. Japan. *54*, 303–308.

Han, Y.K., Kim, M.D., Lee, S.H., Yun, S.H., and Lee, Y.W. (2007). A novel F-box protein involved in sexual development and pathogenesis in *Gibberella zeae.* Mol. Microbiol. *63*, 768–779.

Hanlin, R.T. (1971). Morphology of *Nectria haematococca.* Am. J. Bot. *58*, 105–116.

Heitman, J., Kronstad, J.W., Taylor, J.W., and Casselton, L.A. (eds) (2007). Sex in Fungi: Molecular Determination and evolutionary Implications (ASM Press, Washington, DC), p. 542.

Hong, S.-Y., So, J., Lee, J., Min, K., Son, H., Park, C., Yun, S.-H., and Lee, Y.-W. (2010). Functional analyses of two syntaxin-like SNARE genes, *GzSYN1* and *GzSYN2*, in the ascomycete *Gibberella zeae.* Fungal Genet. Biol. *47*, 364–372.

Hou, Z., Xue, C., Peng, Y., Katan, T., Kistler, H.C., and Xu, J.-R. (2002). A mitogen-activated protein kinase gene (*MGV1*) in *Fusarium graminearum* is required for female fertility, heterokaryon formation, and plant infection. Mol. Plant Microb. Interact. *15*, 1119–1127.

Horn, B.W., Moore, G.G., and Carbone, I. (2009a). Sexual reproduction in *Aspergillus flavus.* Mycologia *101*, 423–429.

Horn, B.W., Ramirez-Prado, J.H., and Carbone, I. (2009b). The sexual state of *Aspergillus parasiticus*. Mycologia *101*, 275–280.

Hoyer, L.L., Fundyga, R., Hecht, J.E., Kapteyn, J.C., Klis, F.M., and Arnold, J. (2001). Characterization of agglutinin-like sequence genes from non-*albicans* Candida and phylogenetic analysis of the ALS family. Genetics *157*, 1555–1567.

Hull, C.M., Boily, M.J., and Heitman, J. (2005). Sex-specific homeodomain proteins Sxi1 alpha and Sxi2a coordinately regulate sexual development in *Cryptococcus neoformans*. Eukaryot. Cell 4,526–535.

Jenczmionka, N.J., Maier, F.J., Lösch, A.P., and Schäfer, W. (2003). Mating, conidiation and pathogenicity of *Fusarium graminearum*, the main causal agent fo the head-blight disease of wheat, are regulated by the MAP kinase gpmk1. Curr. Genet. *43*, 87–95.

Jensen, J.D. (1982). The development of *Gelasinospora reticulospora*. Mycologia 74, 724–737.

Jiang, J., Liu, X., Yin, Y., and Ma, Z. (2011). Involvement of a velvet protein FgVeA in the regulation of asexual development, lipid and secondary metabolism and virulence in *Fusarium graminearum*. PLoS One *6*, e28291. doi:10.1371/journal.pone.0028291.

Jones, A.L., and Aldwinckle, H.S., eds. (1990). Compendium of Apple and Pear Diseases (APS Press, St. Paul, MN), p. 100.

Josefsen, L., Droce, A., Sondergaard, T.E., Sorensen, J.L., Bormann, J., Schafer, W., Giese, H., and Olsson, S. (2012). Autophagy provides nutrients for nonassimilating fungal structures and is necessary for plant colonization but not for infection in the necrotrophic plant pathogen *Fusarium graminearum*. Autophagy *8*, 326–337.

Kerényi, Z., Zeller, K., Hornok, L., and Leslie, J.F. (1999). Molecular standardization of mating type terminology in the *Gibberella fujikuroi* species complex. Appl. Environ. Microbiol. *65*, 4071–4076.

Keszthelyi, A., Jeney, A., Kerényi, Z., Mendes, O., Waalwijk, C., and Hornok, L. (2007). Tagging target genes of the MAT1-1-1 transcription factor in *Fusarium* verticillioides (Gibberella fujikuroi MP-A). Antonie Van Leeuwenhoek *91*, 373–391.

Kim, J.E., Han, K.H., Jin, H.M., Kim, H., Kim, J.C., Yun, S.H., and Lee, Y.W. (2005). Putative polyketide synthase and laccase genes for biosynthesis of aurofusarin in *Gibberella zeae*. Appl. Environ. Microbiol. *71*, 1701–1708.

Kim, J.-E., Lee, H.-J., Lee, J., Kim, K.W., Yun, S.-H., Shim, W.-B., and Lee, Y.W. (2009). *Gibberella zeae* chitin synthase genes, *GzCHS5* and *GzCHS7*, are required for hyphal growth, perithecia formation, and pathogenicity. Curr. Genet. *55*, 449–459.

Kim, H.-K., Lee, T., and Yun, S.-H. (2008). A putative pheromone signaling pathway is dispensable for self-fertility in the homothallic ascomycete *Gibberella zeae*. Fungal Genet. Biol. *45*, 1188–1196.

Kim, H.-K., Lee, T., and Yun, S.-H. (2011). *GzRUM1*, encoding an ortholog of human retinoblastoma Binding Protein 2, is required for ascospore development in *Gibberella zeae*. Plant Pathol. J. *27*, 20–25.

Kim, Y.-T., Lee, Y.-R., Jin, J., Han, K.-H., Kim, H., Kim, J.0C., Lee, T., Yun, S.H., and Lee, Y.-W. (2005). Two different polyketide synthase genes are required for synthesis of zearlenone in *Gibberella zeae*. Mol. Microbiol. *58*, 1102–1113.

Kirk, P.M., Cannon, P.F., David, J.C., and Stalpers, J.A., eds. (2001). Dictionary of the Fungi, 9th edn (CAB International, Oxon, UK).

Klittich, C.J.R., and Leslie, J.F. (1988). Nitrate reduction mutants of *Fusarium moniliforme* (*Gibberella fujikuroi*). Genetics *118*, 417–423.

Lee, J., Chang, I.-Y., Kim, H., Yun, S.-H., Leslie, J.F., and Lee, Y.-W. (2009). Genetic diversity and fitness of *Fusarium graminearum* populations from rice in Korea. Appl. Environ. Microbiol. *75*, 3289–3295.

Lee, J., Lee, T., Lee, Y.-W., Yun, S.-H., and Turgeon, B.G. (2003). Shifting fungal reproductive mode by manipulation of mating type genes: obligatory heterothallism of *Gibberella zeae*. Mol. Microbiol. *50*, 145–152.

Lee, J., Leslie, J.F., and Bowden, R.L. (2008). Expression and function of sex pheromones and receptors in the homothallic ascomycete *Gibberella zeae*. Eukaryot. Cell 7, 1211–1221.

Lee, J., Park, C., Kim, J.-C., Kim, J.-E., and Lee, Y.W. (2010). Identification and functional characterization of genes involved in the sexual reproduction of the ascomycete fungus *Gibberella zeae*. Biochem. Bioph. Res. Co. *401*, 48–52.

Lee, S., Son, H., Lee, J., Min, K., Choi, G.J., Kim, J.-C., and Lee, Y.-W. (2011). Functional Analyses of two acetyl coenzyme A synthetases in the ascomycete *Gibberella zeae*. Eukaryot. Cell *10*, 1043–1052.

Leslie, J.F., Summerell, B.A., and Bullock, S. (2006). The *Fusarium* Laboratory Manual. (Blackwell Publishing, Ames, IA), p. 388.

Lievens, B., Van Baarlen, P., Verreth, C., Van Kerckhove, S., Rep, M., and Thomma, B.P.H.J. (2009). Evolutionary relationships between *Fusarium oxysporum* f. sp. *lycopersici* and *F. oxysporum* f. sp. *radicis-lycopersici* isolates inferred from mating type, elongation factor-1œ and exopolygalacturonase sequences. Mycol. Res. *113*, 1181–1191.

Lin, Y., Son, H., Min, K., Lee, J., Choi, G.J., Kim, J.-C., and Lee, Y.-W. (2012). A putative transcription factor MYT2 regulates perithecium size in the ascomycete *Gibberella zeae*. PLoS One 7, e37859.

Lipke, P.N., and Kurjan, J. (1992). Sexual agglutination in budding yeasts: Structure, function, and regulation of adhesion glycoproteins. Microbiol. Rev. *56*, 180–194.

Liu, W., Xie, S., Zhao, X., Chen, X., Zheng, W., Lu, G., Xu, J.-R., and Wang, Z. (2010). A homeobox gene is essential for conidiogenesis of the rice blast fungus *Magnaporthe oryzae*. Mol. Plant Microb. Interact. *23*, 366–375.

Lord, K.M., and Read, N.D. (2011). Perithecium morphogenesis in *Sordaria macrospora*. Fungal Genet. Biol. *48*, 388–399.

Lysøe, E., Pasquali, M., Breakspear, A., and Kistler, H.C. (2011). The transcription factor FgStuAp influences spore development, pathogenicity, and secondary metabolism in *Fusarium graminearum*. Mol. Plant Microb. Interact. *24*, 54–67.

Ma, L.-J., van der Does, H.C., Borkovich, K.A., Coleman, J.J., Daboussi, M.J., Di Pietro, A., Dufresne, M., Freitag, M., Grabherr, M., Henrissat, B., *et al.* (2010). Comparative genomics reveals mobile pathogenicity chromosomes in *Fusarium*. Nature *464*, 367–373.

Malz S., Grell, M.N., Thrane, C., Maier, F.J., Rosager, P., Felk, A., Albertsen, K.S., Salomon, S., Bohn, L., Schäfer, W., *et al.* (2005). Identification of a gene cluster responsible for the biosynthesis of aurofusarin in the *Fusarium graminearum* species complex. Fungal Genet. Biol. *42*, 420–433.

Martin, T., Lu, S., Tilbeurgh, H., Ripoll, D., Dixelius, C., Turgeon, B.G., and Debuchy, R. (2010). Tracing the origin of the fungal SEX a1 domain places its ancestor in the HMG-box superfamily: implication for fungal mating-type evolution. PLoS One *5*, e15199.

Merhej, J., Urban, M., Dufresne, M., Hammond-Kosack, K.E., Richard-Forget, F., and Barreau, C. (2012). The *velvet* gene, *FgVe1*, affects fungal development and positively regulates trichothecene biosynthesis and pathogenicity in *Fusarium graminearum*. Mol. Plant Pathol. *13*, 363–374.

Meyerowitz, E.M. (2002). Plants compared to animals: The broadest comparative study of development. Science *295*, 1482–1485.

Min, K., Lee, J., Kim, J.-C., Kim, S.G., Kim, Y.H., Vogel, S., Trail, F., and Lee, Y.-W. (2010). novel gene, ROA, is required for proper morphogenesis and discharge of ascospores in *Gibberella zeae*. Eukaryot. Cell *9*, 1495–1503.

Moore-Landecker, E. (1996). Fundamentals of the fungi. Fourth Edition. (Benjamin Cummings Publishing Company, Inc. San Francisco), pp. 574.

Nelson, R.R. (1971). Hormonal involvement in sexual reproduction in the fungi with special reference to F-2, a fungal estrogen. In: Morphological and biochemical events in plant–parasite interaction, Adai, S., and Ouchi, S., eds. (The Phytopathological Society of Japan, Tokyo, Japan), pp. 181–200.

O Donnell, K., Ward, T.J., Geiser, D.M., Kistler, H.C.,and Aoki, T. (2004). Genealogical concordance between the mating type locus and seven other nuclear genes support formal recognition of nine phylogenetically distinct species within the *Fusarium graminearum* clade. Fungal Genet. Biol. *41*, 600–623.

O'Donnell, K., Cigelnik, E., and Nirenberg, H.I. (1998). Molecular systematics and phylogeography of the *Gibberella fujikuroi* species complex. Mycologia *90*, 465–493.

Oide, S., Liu, J., Yun, S.-H., Wu, D., Michev, A., Choi, M.Y., Horwitz, B.A., and Turgeon, B.G. (2010). Histidine kinase two-component response regulator proteins regulate reproductive development, virulence, and stress responses of the fungal cereal pathogens *Cochliobolus heterostrophus* and *Gibberella zeae*. Eukaryot. Cell *9*, 1867–1880.

Park, A.R., Cho, A.-R., Seo, J.-A., Min, K., Son, H., Lee, J., Choi, G.J., Kim, J.-C., and Lee, Y.-W. (2012). Functional analyses of regulators of G protein signaling in *Gibberella zeae*. Fungal Genet. Biol. *49*, 511–520.

Pöggeler, S., Nowrousian, M., and Kück., U. (2006). Fruiting-body development in Ascomycetes. *In:* The Mycota, Growth, differentiation and Sexuality, Kües, U., and Fischer, R., eds. (Springer–Verlag, Berlin, Heidelberg), pp. 325–348.

Proctor, R.H., Butchko, R.A.E., Brown, D.W., and Moretti, A. (2007). Functional characterization, sequence comparisons and distribution of a polyketide synthase gene required for perithecial pigmentation in some *Fusarium* species. Food Addit. Contam. *24*, 1076–1087.

Read, N.D. (1983). A scanning electron microscopic study of the external features of perithecium development in *Sordaria humana*. Can. J. Bot. *61*, 3217–3229.

Read, N.D., (1994). Cellular nature and multicellular morphogenesis of higher fungi. In: Shape and Form in Plants and Fungi, Ingram, D., Hudson, A. eds. (Academic Press, London), pp. 254–271.

Read, N.D., and Beckett, A. (1996). Ascus and ascospore morphogenesis. Mycol. Res. *100*,1281–1314.

Rittenour, W.D., Chen, M., Cahoon, E.B., and Harris, S.D. (2011). Control of glucosylceramide production and morphogenesis by the Bar1 ceramide synthase in *Fusarium graminearum*. PLoS One *6*, e19385.

Schemske, D.W., and Lande, R. (1985). The evolution of self-fertilization and inbreeding depression in plants. II. Empirical observations. Evolution *39*, 41–52.

Schofflemeer, E.A., Klis, F.M., Sietsma, J.H., and Cornelissen, B.J. (1999). The cell wall of *Fusarium oxysporum*. Fungal Genet. Biol. *27*, 275–282.

Shaner, G. (2003). Epidemiology of Fusarium head blight of small grain cereals in North America. In: Fusarium Head Blight of Wheat and Barley, Leonard, K.J., and Bushnell, W.R., eds. (American Phytopathological Society Press, St. Paul, MN), pp. 84–119.

Shim, W.-B., Sagaram, Y.S., Choi, Y.-E, So, J., Wilkinson, H.H., and Lee, Y.-W. (20 06). *FSR1* is essential for virulence and female fertility in *Fusarium verticillioides* and *F. graminearum*. Mol. Plant Microb. Interact. *19*, 725–733.

Shiu, P.K.T., and Glass, N.G. (2000). Cell and nuclear recognition mechanisms mediated by mating type in filamentous ascomycetes. Curr. Opin. Microbiol. *3*, 183–188.

Sikhakolli, U., López-Giráldez, F., Li, N., Common, R., Townsend, J.P., and Trail, F. (2012). Transcriptome analyses during fruiting body formation in *Fusarium graminearum* and *Fusarium verticillioides* reflect species life history and ecology. Fungal Genet. Biol. *49*, 663–673.

Son, H., Min, K., Lee, J., Choi, G.J., Kim, J.-C., and Lee, Y.-W. (2012). Differential roles of pyruvate decarboxylase in aerial and embedded mycelia of the ascomycete *Gibberella zeae*. FEMS Microbiol. Lett. *329*, 123–130.

Son, H., Seo, Y.-S., Min, K., Park, A.R., Lee, J., Jin, J.-M., Lin, Y., Cao, P., Hong, S.-Y., Kim, E.-K., *et al.* (2011) Phenome-based functional analysis of transcription factors in the cereal head bligh fungus, *Fusarium graminearum*. PLoS Pathog. *7*, e1002310.

Spatafora, J.W., Sung, G.-H., Johnson, D., Hesse, C., O.-Rourke, B., Serdani, M., Spotts, R., Lutzoni, G.,

Hofstetter, V., Miadlikowska, J., *et al.* (2006). Five-gene phylogeny of Pezizomycotina. Mycologia *98*, 1018–1028.

Studt, L., Wiemann, P., Kleigrewe, K., Humpf, H.-U., and Tudzynski, B. (2012). Biosynthesis of fusarubins accounts for pigmentation of *Fusarium fujikuroi* perithecia. Appl. Environ. Microbiol. *78*, 4468–4480.

Summerell, B.A., Burgess, L.W., Bullock, S., Backhouse, D., and Nguyen Duc Tri (1998). Occurrence of perithecia of *Gibberella fujikuroi* mating population a (*Fusarium moniliforme*) on maize stubble in Northern Vietnam. Mycologia *90*, 890–895.

Trail, F. (2009). For Blighted Waves of Grain: *Fusarium graminearum* in the post-genomics era. Plant Physiol. *149*, 103–110..

Trail, F. (2007). Fungal cannons: Explosive spore discharge in the Ascomycota. FEMS Microbiol. Lett. *276*,12–18.

Trail, F., and Common, R. (2000). Perithecial development by *Gibberella zeae*: A light microscopy study. Mycologia 92, 130–138.

Trail, F., Gaffoor, I., and Vogel, S. (2005). Ejection mechanics and trajectory of the ascospores of *Gibberella zeae* (anamorph *Fusarium graminearum*). Fungal Genet. Biol. *42*, 528–533.

Trail, F., Xu, H., Loranger, R., and Gadoury, D. (2002). Physiological and environmental aspects of ascospore discharge in *Gibberella zeae*. Mycologia *94*,181–189.

Turgeon, B.G., and Debuchy, R. (2007). *Cochliobolus* and *Podospora*: Mecahnisms of sex determination and the evolution of reproductive lifestyle. In: Sex in Fungi: Molecular determinants and evolutionary implications, Heitman, J., Kronstad, J.W., Taylor, J.W., and Casselton, L.A., eds, (ASM Press, Washington, DC), pp. 93–121.

Urban, M., Mott, E., Farley, T., and Hammond-Kosack, K. (2003). The *Fusarium graminearum MAP1* gene is essential for pathogenicity and development of perithecia. Mol. Plant. Pathol. *4*, 347–359.

Vallejo-Marin, M., and O'Brien, H.E. (2006). Correlated evolution of self-incompatibility and clonal reproduction in *Solanum* (Solanaceae). New Phytol. *173*, 415–421.

VanEtten, H.D. (1978). Identification of additional habitats of *Nectria haematococca* Mating Population VI. Phytopath. *68*,1552–1556.

Wang, C., Zhang, S., Hou, R., Zhao, Z., Zheng, Q., Xu, Q., Zheng, D., Wang, G., Lui, H., Gao, X., *et al.* (2011). Functional analysis of the kinome of the wheat scab fungus *Fusarium graminearum*. PLoS Pathog. *7*, e1002460.

Wang, Y., Liu, W., Hou, Z., Wang, C., Zhou, X., Jonkers, W., Ding, S., Kistler, H.C., and Xu, J.-R. (2011). A novel transcriptional factor important for pathogenesis and ascosporogenesis in *Fusarium graminearum*. Mol. Plant Microb. Interact. *24*, 118–128.

Ward, T.J., Clear, R.M., Rooney, A.P., O'Donnell, K., Gaba, D.O., Patrick, S., Starkey, D.E., Gilbert, J., Geiser, D.M., and Nowicki, T.W. (2008). An adaptive evolutionary shift in Fusarium head blight pathogen populations is driving the rapid spread of more toxigenic *Fusarium graminerarum* in North America. Fungal Genet. Biol. *45*, 473–484.

Wiemann, P., Brown, D.W., Kielgrewe, K., Bok, J.W., Keller, N.P., Humpt, H.-U., and Tudzynski, B. (2010). FfVel1 and FfLae*1*, components of a *velvet*-like complex in *Fusarium fujikuroi*, affect differentiation, secondary metabolism and virulence. Mol. Microbiol. *77*, 972–994.

Wolf, J.C., and Mirocha, C.J. (1973). Regulation of sexual reproduction in *Gibberella zeae* (*Fusarium roseum 'graminearum'*) by F-2 (zearalenone). Can. J. Microbiol. *19*, 725–734.

Yu, H.-Y., Seo, J.-A., Kim, J.-E., Han, K.-H., Shim, W.-B., Yun, S.-H., and Lee, Y.-W. (2008). Functional analyses of heterotrimeric G protein Gœ and Gβ subunits in *Gibberella zeae*. Microbiology *154*, 392–401.

Yun, S.-H., Arie, T., Kaneko, I., Yoder, O.C., and Turgeon, B.G. (2000). Molecualr organization of mating type loci in heterothallic, homothallic and asexual Gibberella/*Fusarium* species. Fungal Genet. Biol. *31*, 7–20.

Zeller, K.A., Bowden, R.L., and Leslie, J.F. (2003). Diversity of epidemic populations of Gibberella zeae from small quadrats in Kansas and North Dakota. Phytopathology 93, 874–880.

Zeller, K.A., Bowden, R.L., and Leslie, J.F. (2004). Population differentiation and recombination in wheat scab populations of Gibberella zeae from the United States. Mol. Ecol. *13*, 563–571.v

Structural Dynamics of *Fusarium* Genomes

3

H. Corby Kistler, Martijn Rep and Li-Jun Ma

Abstract

Fungi in the genus *Fusarium* have a great negative impact on the world economy; yet also hold great potential for answering many fundamental biological questions. The advance of sequencing technologies has made possible the connection between phenotypes and genetic mechanisms underlying the acquisition and diversification of such traits with economic and biological significance. This chapter provides a historical view of our understanding of genomic structural variation among *Fusarium* species. Prior to the genomic era, chromosomal variation was observed between *Fusarium* species and among isolates of *F. oxysporum* and *F. solani* (teleomorph *Nectria haematococca*). Such observations led to the discovery of supernumerary chromosomes in *Nectria haematococca* MPVI and have established their role in fungus–plant interactions. Contemporary comparative genomic studies not only have confirmed the existence of supernumerary chromosomes in the *F. oxysporum* and *F. solani* genomes, but also have provided strong evidence for the horizontal transmission of these chromosomes and their role as genetic determinants of host specific virulence. Overall, knowledge of the highly dynamic *Fusarium* genomes establishes them as eukaryotic models allowing greater understanding of genome plasticity and adaptive evolution to ecological niches.

Introduction

As significant plant pathogens that cause disease on nearly every agriculturally important plant, *Fusarium* species have been the subject of applied and basic biological research related to disease management, mycotoxin production and evolutionary biology. Basic research on *Fusarium solani* (teleomorph *Nectria haematococca* MPVI) led to the pioneering discovery of conditionally dispensable (CD) or supernumerary chromosomes in fungi and demonstrated their functional implication for virulence (Miao *et al.*, 1991a). Naturally, *Fusarium* genome sequencing projects were initiated among a few important fungal species, immediately after the completion of the human genome project and the establishment of the principles of whole-genome shotgun sequencing (Galagan *et al.*, 2005). Utilizing genomic information, the *Fusarium* research community have collectively produced wide-ranging contributions to the understanding of plant pathogenicity, such as the importance of effectors of *F. oxysporum* for virulence and plant immunity (Takken and Rep, 2010) and mycotoxins of *F. graminearum* as pathogenicity determinants and food contaminants (Goswami and Kistler, 2004; Alexander *et al.*, 2009). At the same time, rapid progress in *Fusarium* genomic research has advanced our understanding of the genome dynamics of this group of organisms and established them as evolutionary models to study genome plasticity as it relates to adaptation to ecological niches (Cuomo *et al.*, 2007; Coleman *et al.*, 2009; Ma *et al.*, 2010). In this chapter, we will provide a historic view of our understanding of the structural genome dynamics of this group of organisms, including the development of *Fusarium* genome projects and advances enabled by whole-genome sequencing.

Cytological and electrophoretic karyotypes of *Fusarium* species

Prior to the advent of large-scale genomic sequencing projects, both cytological and electrophoretic approaches were used to determine the chromosomal content and genome size of *Fusarium* species. Early cytological estimates employed non-fluorescent staining of meiotic (pachytene) or mitotic metaphase nuclei. These studies yielded chromosome counts that greatly underestimated the actual chromosome number due to the technical limitations of the methods (reviewed for *F. solani* by Taga *et al.*, 1998).

With the introduction of pulsed field gel electrophoresis in the 1980s (Schwartz and Cantor, 1984), it became possible to separate and visualize large DNA molecules in the size range of fungal chromosomes (100 kb to 6 Mb). Using appropriate molecular markers, the size of each chromosome was inferred and an estimate of total genome size was calculated. These electrophoretic karyotypes (EKs) became a standard method for defining the genomic makeup of many fungi. For fungi with small genomes such as yeasts (Carle and Olson, 1985; Magee and Magee, 1987) or smuts (McCluskey and Mills, 1990; McCluskey *et al.*, 1994), all chromosomes can be fully resolved by electrophoretic karyotyping.

Numerous strains of *Fusarium* have been analysed for EK. Because many *Fusarium* genomes have larger chromosomes and more complex genome structure than yeasts and smuts, this approach has been somewhat less successful. For example, all chromosomes of *F. graminearum* are > 8 Mb (Gale *et al.*, 2005) and thus are too large to be resolved by PFGE. For *F. oxysporum* and *F. solani* genomes, most chromosomes can be separated by PFGE, but many chromosomal bands are compressed at the top of each gel, which makes clear resolution of the larger chromosomes problematic. Additionally, for strains that contain more than one chromosome of similar size, ambiguous 'doublet' or 'triplet' bands may occur and create inaccuracy in genome size and structure prediction (Boehm *et al.*, 1994). Despite these limitations, several important observations have been made by examination of *Fusarium* EKs.

First, there is pervasive chromosomal polymorphism among strains of several *Fusarium* species including *F. oxysporum* (Fig. 3.1) and *F. solani*. Strains of these species often contain chromosome length polymorphisms or even differences in apparent chromosome number (Boehm *et al.*, 1994; Miao *et al.*, 1991b; Migheli *et al.*, 1995; Rosewich *et al.*, 1999; Taga *et al.*, 1999; Temporini and VanEtten, 2002). Size polymorphisms may be so great that homologous chromosomes are difficult to assign based on size alone. Evidence suggests that these polymorphisms may occur due to insertion, deletion, or nonreciprocal translocation (Teunissen *et al.*, 2002) between dispersed repetitive elements (Daviere *et al.*, 2001), or by aneuploidy (Garmaroodi and Taga, 2007) or segmental duplication (Kistler *et al.*, 1995). Despite widespread chromosomal polymorphism, clonally derived strains of *F. oxysporum* vary less than strains from separate clonal lineages of the species complex (Boehm *et al.*, 1994; O'Donnell *et al.*, 1998). In contrast, isolates of some *Fusarium* species, such as *F. verticillioides*, show little polymorphism revealed by electrophoretic karyotyping, yet chromosome polymorphisms clearly distinguish *F. verticillioides* from related species (Xu *et al.*, 1995).

Another source of chromosome polymorphism revealed by EKs is the presence of supernumerary chromosomes (Covert, 1998). First observed as containing the pea pathogenicity determinants in *F. solani* (Miao *et al.*, 1991a; Han *et al.*, 2001), supernumerary chromosomes also encode genes for pathogenicity towards other hosts (Covert *et al.*, 1996) and for rhizosphere

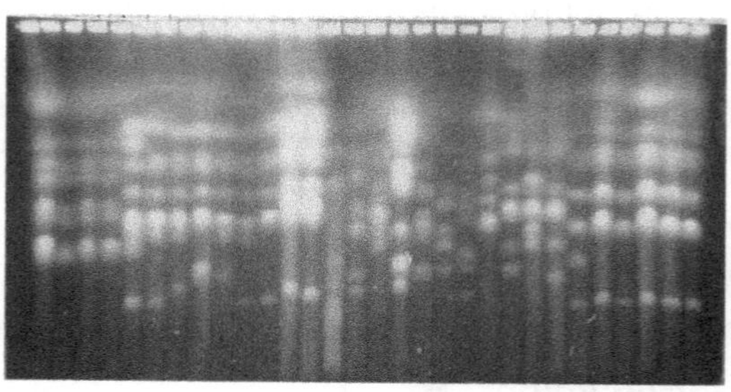

Figure 3.1 Electrophoretic karyotype variation among 30 strains of *F. oxysporum*.

niche adaptation (Rodriguez-Carres *et al.*, 2008). These supernumerary chromosomes may be lost during meiosis (Miao *et al.*, 1991a), concomitant with parasexual recombination (Teunissen *et al.*, 2002) or upon genetic transformation (Wasmann and VanEtten, 1996). Loss of supernumerary chromosomes can be induced by treatment with the benzimidazole fungicide benomyl, that inhibits microtubule formation (VanEtten *et al.*, 1998), and supernumerary chromosomes have been targeted for site-directed chromosome breakage (Kistler *et al.*, 1996). As the presence or absence of supernumerary chromosomes does not affect fitness of the organism in complete medium, they are also referred to as conditionally dispensable chromosomes. It is unclear if supernumerary chromosomes are more prone to loss or alteration compared to conserved chromosomes, or if changes in supernumerary chromosomes can be fully explained by their dispensable content.

Microscopic visualization of *Fusarium* mitotic chromosomes using fluorescent probes and the 'germ tube burst method' (Taga and Murata, 1994) has improved accuracy of chromosome counts and revealed new insights into whole genome arrangements. Fluorescence *in situ* hybridization probes directed towards DNA sequences isolated from supernumerary chromosomes have shown the unique content of these chromosomes compared to the rest of the genome (Taga *et al.*, 1999). This same technique has been used to distinguish monosomic from polysomic supernumerary chromosomes (Garmaroodi and Taga, 2007) and has confirmed chromosome number counts based on genetic and physical maps (Gale *et al.*, 2005).

Microscopic visualization of restriction-enzyme digested and fluorescently labelled chromosomes from yeast showed the promise for creating large-scale, ordered restriction maps from whole genomes (Schwartz *et al.*, 1993). This single-molecule based technique, known as optical mapping, first was used to create *de novo* restriction maps from bacterial genomes (Lin *et al.*, 1999; Lim *et al.*, 2001) and later was successfully applied to other eukaryotic organisms (Lai *et al.*, 1999) including fungi. Electronically generated restriction maps obtained from DNA sequence assemblies can be matched to optical maps to create a consolidated physical map of DNA scaffolds, in many cases corresponding to whole chromosomes. Optical maps currently have been created for the genomes of *F. solani* (Coleman *et al.*, 2009) and several strains of *F. oxysporum* (L.J. Ma, 2010, unpublished).

Fusarium genome sequences

Genome sequencing provides the ultimate solution to understand genome structure, as it reveals the primary DNA sequence. The first sequenced *Fusarium* genome was of *F. graminearum* (Cuomo *et al.*, 2007), funded in 2002 by the National Research Initiative within the US Department of Agriculture's Cooperative State Research Education and Extension Service (now known as the National Institute of Food and Agriculture) through their Microbial Genome Sequencing Project. Continuing this effort, a *Fusarium* comparative project targeted two additional *Fusarium* species, *F. oxysporum f.* sp. *lycopersici* and *F. verticillioides*, and was funded through the same agency (Ma *et al.*, 2010). At the same time, a fourth *Fusarium* genome, that of *Fusarium solani* f. sp. *pisi* was sequenced under the auspices of the US Department of Energy's Office of Science, Biological and Environmental Research Program (Coleman *et al.*, 2009).

Efforts to define the genome structure of *F. graminearum* and *F. verticillioides* were straightforward, benefiting from the existing genetic maps of both genomes (Xu and Leslie, 1996; Gale *et al.*, 2005) and the fact that these two genomes contain minimal repeat sequences (Cuomo *et al.*, 2007; Ma *et al.*, 2010). In contrast, no meiotic linkage map could be created for the asexual *F. oxysporum* genome, and so therefore an optical map for the sequenced strain was constructed. For *F. solani*, an optical map was also generated to assist the assembly process. The first surprise revealed by comparing genome maps was that the four sequenced *Fusarium* genomes differ significantly in genome size, ranging from 36 Mb of *F. graminearum* to >60 Mb of *F. oxysporum* and in number of chromosomes, from 4 to 17 (Table 3.1).

Table 3.1 Sequenced *Fusarium* genomes*

Species	Strain	Length (Mb)	Chromosomes	Repeats (%)	Genes
F. graminearum	PH-1	36	4	0.67	13,332
F. verticillioides	7600	42	11	1.76	14,179
F. oxysporum	4287	60	15	27.4	17,735
F. solani	77-13-4	51	17	15.1	15,707

*The genomic data for *F. graminearum*, *F. verticillioides* and *F. oxysporum* can be accessed at: http://www.broadinstitute.org/annotation/genome/*Fusarium*_group/MultiHome.html. The genomic data for *F. solani* can be accessed at: http://genome.jgi-psf.org/Necha2/Necha2.home.html

Some *Fusarium* species contain unique supernumerary chromosomes

The assembly of the *F. solani* genome confirmed the presence of supernumerary chromosomes reported previously (Miao *et al.*, 1991a; Coleman *et al.*, 2009). Compared with the rest of the genome, the supernumerary chromosomes of *F. solani* are more repetitive, lower in G+C content, and enriched for duplicated genes and genes without homologues in other *Fusarium* genomes (Coleman *et al.*, 2009). A comparative study between the genome sequences of *F. graminearum*, *F. verticillioides* and *F. oxysporum* also revealed the presence of the supernumerary chromosomes in *F. oxysporum* (Ma *et al.*, 2010). As these supernumerary chromosomes are only observed in the *F. oxysporum* lineage, they are also referred to as lineage-specific (LS) chromosomes. The sequenced *F. oxysporum* f. sp. *lycopersici* genome contains four LS chromosomes when compared to the other two *Fusarium*, as well as lineage-specific extensions to its two largest chromosomes. These LS regions together make up more than one-quarter of the genome and account for most of the difference in genome length.

F. verticillioides also may have a supernumerary chromosome. The smallest, 600 kb chromosome as defined by the *F. verticillioides* genetic map (Xu and Leslie, 1996), was not identified in the genome assembly. One simple explanation could be that there are insufficient marker sequences to link contigs to this chromosome. However, this mini-chromosome was sometimes observed to be dispensable during culturing in the lab, a characteristic similar to supernumerary chromosomes described in *F. solani* (L.R. Xu, personal communication). It is unclear whether this mini-chromosome was actually present in the sequenced strain and, if so, whether the sequence of the smallest chromosome was captured in the current assembly. At present, much of the genome assembly is anchored to the 11 chromosomes of the genetic map with the exception of 10 scaffolds that, perhaps by coincidence, total 680 kb in size.

Pathogenicity chromosomes

The genetic determinants of host specific virulence for both *F. solani* and *F. oxysporum* are localized on supernumerary chromosomes. The 1.6 Mb chromosome 14 of *F. solani* contains the pea pathogenicity (PEP) locus, which encodes several genes that influence lesion size on epicotyls of the host *Pisum sativum* (Han *et al.*, 2001). This chromosome also carries genes for utilization of homoserine, an amino acid particularly enriched in the pea rhizosphere. A strain of *F. solani* containing the homoserine utilization (HUT) locus on chromosome 14 was more competitive in the pea rhizosphere than a HUT^- mutant (Rodriguez-Carres *et al.*, 2008). Chromosome 14 thus is involved in several aspects of niche adaption that may confer fitness in association with a particular host species. A chromosome in *F. oxysporum*, coincidentally also named chromosome 14, also contains genes that define the pathogenic relation with its host species, tomato. Most genes for small proteins secreted into the xylem (*SIX* genes) are located on chromosome 14 (Ma *et al.*, 2010) which include *SIX3* (*AVR2*) and *SIX1* (*AVR3*) that encode effectors that induce resistance to tomato wilt conferred by tomato genes *I-2* and *I-3*, respectively (Takken and Rep, 2010; Ma *et al.*, 2010).

In addition to the known virulence factors encoded on pathogenicity chromosomes in *F. solani* and *F. oxysporum*, these chromosomes also are enriched for repetitive sequences including full length, potentially active transposable elements and are deficient in genes for housekeeping functions (Coleman *et al.*, 2009; Ma *et al.*, 2010). Secreted effectors and transcription factors also are enriched on *F. oxysporum* pathogenicity chromosomes (Table 3.2) (Ma *et al.*, 2010).

Repeat content of supernumerary chromosomes

Unlike most sequenced fungal genomes, genomes of *F. graminearum* and *F. verticillioides* have very low repeat content with less than 2% of their genome as repetitive sequence. In contrast, both *F. solani* and *F. oxysporum* have much higher levels of repeats, calculated to be 15% and 28% of their overall genome, respectively (Coleman *et al.*, 2009; Ma *et al.*, 2010, Fig. 3.2). The repeats in *F. oxysporum* genome are highly similar, as a majority of the repeat families have >95% sequence identity (Fig. 3.2). Much of the highly identical repeated

Table 3.2 Predicted functions of genes assigned to *F. oxysporum* f. sp. *lycopersici* chromosome 14 (manual inspection)

Predicted functions	Number of genes/elements
DNA/RNA	19
Transcription factors	9
Chromatin-related functions	4
Nucleic acid-related functions	6
Nutrition/protection/secondary metabolism	47
Carbohydrate-active enzymes	7 (of which 2 secreted)
Proteases	5 (of which 1 secreted)
Cytochrome P450s	6
Secreted enzymes (other)	6
Intracellular enzymes (other)/metabolism	14
Methyl transferases	6
Transporters	3
Secreted proteins other than enzymes	20
Small secreted proteins (<300 residues)	16 (of which 9 SIX genes)
Larger secreted proteins with unknown function	4
Other functions	17
Protein kinases	2
Haemolysin-like	2
Vegetative incompatibility-like	4
Other functions (e.g. GTPase, cyclins)	9
No function assigned	142
Related sequences in other fungi	69
Homologues only in Fol genome	9
No similarities	64
Transposable elements	>243
Annotated transposable elements	47
Unannotated transposable elements	>196

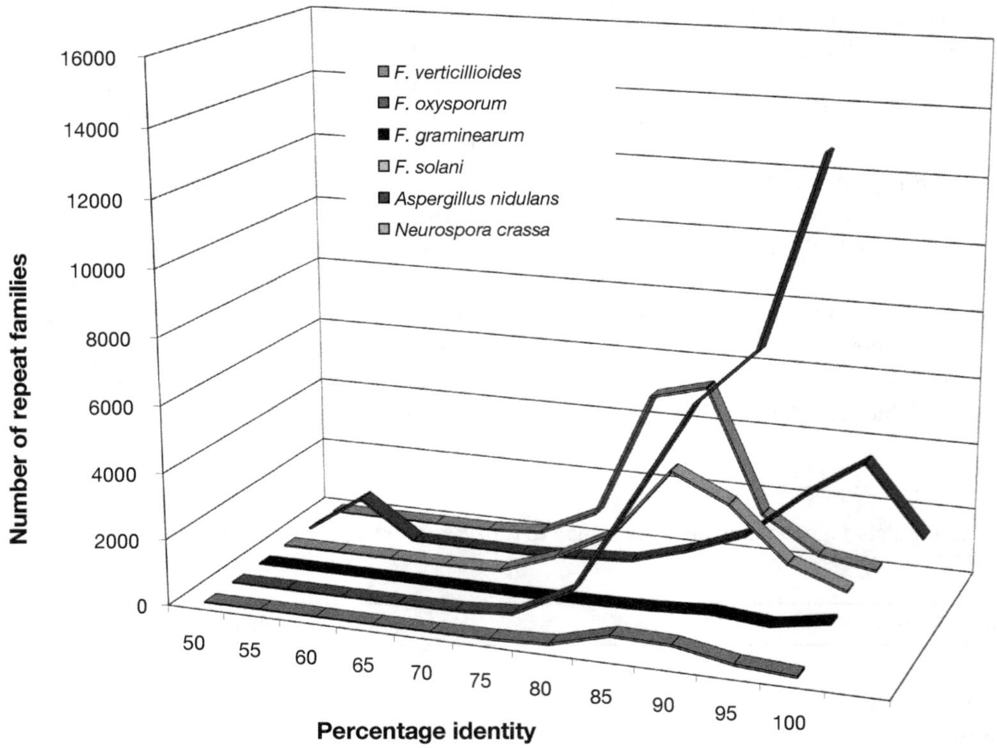

Figure 3.2 Degree of sequence identity for repeated sequences within six fungal species. A colour version of this figure is located in the plate section at the back of the book.

sequence in *F. oxysporum* is due to several large scale inter- and intra-chromosomal segmental duplications that share ~99% sequence identity. The repetitive sequences in the *F. oxysporum* genome also include highly identical transposable elements. Overall, transposable elements (TEs) in *F. oxysporum* account for approximately 4% of the genome and mostly are restricted to LS regions, which contain 95% of the DNA transposons (Ma *et al.*, 2010). Supernumerary chromosomes of *F. solani* also are enriched in transposons with chromosomes 14 and 15 having the majority of mapped DNA transposons (Coleman *et al.*, 2009). The supernumerary chromosomes of *F. solani* also are enriched in duplicated sequences but no large-scale regions of segmental duplication are found. Most repeats in *F. solani* do not have a high degree of identity (Fig. 3.2), a fact that has been attributed to rapid divergence of its repeated sequences by way of a fungal pre-meiotic mechanism for repeat-induced point mutation (RIP) in *F. solani* (Coleman *et al.*, 2009). Evidence for RIP among highly repeated sequences is not found in *F. oxysporum*, likely due to its lack of meiosis and sexual reproduction (Ma *et al.*, 2010). However, the laboratory strain of *F. solani* that was chosen for sequencing was derived from ascospores of several generations of genetic crosses in the laboratory (Kistler and VanEtten, 1984). This particular species of *Fusarium* in fact is almost never found to reproduce sexually in nature (VanEtten and Kistler, 1988) and so it would be interesting to see if a strain freshly isolated from the field would exhibit higher levels of repetitive sequence identity than the sequenced laboratory strain. Both *F. graminearum* and *F. verticillioides* have active RIP mechanisms (Clutterbuck, 2011; Cuomo *et al.*, 2007; Ma *et al.*, 2010) and routinely reproduce sexually in nature; we speculate that this could be responsible for each genome having a very low level of repeated sequence (Fig. 3.2). Recent work (Van de Wouw *et al.*, 2010) suggests

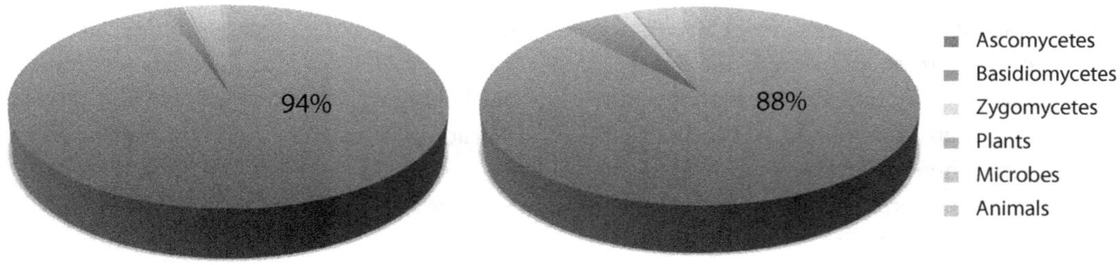

Figure 3.3 Homologous profile of *F. oxysporum* genes encoded in the conserved versus LS regions comparing to proteomes across different kingdoms. The *Fol* proteins were searched using BLASTP (1e-20) against the NCBI metazoan, plant, microbial gene sets available at ftp://ftp.ncbi.nlm.nih.gov/gene/DATA/GENE_INFO (21 February 2008 version) and the non-*Fusarium* fungal database including the protein sets from Ascomycete: two fungal genomes from each subphylum Sordariomycetes (*Magnaporthe grisea*, *Neurospora crassa*), subphylum Leotiomycetes (*Botrytis cinerea*, *Sclerotinia sclerotiorum*), and Eurotiomycetes (*Aspergillus fumigatus*, *A. oryzae*); Basidiomycete: *Ustilago maydis*, *Coprinus cinereus*, *Cryptococcus neoformans* serotype A; and a zygomycete fungal *Rhizopus oryzae* protein set. A colour version of this figure is located in the plate section at the back of the book.

that proximity of fungal effector genes to repetitive elements such as transposons may facilitate their mutation via RIP and hasten their adaptation to new host genotypes in response to selective pressure. Thus the transposon rich environment of *Fusarium* supernumerary chromosomes may be conducive to genome plasticity required for nimble genetic changes associated with plant – fungal interactions.

Horizontal transfer of supernumerary chromosomes

Several lines of evidence suggest that supernumerary chromosomes of both *F. oxysporum* and *F. solani* originated by lateral transfer from a foreign fungal source (Coleman *et al.*, 2009; Ma *et al.*, 2010). Besides being enriched in mobile elements such as transposons, these chromosomes contain largely unique genes, found only in the source strain and have characteristics such as G+C content or codon usage significantly divergent from the core genome. A comparative study of the genes on *F. oxysporum* supernumerary chromosomes that match sequences in the GenBank NR database revealed that a significantly smaller proportion of these genes have homologous sequence matches (67% compared to 91% in the core region of the genome). However, the genes from these two partitions of the genome still share a similar distribution profile of homologues among kingdoms of life (Fig. 3.3) suggesting that the source of the transfer in the *F. oxysporum* lineage is fungal. Phylogenetic analysis using all the genes (362) in LS regions that share homologues in seven selected ascomycete genomes, including the four *Fusarium* genomes, *M. grisea, N. crassa* and *A. nidulans,* indicates a distinct evolutionary origin of these genes, near the root of the *Fusarium* genus (Ma *et al.*, 2010). A comparative study using both Illumina sequence of *F. oxysporum* strain Fo5176, a pathogen of *Arabidopsis* (Thatcher *et al.*, 2009), and EST sequences from *F. oxysporum* f. sp. *vasinfectum* (Dowd *et al.*, 2004), a pathogen of cotton, illustrated that the *F. oxysporum* LS regions also differ considerably in sequence between strains within the *F. oxysporum* species complex (Ma *et al.*, 2010). It is clear that these supernumerary chromosomes were not vertically transmitted from the last common ancestor of the *Fusarium* genus, but instead were likely horizontally acquired. The high diversity of LS sequences among different lineages suggests this may be an on-going process.

Experiments conducted at the University of Amsterdam demonstrated the transfer of entire LS chromosomes through simple co-incubation between two otherwise genetically isolated strains of *F. oxysporum* (Ma *et al.*, 2010). Thus, in a single event, an entire suite of genes required for host compatibility can be transferred to a different genetic lineage. If the recipient lineage had an environmental adaptation different from the donor, transfer could increase the overall incidence of disease in the host by introducing pathogenicity in a genetic background pre-adapted and more fit in a local environment. Alternatively, (re)combination with already present LS regions may lead to emergence of strains with new characteristics, such as pathogenicity towards new hosts.

Partitioning of diversity in genomes without supernumerary chromosomes

While the *F. graminearum* genome has few repetitive elements and no supernumerary chromosomes, it contains regions with elevated polymorphism (SNP) density, as well as increased rates of recombination (Gale *et al.*, 2005; Cuomo *et al.*, 2007). These high diversity regions are found not only at all subtelomeric locations but also at broad interstitial chromosomal sites. *F. graminearum* and relatives are peculiar in that they contain four large chromosomes instead of 11 or more smaller ones found in other *Fusarium* species (e.g. Boehm *et al.*, 1994; Xu *et al.*, 1995; Taga *et al.*, 1998). Comparison of *F. graminearum* with *F. verticillioides* and *F. oxysporum* indicates that the highly variable regions of *F. graminearum* are at locations where chromosomes may have fused end-to-end in an ancestor of the *F. graminearum* clade (Fig. 3.4). Maintenance of high diversity at these formerly subtelomeric regions of the genome is a mystery. Interestingly, these high-diversity regions are enriched for *F. graminearum*-specific sequences, genes for predicted secreted proteins and genes specifically expressed *in planta*. Therefore, while this fungus has no supernumerary chromosomes, the genes that likely impact the plant–fungus interaction are nonetheless found at relatively dynamic locations of the fungal genome.

Conclusions

Overall, comparative genomic studies suggest that the ancestral form of the *Fusarium* clade could be a heterothallic species with ~11 core chromosomes. Certain lineages including the *F. graminearum*

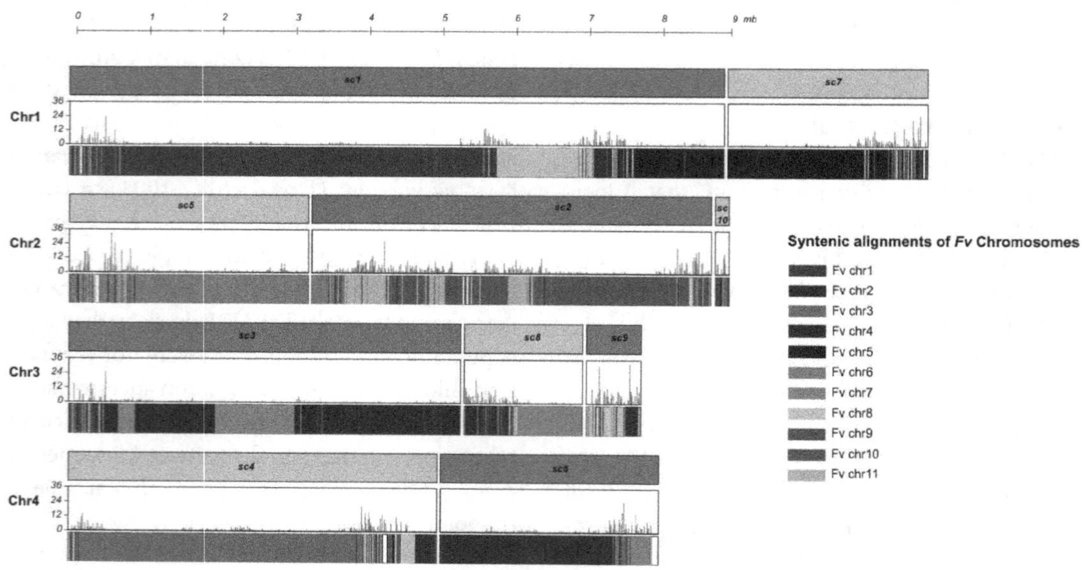

Figure 3.4 Whole genome alignments of *F. graminearum* with *F. verticillioides* illustrating chromosomal fusion events in FG. The alignments display end-to-end synteny in large blocks with the exception of *F. verticillioides* chromosome ends and reveal multiple chromosome fusions in *F. graminearum*. The previously described highly polymorphic and recombinogenic regions of *F. graminearum* (Cuomo *et al.*, 2007) correspond to *F. verticillioides* chromosome ends, including the implied interstitial fusion sites. A colour version of this figure is located in the plate section at the back of the book.

species complex evolved a more compact genomic structure through chromosomal fusions and rigorous control of the proliferation of repetitive sequences. In this lineage, a homothallic sexual reproduction lifestyle emerged in *F. graminearum* and related species, perhaps to satisfy a requirement for ascospores to initiate the life cycle of the fungus by infection of its host (Goswami and Kistler, 2004). At the other end of the spectrum, *F. oxysporum* and possibly *F. solani* developed mechanisms that enable the acquisition of novel genetic material by horizontal transfer that provides some advantage for the organism, such as the ability to broaden the range of host plants it may infect. Horizontal transfer may also allow for non-Mendelian genetic recombination that could compensate for the accumulation of deleterious mutations in these organisms in the absence of sexual recombination. More likely, we speculate, mobile supernumerary chromosomes may function as selfish elements that serve as lifeboats for gene clusters with joint adaptive function that may escape from an asexual genetic lineage destined for extinction by Muller's Ratchet (Andersson and Hughes, 1996), to a more fit genetic lineage.

There are probably both costs and benefits to maintaining a highly flexible genome such as in *F. oxysporum*. With flexibility may come instability, which may increase the chance of organism malfunction – genomic destabilization is, after all, a hallmark feature of nearly all cancer cells (Anderson *et al.*, 2001). On the other hand, genetic plasticity achieved through genomic flexibility can play an important role in adaptation to diverse environments and challenges. For instance, uneven evolutionary rates were observed within the genome of strains of the Irish potato famine pathogen *Phytophthora infestans*, with genes in repeat-rich regions showing higher rates of structural polymorphisms and positive selection. These loci are enriched in genes induced *in planta*, connecting host adaptation to genome flexibility (Raffaele *et al.*, 2010; Haas *et al.*, 2009).

Clearly each organism needs to strike a balance between genome stability and genome flexibility. Interestingly, the *Fusarium* genomes represent different 'solutions', with *F. graminearum* and *F. verticillioides* showing characteristics of relative genome 'streamlining', while *F. oxysporum* and *F. solani* are quite flexible in chromosome content and arrangement of genetic material. Collectively, therefore, these genomes have the potential to enhance our understanding of the relation between genome dynamics and evolution. Specifically, horizontal gene or chromosome transfer (HGT or HCT) spreads genetic diversity by instantly introducing new genes into existing genomes, accelerating genome innovation as an alternative to gene duplication, whole genome duplication or hybridization between related species.

Since completion of the reference human genome sequence (Lander, 2011), great advancement in transcriptomics, epigenomics, proteomics and network modelling have already illustrated the power of the 'post genomic era.' Applying the full range of these technologies to investigate the *Fusarium* genus will generate exciting novel insight, for the benefit of agriculture as well as our understanding of evolutionary processes acting on genomes.

Acknowledgements

This chapter is dedicated to Hans VanEtten, a pioneer of *Fusarium* biology and genetics and an inspirational mentor and colleague. H.C.K. and L.J.M. are grateful for the support of United States Department of Agriculture, National Institute of Food and Agriculture Grant awards 2008-35604-18800 and 2008-35600-04691.

References

Alexander, N.J., Proctor, R.H., and McCormick, S.P. (2009). Genes, gene clusters, and biosynthesis of trichothecenes and fumonisins in *Fusarium*. Toxin Rev. *28*, 198–215.

Anderson, G.R., Brenner, B.M., Swede, H., Chen, N., Henry, W.M., Conroy, J.M., Karpenko, M.J., Issa, J.P., Bartos, J.D., Brunelle, J., *et al.* (2001). Intrachromosomal genomic instability in human sporadic colorectal cancer measured by genome-wide allelotyping and inter-(simple sequence repeat) PCR. Cancer Res. *61*, 8274–8283.

Andersson, D.I., and Hughes, D. (1996). Muller's ratchet decreases fitness of a DNA-based microbe. Proc. Natl. Acad. Sci. U.S.A. *93*, 906–907.

Bartlett, J.M., and Stirling, D. (2003). A short history of the polymerase chain reaction. Methods Mol. Biol. *226*, 3–6.

Boehm, E.W.A., Ploetz, R.C., and Kistler, H.C. (1994). Statistical-analysis of electrophoretic karyotype variation among vegetative compatibility groups of

Fusarium oxysporum f. sp. *cubense*. Mol. Plant Microbe Interact. *7*, 196–207.

Carle, G.F., and Olson, M.V. (1985). An electrophoretic karyotype for yeast. Proc. Natl. Acad. Sci. U.S.A. *82*, 3756–3760.

Clutterbuck, A.J. (2011). Genomic evidence of repeat-induced point mutation (RIP) in filamentous ascomycetes. Fungal Genet. Biol. *48*, 306–326.

Coleman, J.J., Rounsley, S.D., Rodriguez-Carres, M., Kuo, A., Wasmann, C.C., Grimwood, J., Schmutz, J., Taga, M., White, G.J., Zhou, S., *et al.* (2009). The genome of *Nectria haematococca*: Contribution of supernumerary chromosomes to gene expansion. PLoS Genet. *5*, e1000618.

Covert, S.F. (1998). Supernumerary chromosomes in filamentous fungi. Curr. Genet. *33*, 311–319.

Covert, S.F., Enkerli, J., Miao, V.P.W., and VanEtten, H.D. (1996). A gene for maackiain detoxification from a dispensable chromosome of *Nectria haematococca*. Mol. Gen. Genet. *251*, 397–406.

Cuomo, C.A., Güldener, U., Xu, J.R., Trail, F., Turgeon, B.G., Di Pietro, A., Walton, J.D., Ma, L.J., Baker, S.E., Rep, M., *et al.* (2007). The *Fusarium graminearum* genome reveals a link between localized polymorphism and pathogen specialization. Science *317*, 1400–1402.

Daviere, J.M., Langin, T., and Daboussi, M.J. (2001). Potential role of transposable elements in the rapid reorganization of the *Fusarium oxysporum* genome. Fungal Genet. Biol. *34*, 177–192.

Dowd, C., Wilson, L.W., and McFadden, H. (2004). Gene expression profile changes in cotton root and hypocotyl tissues in response to infection with *Fusarium oxysporum* f. sp *vasinfectum*. Mol. Plant Microbe Interact. *17*, 654–667.

Galagan, J.E., Henn, M.R., Ma, L.J., Cuomo, C.A., and Birren, B. (2005). Genomics of the fungal kingdom: insights into eukaryotic biology. Genome Res. *15*, 1620–1631.

Gale, L.R., Bryant, J.D., Calvo, S., Giese, H., Katan, T., O'Donnell, K., Suga, H., Taga, M., Usgaard, T.R., Ward, T.J., *et al.* (2005). Chromosome complement of the fungal plant pathogen *Fusarium graminearum* based on genetic and physical mapping and cytological observations. Genetics *171*, 985–1001.

Garmaroodi, H.S., and Taga, M. (2007). Duplication of a conditionally dispensable chromosome carrying pea pathogenicity (PEP) gene clusters in *Nectria haematococca*. Mol. Plant Microbe Interact. *20*, 1495–1504.

Goswami, R.S., and Kistler, H.C. (2004). Heading for disaster: *Fusarium graminearum* on cereal crops. Mol. Plant Pathol. 5.

Haas, B.J., Kamoun, S., Zody, M.C., Jiang, R.H.Y., Handsaker, R.E., Cano, L.M., Grabherr, M., Kodira, C.D., Raffaele, S., Torto-Atalibo, T., *et al.* (2009). Genome sequence and analysis of the Irish potato famine pathogen *Phytophthora infestans*. Nature *461*, 393–398.

Han, Y., Liu, X.G., Benny, U., Kistler, H.C., and VanEtten, H.D. (2001). Genes determining pathogenicity to pea are clustered on a supernumerary chromosome in the fungal plant pathogen *Nectria haematococca*. Plant J. *25*, 305–314.

Kistler, H.C., Benny, U., Boehm, E.W.A., and Katan, T. (1995). Genetic duplication in *Fusarium oxysporum*. Curr. Genet. *28*, 173–176.

Kistler, H.C., Meinhardt, L.W., and Benny, U. (1996). Mutants of *Nectria haematococca* created by a site-directed chromosome breakage are greatly reduced in virulence toward pea. Mol. Plant Microbe Interact. *9*, 804–809.

Kistler, H.C., and VanEtten, H.D. (1984). Three non-allelic genes for pisatin demethylation in the fungus Nectria haematococca. J. Gen. Microbiol. *130*, 2595–2603.

Lai, Z., Jing, J., Aston, C., Clarke, V., Apodaca, J., Dimalanta, E.T., Carucci, D.J., Gardner, M.J., Mishra, B., Anantharaman, T.S., *et al.* (1999). A shotgun optical map of the entire *Plasmodium falciparum* genome. Nat. Genet. *23*, 309–313.

Lander, E.S. (2011). Initial impact of the sequencing of the human genome. Nature *470*, 187–197.

Lim, A., Dimalanta, E.T., Potamousis, K.D., Yen, G., Apodoca, J., Tao, C., Lin, J., Qi, R., Skiadas, J., Ramanathan, A., *et al.* (2001). Shotgun optical maps of the whole *Escherichia coli* O157,H7 genome. Genome Res. *11*, 1584–1593.

Lin, J., Qi, R., Aston, C., Jing, J., Anantharaman, T.S., Mishra, B., White, O., Daly, M.J., Minton, K.W., Venter, J.C., *et al.* (1999). Whole-genome shotgun optical mapping of *Deinococcus radiodurans*. Science *285*, 1558–1562.

Ma, L.J., van der Does, H.C., Borkovich, K.A., Coleman, J.J., Daboussi, M.J., Di Pietro, A., Dufresne, M., Fretag, M., Grabherr, M., Henrissat, B., *et al.* (2010). Comparative genomics reveals mobile pathogenicity chromosomes in *Fusarium*. Nature *464*, 367–373.

Magee, B.B., and Magee, P.T. (1987). Electrophoretic karyotypes and chromosome numbers in *Candida* species. J. Gen. Microbiol. *133*, 425–430.

Maxam, A.M., and Gilbert, W. (1977). New method for sequencing DNA. Proc. Natl. Acad. Sci. U.S.A. *74*, 560–564.

McCluskey, K., Agnan, J., and Mills, D. (1994). Characterization of genome plasticity in *Ustilago hordei*. Curr. Genet. *26*, 486–493.

McCluskey, K., and Mills, D. (1990). Identification and characterization of chromosome length polymorphisms among strains representing 14 races of *Ustilago hordei*. Mol. Plant Microbe Interact. *3*, 366–373.

Mendel, G. (1866). Experiments in plant hybridization. J. Royal Horticult. Soc. 26, 1–32.

Miao, V.P., Covert, S.F., and VanEtten, H.D. (1991a). A fungal gene for antibiotic-resistance on a dispensable (B) chromosome. *Science 254*, 1773–1776.

Miao, V.P.W., Matthews, D.E., and VanEtten, H.D. (1991b). Identification and chromosomal locations of a family of cytochrome P-450 genes for pisatin detoxification in the fungus *Nectria haematococca*. Mol. Gen. Genet. *226*, 214–223.

Migheli, Q., Berio, T., Gullino, M.L., and Garibaldi, A. (1995). Electrophoretic karyotype variation among pathotypes of *Fusarium oxysporum* f.sp. *dianthi*. Plant Pathol. *44*, 308–315.

O'Donnell, K., Kistler, H.C., Cigelnik, E., and Ploetz, R.C. (1998). Multiple evolutionary origins of the fungus causing Panama disease of banana: Concordant evidence from nuclear and mitochondrial gene genealogies. Proc. Natl. Acad. Sci. US95, A., 2044–2049.

Raffaele, S., Farrer, R.A., Cano, L.M., Studholme, D.J., MacLean, D., Thines, M. *et al.* (2010). Genome evolution following host jumps in the Irish potato famine pathogen lineage. Science *330*, 1540–1543.

Rodriguez-Carres, A., White, G., Tsuchiya, D., Taga, M., and VanEtten, H.D. (2008). The supernumerary chromosome of *Nectria haematococca* that carries pea-pathogenicity-related genes also carries a trait for pea rhizosphere competitiveness. Appl. Environ. Microbiol. *74*, 3849–3856.

Rosewich, U.L., Pettway, R.E., Katan, T., and Kistler, H.C. (1999). Population genetic analysis corroborates dispersal of *Fusarium oxysporum* f. sp *radicis-lycopersici* from Florida to Europe. Phytopathology *89*, 623–630.

Sanger, F., Air, G.M., Barrell, B.G., Brown, N.L., Coulson, A.R., Fiddes, J.C. *et al.* (1977). Nucleotide-sequence of bacteriophage phi X174 DNA. Nature *265*, 687–695.

Schwartz, D.C., and Cantor, C.R. (1984). Separation of yeast chromosome-sized DNAs by pulsed field gradient gel-electrophoresis. Cell *37*, 67–75.

Schwartz, D.C., Li, X., Hernandez, L.I., Ramnarain, S.P., Huff, E.J., and Wang, Y.K. (1993). Ordered restriction maps of *Saccharomyces cerevisiae* chromosomes constructed by optical mapping. Science *262*, 110–114.

Taga, M., and Murata, M. (1994). Visualization of mitotic chromosomes in filamentous fungi by fluorescence staining and fluorescence *in situ* hybridization. Chromosoma *103*, 408–413.

Taga, M., Murata, M., and Saito, H. (1998). Comparison of different karyotyping methods in filamentous ascomycetes – a case study of *Nectria haematococca*. Mycol. Res. *102*, 1355–1364.

Taga, M., Murata, M., and VanEtten, H.D. (1999). Visualization of a conditionally dispensable chromosome in the filamentous ascomycete *Nectria haematococca* by fluorescence *in situ* hybridization. Fungal Genet. Biol. *26*, 169–177.

Takken, F., and Rep, M. (2010). The arms race between tomato and *Fusarium oxysporum*. Mol. Plant Pathol. *11*, 309–314.

Temporini, E.D., and VanEtten, H.D. (2002). Distribution of the pea pathogenicity (*PEP*) genes in the fungus *Nectria haematococca* mating population VI. Curr. Genet. *41*, 107–114.

Teunissen, H.A.S., Verkooijen, J., Cornelissen, B.J.C., and Haring, M.A. (2002). Genetic exchange of avirulence determinants and extensive karyotype rearrangements in parasexual recombinants of *Fusarium oxysporum*. Mol. Genet. Genomics *268*, 298–310.

Thatcher, L.F., Manners, J.M., and Kazan, K. (2009). *Fusarium oxysporum* hijacks COI1-mediated jasmonate signaling to promote disease development in *Arabidopsis*. Plant J. *58*, 927–939.

Van de Wouw, A.P., Cozijnsen, A.J., Hane, J.K., Brunner, P.C., McDonald, B.A., Oliver, R.P., and Howlett, B.J. (2010). Evolution of linked avirulence effectors in *Leptosphaeria maculans* is affected by genomic environment and exposure to resistance genes in host plants. PLoS Pathog. *6*, e1001180.

VanEtten, H., Jorgensen, S., Enkerli, J., and Covert, S.F. (1998). Inducing the loss of conditionally dispensable chromosomes in *Nectria haematococca* during vegetative growth. Curr. Genet. 33, 299–303.

VanEtten, H.D., and Kistler, H.C. (1988). *Nectria haematococca* mating populations I and VI. In Genetics of Plant Pathogenic Fungi, Sidhu, G.S., ed. (Academic Press, San Diego), pp. 189–206.

Wasmann, C.C., and VanEtten, H.D. (1996). Transformation-mediated chromosome loss and disruption of a gene for pisatin demethylase decrease the virulence of *Nectria haematococca* on pea. Mol. Plant Microbe Interact. *9*, 793–803.

Watson, J.D., and Crick, F.H. (1953). Molecular structure of nucleic acids: A structure for deoxyribose nucleic acid. Nature *248*, 765.

Xu, J.R., and Leslie, J.F. (1996). A genetic map of *Gibberella fujikuroi* mating population A (*Fusarium moniliforme*). Genetics *143*, 175–189.

Xu, J.R., Yan, K., Dickman, M.B., and Leslie, J.F. (1995). Electrophoretic karyotypes distinguish the biological species of *Gibberella fujikuroi* (*Fusarium* section Liseola). Mol. Plant Microbe Interact. *8*, 74–84.

Molecular Genetics and Genomic Approaches to Explore *Fusarium* Infection of Wheat Floral Tissue

4

Martin Urban and Kim E. Hammond-Kosack

Abstract

The most destructive phase of the wheat–*Fusarium* interaction commences at anthesis and results in lower grain yields, reduced grain quality and the contamination of grain with harmful mycotoxins. Current control strategies are often inadequate. Globally, *F. graminearum* is the most problematic species. A recent microscopic study has revealed a hitherto unsuspected latent phase where hyphae symptomlessly advance the infection through living wheat floral tissues prior to host cell death. Various forward and reverse genetics methods have been developed to explore the repertoire of *Fusarium* genes contributing to disease formation, mycotoxin production and sporulation. At the time of writing this chapter, 159 genes are known to contribute to virulence. A newly devised seven-stage floral disease assessment key is described to assist in the inter-comparison of mutant phenotypes.

Various innovative bioinformatics approaches are currently being used to predict additional virulence components, by taking advantage of the wealth of genomic, transcriptomic, metabolomic and phenotypic knowledge available. These include (1) InParanoid analyses to infer gene function by using the phenomics datasets available for ~100 pathogenic species in the Pathogen-Host Interaction database, (2) the prediction of protein–protein interaction networks, and (3) statistical analysis of the spatial distribution of specific gene types within the genomic landscape and via comparative phytopathogen genome analyses. Soon data arising from various next generation sequencing approaches will increase the precision of both experimental and predictive studies.

Introduction

A global disease problem

Wheat accounts for 32% of global cereal production (www.faostat.fao.org, 2009) and is the world's fourth highest provider of calories for human consumption. Fusarium head blight disease (FHB), also called Fusarium ear blight (FEB) or Fusarium head scab (http://scabusa.org), occurs in all major wheat-growing areas of the world. The disease is caused by up to 17 species of *Fusarium*, but those of greatest concern are *F. avenaceum*, *F. culmorum*, *F. graminearum* and more recently *F. langsethiae* and *Microdochium nivale*. Most of these floral-infecting species can also infect the wheat stem base and the root system. Other *Fusarium* species including *F. pseudograminearum*, due to local climatic conditions, predominantly infect at the base of the wheat plant and cause crown rot disease.

FHB can introduce several mycotoxins into the food chain, including the trichothecenes deoxynivalenol (DON), its acetylated derivatives 3A-DON and 15A-DON and nivalenol as well as zearalenone, and decrease cereal grain yield and grain quality. Globally, losses in yield in wheat crops due to FHB can be up to 50%. In the USA, direct and secondary economic losses due to FHB for all crops were estimated to be $2.7 billion from 1998 to 2000 alone (Nganje *et al.*, 2004).

The other major wheat-producing countries where FHB is of major concern are China, Russia, India and France. However, there is considerable seasonal and regional variability in the severity of epidemics and hence the amount of mycotoxin in the harvested grain. Regulations set by the USA and EU to minimize human exposure to mycotoxins now limits DON concentrations in finished grain products for human consumption to 1 ppm in the USA and 0.75 ppm in the EU (European Commission Regulation no. 856/2005; www.scabusa.org). In wheat-growing regions of the world where climatic change is predicted to increase temperatures, this shift is predicted to result in increased *Fusarium* mycotoxin contamination. This is because *Fusarium* species producing lower levels of mycotoxin are gradually being replaced by more aggressive *Fusarium* species producing higher levels of mycotoxin (Paterson *et al.*, 2010). In addition, these warmer conditions are also resulting in a greater frequency of farmers growing maize in field crop rotations. *Fusarium*-infected maize stubble are more persistent than wheat stubble and thereby the overall inoculum load within these geographical regions will increase over time.

In wheat (*Triticum aestivum*), several resistance quantitative trait loci (QTL) have been identified in non-elite germplasm which are partially effective against *Fusarium* infections in floral tissue (Buerstmayr *et al.*, 2009). However, a recent study has shown that the effectiveness of the resistance conferred by the major QTLs on chromosomes 3BS (locus *FHB1*) and 5A either singly or in combination is strongly influenced by the overall wheat genetic background (Salameh *et al.*, 2010). In the higher yielding wheat germplasm, the resistance conferred by the *FHB1* locus was less effective. The use of fungicides to control FHB reduces visible disease symptoms but may be less effective at limiting mycotoxin contamination. For example, a treatment with sublethal dose of prothioconazole and fluoxastrobin fungicides increased mycotoxin production on *F. graminearum* inoculated wheat spikes compared with untreated control plants (Audenaert *et al.*, 2010). Combining the best resistant germplasm and best fungicide only provides 50–60% disease control. Currently, the most successful ways to control FHB is (1) to limit the amount of inoculum from one growth season to the next by rotating wheat with crops other than maize, (2) ploughing in infected stubble residues (Parry *et al.*, 1995) and (3) by the application of azole-based fungicides to floral tissue (Nicholson *et al.*, 2003). New strategies to control this destructive disease of cereals are urgently required.

Overview of the infection process in susceptible wheat heads

Infection cycle

Fusarium head blight is a monocyclic disease occurring only once during a cropping season. Infection of wheat heads is initiated during the short period of flowering or anthesis, which in a typical field, occurs over a 14-day window. Rain-splash and wind disseminate spores that arise from the fungus growing saprophytically on crop stubble. Unlike *F. culmorum*, *F. graminearum* also produces sexual structures, called perithecia, and forcibly discharges ascospores into the air (Trail and Common, 2000). These dispersed spores become deposited on different wheat tissues including the upper canopy leaves, the extruded anthers, the outer glumes and within the bracts. Earlier studies have demonstrated that *F. culmorum* and *F. graminearum* invade floral tissues either passively through stomata or by directly penetrating floral tissues (Kang and Buchenauer, 2000; Pritsch *et al.*, 2000). Successful infections are primarily established through saprophytic colonization of the senescing anthers, which provide an easily accessible source of nutrients and growth-stimulatory factors such as choline and betaine (Strange and Smith, 1978). Anther extrusion and the extent of flower opening during anthesis contribute to disease susceptibility (Parry *et al.*, 1995). The fungus grows first ectopically on pollen and anther tissue, and then down the filaments to colonize the stigma tissue (Beacham *et al.*, 2009). Hyphae produce mycotoxins as soon as the host tissue is colonized (Kang and Buchenauer, 1999). The fungus does not differentiate any obvious pre-infection structures such as an appressorium; however, swollen structures during hyphal tip invasion have sometimes been reported (Boenisch and Schäfer, 2011; Jansen *et al.*, 2005; Rittenour and Harris, 2010). Once a critical biomass is reached, infection hyphae penetrate the exposed floral tissues including the softer layers of the bract tissues, the base of the ovary and the developing embryo. Then within the wheat tissue, the fungal hyphae advance in the intercellular spaces between the host cells. The infected wheat head at first appears healthy. However, once a single visibly diseased spikelet has been observed up to a third of a susceptible ear may already have been colonized (Fig. 4.1). Therefore, during *Fusarium* colonization of susceptible wheat heads there is a considerable latent phase between initial infection and the onset of visible symptoms (Brown *et al.*, 2010). *F. graminearum* is homothallic and can develop perithecia on fully colonized wheat heads under high-humidity conditions. In NW Europe, this stage is rarely observed, presumably owing to unfavourable climatic conditions. Perithecia, ascospores and vegetative conidia can be formed on infected wheat heads and in the crop stubble to propagate the disease, while chlamydospores and mycelia allow survival in the soil between crops (Parry *et al.*, 1995).

Symptomless floral infections

Recently, the infection biology of *F. graminearum* on wheat heads was investigated after the entry of hyphae into the rachis (Brown *et al.*, 2010). This revealed several new features of the infection cycle, including the presence of a substantial latent phase of infection where the fungal colonization remained macroscopically symptomless for several days. This symptomless phase was retained throughout *F. graminearum* colonization of the wheat head and visible disease symptoms were only evident several centimetres behind the extending hyphal front. For example, as early as 5 days post infection of a susceptible genotype, a third of the wheat head may already be colonized yet appear healthy. Within the wheat tissue, the *Fusarium* hyphae advance initially in the intercellular spaces between live wheat cells (Fig. 4.1F) (Brown *et al.*, 2010). In the early phases of infection, this symptomless intercellular growth throughout the cortical tissue represents the majority of the mycelium colonizing the host. Subsequently, the phloem and xylem tissues were invaded by intracellular hyphae and phloem cell integrity was lost. Within the rachis considerable intracellular cortex colonization occurred coincident with the entire loss of the host cell contents (Fig. 4.1E). No host cell death was evident ahead of the infection, except the highly photosynthetic chlorenchyma cells just below the rachis surface.

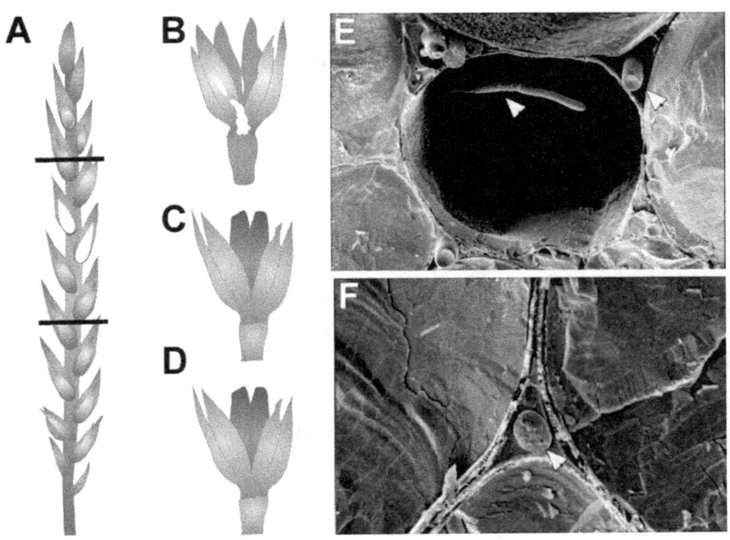

Figure 4.1 Symptomless infection at the advancing hyphal front during plant infection. (A) Wheat head point-inoculated at 5 days post inoculation (dpi). *F. graminearum* inoculated spikelets are bleached, indicated by a white oval. The region between the two black lines has been confirmed to be infected using microscopy. (B) The inoculated spikelet with visible mycelium and disease symptoms. (C and D) Sequential spikelets below, infected but without symptoms. (E) Penetration of dead wheat cell coincides with disease symptoms. (F) Colonization of intercellular spaces during symptomless infection. White arrow=hyphae.

Overall, the *Fusarium* hyphal colonization was found to advance more rapidly below than above the initial point of infection. The neighbouring spikelets were reached by growing in between the thin-walled cortical cells surrounding the vascular tissues as it passes through the gap in the sclerenchyma tissue. Alternatively, hyphae were found to grow directly through the numerous pit fields in the heavily lignified cell walls of the sclerenchyma cells, where repeated constriction and then expansion in hyphal diameter was evident. Behind the infection front, both the inter- and intracellular hyphae became abundant, resulting in the collapse of the non-lignified cell types. In this middle zone of infection, hyphal diameters were considerably enlarged compared to those leading the infection through the intercellular spaces of the cortex and then through the xylem vessels. Later on, far behind the infection front, both the inter- and intracellular hyphae were devoid of contents and had often collapsed. These 'ghosts' become the most abundant hyphal type in the centre of the infection. At late stages of infection, the pathogen switched from predominantly vertical to lateral growth and a mass of highly branched mycelium accumulated below the surface of the rachis. Here the lignified host cell walls became heavily degraded and hyphae ruptured the epidermis and produced masses of aerial mycelium. Under favourable conditions in the field, the formation of abundant asexual spores is typically observed in these regions of the wheat head and results in the appearance of the characteristic pink disease symptoms at the base of each bleached spikelet.

Overall, this recent study has revealed that during all phases of the wheat head infection process, *F. graminearum* maintains a zone where the most advancing hyphae are surrounded by living host cells. This finding has important implications for the characterization of genes and proteins required for the establishment and maintenance of fungal pathogenicity. Recently, the same group has shown that at the advancing hyphal front, the expression of various *TRI* genes required for mycotoxin production (trichothecene) is maximal compared with expression within the hyphae residing in the neighbouring, previously colonized rachis tissue (Brown *et al.*, 2011).

The role of mycotoxins during virulence

Considerable efforts have been devoted to understand the molecular basis of mycotoxin biosynthesis. The biosynthetic pathway for trichothecene production is now well defined and the pathway from primary isoprenoids to DON, 3A-DON, 15A-DON and NIV requires ~15 biochemical steps (Brown *et al.*, 2002; Desjardins, 2006; Kimura *et al.*, 2003). The initial committing step in trichothecene biosynthesis is catalysed by the enzyme trichodiene synthase, encoded by the *TRI5* gene, to produce the intermediate trichodiene (Hohn *et al.*, 1998; Proctor *et al.*, 1995b; Tag *et al.*, 2000). *Fusarium* strains lacking a functional *TRI5* are trichothecene non-producing. The trichothecenes exert their biological influence by binding to the peptidyl transferase protein in the 60S ribosomes of eukaryotic cells and inhibiting protein translation (Cundliffe *et al.*, 1974). Three key observations were made following the inoculation of *TRI5* mutant strains onto wheat heads. Firstly, the ability to synthesize trichothecenes is critical for the virulence of *F. graminearum* (Proctor *et al.*, 1995a, 1997). Secondly, trichothecenes are required for successful colonization through the wheat rachis nodes and are hypothesized to suppress plant defence responses (Jansen *et al.*, 2005). Thirdly, in the absence of trichothecenes, *F. graminearum* only produces eye shaped lesions following spray inoculations on the wheat glumes (Fig. 4.2) (Cuzick *et al.*, 2008). These lesions resemble the disease symptoms produced by the non-trichothecene-producing species *M. nivale* (Bai *et al.*, 2002). Trichothecenes are not required for *F. graminearum* colonization of other host plant species, for example either maize cobs (Maier *et al.*, 2006) or *Arabidopsis* floral tissues (Cuzick *et al.*, 2008).

A detailed analysis of the distribution of DON throughout the entire wheat head during infection detected measurable amounts after four days using an enzyme-linked immunosorbent assay (ELISA) (Savard *et al.*, 2000). Spikelets below the inoculation point reached 500–600 ppm while the corresponding internodes of the rachis contained 1000–1200 ppm. Much lower amounts of DON were found in spikelets and rachis above the

Figure 4.2 *F. graminearum* mutants inoculated on wheat heads. The disease symptoms caused by various single gene deletion mutants of *F. graminearum* produced in the strain PH-1 when inoculated on fully susceptible wheat heads. The *map1* mutant hyphae only colonizes the anthers, the *tri5* mutant causes eye shaped lesions on the glumes, the *pkar* mutant colonizes and bleaches the two inoculated spikelets, the *snf1* mutant colonizes both inoculated spikelets then enters the rachis but fails to infect the neighbouring spikelets, whereas the wild-type strain causes disease throughout the wheat ear by 16 days.

inoculation point. In an electron microscopical study of *F. culmorum* infected wheat heads using immunogold labelling of DON and its acetylated derivatives, the toxins were detected as early as 36 h after inoculation in wheat parenchyma cells (Kang and Buchenauer, 1999). The toxins diffused into most plant cell compartments including the cytoplasm and were found to be translocated upwards and downwards in the rachis through phloem and xylem tissues.

Definitions of pathogenicity and virulence factors

Pathogenicity factors are molecules absolutely required to cause disease. Virulence factors are molecules produced by pathogens that contribute to the ability of pathogens to cause disease. Essential-for-life molecules, which are required for basic survival and growth are not considered virulence factors (Korves and Colosimo, 2009). However, a grey area exists where the mutation of a gene or genes leads to reduced growth *in vitro* under certain media conditions and, by extension, reduced virulence *in planta*. A function in pathogenicity and/or virulence is often reported in the literature when the growth reduction *in vivo* is considerably greater than the growth reduction observed *in vitro*. Throughout this chapter, these mutants, although not affected in bona fide virulence factors, will also be called virulence mutants. Perhaps a more pragmatic viewpoint is, that these genes/proteins are not 'essential-for-life', which were the primary targets for fungicide development in the past. However, with the need for discovery of novel fungicide target sites all molecules contributing to *in planta* growth can be considered potential intervention targets.

Scope and objective

The objective of this chapter is to summarize and discuss the ways in which *Fusarium* pathogenicity and virulence genes have so far been discovered through molecular genetics analyses, by interrogation of the sequenced *F. graminearum* genome and through comparative genome analyses with other fungal pathogens. We then explore the use of various *in silico* predictive techniques to guide the selection of *F. graminearum* genes for future functional analysis and describe the development of the first protein–protein interaction networks containing experimentally proven virulence proteins. Topics which are beyond the scope of this chapter and are discussed elsewhere include the formation of crown rot disease (Mudge *et al.*, 2006), the known target sites of the fungicides used to control *Fusarium* floral infections on wheat (www.frac.info) and the development of species-specific diagnostic tools to distinguish the numerous wheat infecting *Fusarium* species (O'Donnell *et al.*, 2004).

Molecular genetics approaches to analyse *F. graminearum* virulence

By characterizing the key genes, and then the pathways and networks contributing to *Fusarium* virulence in wheat floral tissue, the essential proteins and enzymes can be defined and compared with the requirements of other important plant pathogens infecting wheat, other cereal species and/or non-cereal hosts. Studies of this type should also identify a sub-set of key proteins, pathways and networks as possible chemical control targets. These novel target intervention sites may be utilized in three ways: (1) for the development of novel classes of fungicides, (2) to find markers to assist molecular breeding approaches, and (3) for the design of genetically modified crop genotypes.

Forward genetics approaches

Random insertional mutagenesis followed by *in planta* phenotyping is for all pathogenic species a comparatively unbiased method to scan the genome for the presence of virulence genes that function in a non-redundant manner but are not essential for life. A library of random insertion mutants for *F. graminearum* can be generated by transformation, mediated either (1) by polyethylene glycol (PEG) or (2) by *Agrobacterium tumefaciens* with a plasmid containing the *hph* gene conferring resistance to the antifungal agent hygromycin B (Baldwin *et al.*, 2010b; Han *et al.*, 2004; Seong *et al.*, 2005). The later approach was highly efficient in *F. pseudograminearum* but less effective in *F. culmorum* and *F. graminearum* (Malz *et al.*, 2005). Both methods are suitable for targeted gene insertions in all three species and replacement and overexpression vectors have recently been developed for *A. tumefaciens* and successfully tested in *F. graminearum* (Frandsen *et al.*, 2008). A further approach involved the use of a transposable element from *F. oxysporum* that is active in *F. graminearum* (Dufresne *et al.*, 2008).

The analysis of mutants obtained in forward genetics screens, is in most cases essentially straightforward. However, in some studies, gene identification was impeded by non-random integration events due to the presence of insertion hotspots or the introduction of small chromosomal deletions (Baldwin *et al.*, 2010a; Dufresne *et al.*, 2008). In the latter case, Baldwin *et al.* (2010a) identified a disease attenuated *Fusarium* insertional mutant, called *daf10* in which the plasmid insertion event into the end region of chromosome 1 induced a large chromosomal deletion spanning 146 predicted genes including the *TRI1* gene encoding a cytochrome P450 monooxygenase. The *TRI1* protein is required for the conversion of calonectrin to 8-hydroxycalonectrin, which is an essential step during DON trichothecene biosynthesis. Despite the loss of 145 other genes, *in vitro* growth and sporulation is minimally affected in the *daf10* mutant.

Reverse genetics approaches

Targeted gene insertion and deletion methods to study the function of candidate virulence genes were made possible initially by the availability of an effective calcium chloride/PEG mediated protoplast transformation protocol. This technique was first established for *F. graminearum* to characterize the trichodiene synthase (*TRI5*) gene (Proctor *et al.*, 1995a). In this study, the *TRI5* gene was cloned using degenerate PCR primers complementary to the orthologous gene in *F. sporotrichioides*. Subsequently, gene targeting studies in *F. graminearum* were made significantly easier with the release of the 10× coverage genomic sequence by the Broad Institute (http://www.broad.mit.edu) (Cuomo *et al.*, 2007). The sequenced strain, called PH-1, is assigned to *F. graminearum* (formerly genetic lineage 7) based on sequence diversity in 11 nuclear genes (O'Donnell *et al.*, 2004). PH-1 produces relatively high levels of DON and 15A-DON, forms abundant perithecia and ascospores on carrot agar medium and is highly virulent on wheat, barley and maize. However, due to strict international plant pathogen quarantine regulations, *Fusarium* researchers in several parts of the world are unable to work with this North American strain and use local *F. graminearum* isolates instead.

Targeted gene insertion mutants can be created by amplifying as little as ~500 bp regions of coding DNA by PCR and cloning the product into a suitable transformation vector capable of conferring antibiotic resistance (Maier *et al.*, 2005). A single homologous recombination event integrates the entire vector at the target site thereby splitting the gene into two parts. In most cases insertion into the target gene results in a complete loss of gene function. The generation of targeted insertion mutants is potentially an efficient way to identify and functionally characterize numerous fungal genes. However, this approach has two drawbacks. Firstly, the transformation vector may integrate into the genome at non-homologous locations (ectopic). This process is believed to be mediated by the non-homologous end-joining (NHEJ) pathway, which competes with the pathway for homologous recombination (Meyer, 2008; Watson *et al.*, 2008). Secondly, subsequent recombination events can potentially excise the integrated foreign DNA in the mutant in succeeding generations. This may potentially restore wild-type gene function and the loss of antibiotic resistance.

Gene replacement mutants can be created by a double homologous recombination process. Here, the target coding gene sequence is replaced by DNA conferring antibiotic resistance after two homologous recombination events. A major factor in the targeting efficiency is the length of the flanking regions available for homologous recombination (Weld *et al.*, 2006). For successful *F. graminearum* transformations, typically 1kb up- and downstream sequence flanking the target gene are required to achieve high efficiency (Turgeon *et al.*, 2010). Other taxonomically closely related phytopathogenic fungi including the ascomycetes *Magnaporthe oryzae*, *Cochliobolus carbonum*, *Botrytis cinerea* and *Stagonospora nodorum* also require similar length flanking sequence for efficient double homologous integration. This is in contrast to the yeast *Saccharomyces cerevisiae* which requires as little as 50 bp in each flank.

The frequency of double homologous recombination in *F. graminearum* can also vary significantly depending on the target sequence. For example, the GPMK1/MAP1 gene was partially deleted with an efficiency of 76% (Urban *et al.*, 2003) while *TRI14*, a gene within the trichothecene gene cluster, was deleted with an efficiency of 6% (Dyer *et al.*, 2005).

Recently several PCR-based methods have been successfully implemented, which eliminate the need for cloning. Instead, two or three rounds of PCR are used to synthesize DNA fragments for targeted gene deletion. These methods include the split marker (Fig. 4.3) and the double-joint PCR technique (Catlett *et al.*, 2003; Yu *et al.*, 2004), respectively.

Another approach to explore gene function in filamentous fungi has been through RNA silencing (RNA_i) (Hammond and Keller, 2005; Kadotani *et al.*, 2003). In the latter case, the effective silencing of the *TRI6* regulatory gene in *F. graminearum* eliminated DON production and severely compromised FHB symptom formation (McDonald *et al.*, 2005).

Gene expression

The study of fungal gene expression under diverse conditions including wheat infection can be instrumental in the identification of fungal virulence genes, which are only induced *in planta*. A valuable tool to study the transcriptome is the *F. graminearum* Affymetrix Gene Chip (Guldener *et al.*, 2006a). A variety of *F. graminearum* gene expression datasets and expressed sequence tags (ESTs) are already publicly available at PLEXdb (Wise *et al.*, 2007) and COGEME (Soanes *et*

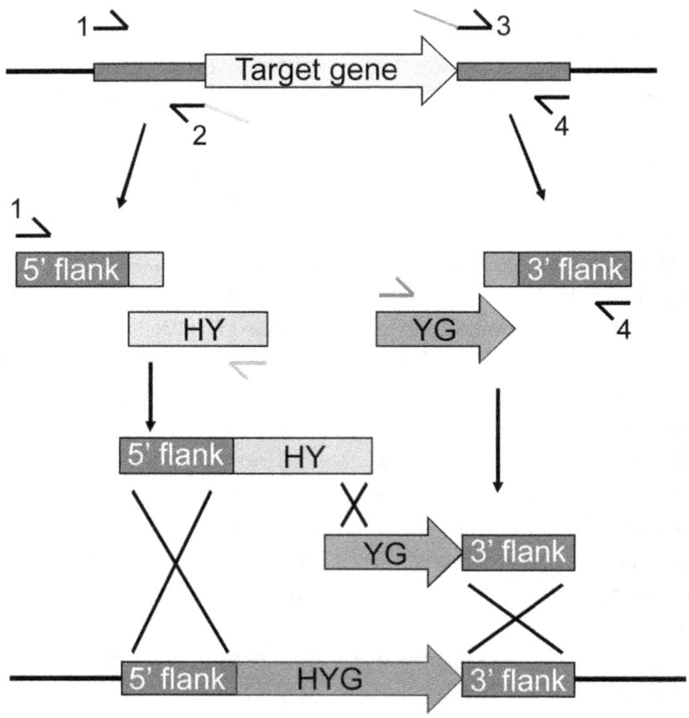

Figure 4.3 Split marker gene deletion method established in *F. graminearum* (Catlett *et al.*, 2003). Regions either side of the target gene are amplified by PCR using primers 1–4. During the second round PCR, each flank is then fused to an incomplete version of the hygromycin resistance gene (HYG), referred to as 'HY' and 'YG' respectively. The two amplicons are then transformed into the fungus. A triple homologous recombination event excises the target gene and reconstitutes a functional HYG sequence.

al., 2002), respectively. The datasets available at PLEXdb include experiments with different infected plant tissues, experiments on conidial germination, sexual development, and different media/chemicals (Table 4.1). Other experiments measure the effects of specific gene mutations.

The identified *in planta* induced genes were mapped to each chromosome and found to be predominantly located in the subteleomeric regions of the chromosome and other regions associated with a high rate of recombination (Fig. 4.4) (Cuomo *et al.*, 2007). These genes were suggested to contribute to virulence and one-third of these genes are predicted to code for secreted proteins.

Virulence assays on wheat and barley heads

Wheat and barley are generally inoculated during mid-anthesis, when the flowering cereal head is most susceptible. *F. graminearum* inoculum for virulence assays can easily be produced *in vitro*. Conidia can be grown on certain complete or minimal media that support sporulation either in liquid cultures or on agar plates (Cappellini and Peterson, 1965; Leslie and Summerell, 2006). Perithecia bearing ascospores can be produced on carrot agar medium (Leslie and Summerell, 2006).

The point-inoculation method uses 5 to 10 μl droplets of conidial suspension, which are placed into one or two spikelets located in the upper third or near the centre of the head. When droplets are placed into a spikelet at the base of a wheat head, disease development is quicker and this results in more pronounced FHB symptoms. However, these symptoms may be due to the fungus disrupting the vascular system, resulting in reduced transport of water into the head. Therefore macroscopic symptom development may not reflect the speed with which the fungus

Table 4.1 Publicly available *F. graminearum* microarray gene expression datasets

Experiment	PLEXdb accession*	Reference
In planta		
Infected barley and wheat spikes	FG1, FG15	Guldener *et al.* (2006b), Lysoe *et al.* (2011b)
Wheat stem post anthesis	FG16	Guenther *et al.* (2009)
In vitro		
Perithecial development	FG5	Hallen *et al.* (2007)
Asexual spore germination	FG7	Seong *et al.* (2008)
Media experiments, trichodiene treatment, DON induction	FG2, FG10, FG14	Gardiner *et al.* (2009), Guldener *et al.* (2006b), Seong *et al.* (2009)
Mutant characterization	FG6, FG11, FG13	Hallen and Trail (2008), Lysoe *et al.* (2011a), Seong *et al.* (2009)

*www.PLEXdb.org

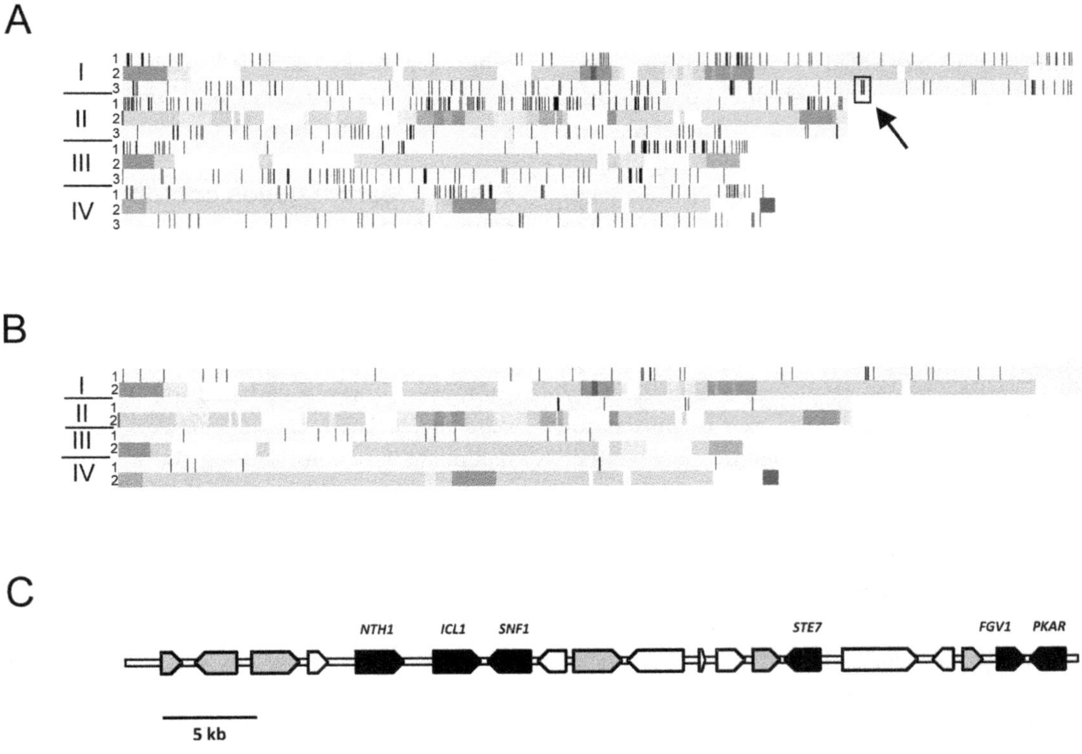

Figure 4.4 Genetic and gene information displayed across the four *F. graminearum* chromosomes. (A) Roman numerals indicate the four chromosomes of 11.9, 9.0, 7.7 and 8.1 Mbp in length. Row 1: The *in planta* induced *F. graminearum* genes during barley spike infection (Boddu *et al.*, 2006). Genes are indicated by black vertical lines. Row 2: The frequency of recombination in a cross between the sequenced strain PH-1 and strain MN00–676 (Gale *et al.*, 2005). The light grey to dark grey shading indicates the increasing rates of recombination frequency detected (overall range 0 to 8 cM/27kbp). Row 3: Distribution of virulence gene homologues from the plant and animal pathogenic fungal species stored in the PHI-base database Ver. 3.1 with significant homology (BLAST score value >1 e-100). The black arrow and box indicate a micro-region, where the non–random clustering of virulence gene homologues was observed (Beacham, 2011). (B) Distribution of the 159 experimental verified virulence genes listed in Table 4.2 and 4.3. (C) Gene content of the micro-region (FGSG_09891 to FGSG_09908) on chromosome 1 indicated by an arrowed box in panel A. Genes essential for full disease development are depicted in black. Genes with no role in virulence are shown in grey (functional annotation available) or white (conserved hypothetical protein) (Beacham, 2011).

grows through the flowering tissue. Point inoculations in spikelets located towards the middle or top of the wheat head are preferable. Spore droplets may contain from 5×10^4 to 1×10^6 conidia/ml. Some experimenters include a weak detergent into spore suspensions, for example 0.01% Tween 20 or Triton 60. An alternative option for point-inoculation, when a test strain fails to produce spores, is the agar plug inoculation method. Here a small agar plug containing a mixture of spores and mycelic fragments is taken from the advancing edge of a relatively young colony grown on minimal agar using a Pasteur pipette and placed directly into a floret.

Spray-inoculation may also serve as an infection method to test for loss of virulence. Solutions with a spore concentration of 5×10^4 are sprayed with a small hand-held sprayer until the cereal head is evenly covered with fine droplets of the spore solution. However, only *Fusarium* mutants highly reduced in virulence will show any obvious reduction in the development of disease symptoms in such an assay due to the high probability for multiple infection sites per head (Cuzick *et al.*, 2008).

After the inoculum has been applied, wheat heads are usually covered with small plastic bags or the whole plants are placed in high humidity conditions using humidity boxes or misters for 2–3 days. These conditions mimic rainy weather conditions in the field and favour early disease development. Plants are then transferred to normal cereal growing conditions and visible disease symptoms are monitored on a regular basis over the next 2–3 weeks. Disease development is quantitatively reported by using several methods including the following: (1) counting the total number of bleached spikelets, (2) counting only the number of bleached spikelets below the inoculation point or (3) counting the total number of heads infected. In some studies, researchers have also dissected the wheat heads from a late infection time point and assessed grain development (Baldwin *et al.*, 2010a; Voigt *et al.*, 2005). The amount of mycotoxin production in inoculated spikes was shown to be correlated with virulence (Maier *et al.*, 2006). For toxin diagnostics the inoculated spikes are harvested when the plant tissue shows visible necrosis. The plant tissue can then be analysed using a range of toxin diagnostics techniques including GCMS, LCMS or commercially available ELISA kits (Lowe *et al.*, 2011).

Frequently, *F. graminearum* infections are assessed under field conditions in an effort to identify QTL based resistance in wheat and barley cultivars using natural and artificial inoculation (Buerstmayr *et al.*, 2009). By contrast, field evaluation of the virulence of transgenic *Fusarium* strains have rarely been conducted. A trichothecene non-producing strain (*TRI5*) was assessed in three controlled field experiments using the droplet inoculation method. The results were consistent with parallel experiments in environmentally controlled growth rooms and showed that virulence is reduced in the trichothecene non-producing strains (Desjardins *et al.*, 1996). The testing of other genetically modified strain under field conditions has so far not been reported in the peer-reviewed literature.

F. graminearum genes contributing to virulence

Disease formation is a complex procedure, where the fungus needs to adapt to nutrient availability and interacts with its host using a variety of signalling and regulatory networks and metabolic pathways. However, certain parts of the fungal cell's molecular interaction networks may have redundant functions. During the last 15 years, using the approaches outlined above, an increasing number of *F. graminearum* genes have been functionally characterized. Single gene modification experiments so far reported have identified 159 genes contributing to virulence (Tables 4.2 and 4.3). A further 28 genes when tested through deletion/disruption experiments were shown not to affect virulence. The linear distribution of these virulence genes when displayed across the four chromosomes reveals a non-random distribution pattern (Beacham, 2011). Interestingly, most genes reside within genomic regions which exhibit either a low recombination frequency or no recombination and are situated away from the subtelomeric regions (Fig. 4.4B). The exception to this pattern are the *TRI* genes involved in DON mycotoxin production and the gene FGSG_00007 which is highly expressed under DON inducing conditions *in vitro* (Gardiner *et al.*, 2009).

The vast majority of the identified virulence genes (90%) have been published since 2005 owing to the availability of the *F. graminearum* genome sequence (www.broadinstitute.org). In 2011, two high-throughput studies reported on the analysis of all putative transcription factors (709 genes) (Son *et al.*, 2011b) and the complete kinome in *F. graminearum* (116 protein kinases) (Wang *et al.*, 2011a). These two studies resulted in the identification of 96 virulence genes (Table 4.3).

Functional categories of virulence genes

The virulence genes can be grouped by functional annotation into 10 categories (Fig. 4.5). The annotation scheme employed is the functional classification system for proteins (FunCat) developed by the Munich Information Centre for Protein Sequences (MIPS) for yeast (Ruepp *et al.*, 2004; http://mips.helmholtz-muenchen.de/genre/proj/FGDB/). Most of the virulence genes reported in Table 4.2 (i.e. excluding the two high-throughput studies) belong to four categories: (1) metabolism (24%); (2) cellular communication/signal transduction mechanisms (22%); (3) cell rescue, defence and virulence/adaptation to nutrient conditions (13%); and (4) transcription (10%).

The first MIPS category, 'metabolism', is highly overrepresented and this probably reflects the fact that fungi need to adapt their primary metabolism to nutrient conditions encountered during host invasion. In genome-wide mutagenesis screens, several genes were identified which have a role in nutritional function. Whilst these mutants still can grow on artificial media containing appropriate supplements, virulence is clearly affected. For example, a number of genes affecting amino acid metabolism [e.g. methionine (Han *et al.*, 2004; Seong *et al.*, 2005), adenine and arginine biosynthesis (Kim *et al.*, 2007)] have been identified indicating their critical importance in disease. In another case, the importance of fatty acid metabolism in disease was revealed only after the creation of a double mutant (*icl1mcl1*) affecting the glyoxylate and methylcitrate cycle simultaneously. The single gene deletion mutants were not affected in virulence (Lee *et al.*, 2009a). Finally, two secreted lipase genes, *FGL1* and *LIP1*, have been functionally characterized. *FGL1* is required for virulence and speculated to suppress wheat callose synthesis during plant defence responses (Voigt, 2011; Voigt *et al.*, 2005). In contrast, *LIP1* does not contribute to virulence (Feng *et al.*, 2005).

At present, no additional secondary metabolites, other than DON and related trichothecenes, are known to play a role in disease formation. Polyketide synthases (PKSs) are a family of enzymes involved in the biosynthesis of the mycotoxins zearalenone, aurofusarin, fusarin C and a purple perithecial pigment (See Chapter Brown). Unexpectedly, single gene mutants of the complete set of 15 PKSs in *F. graminearum* showed wild-type virulence on wheat (Gaffoor *et al.*, 2005; Gaffoor and Trail, 2006; Malz *et al.*, 2005). Experiments are currently under way to test the virulence of mutant strains in which multiple PKS genes are non-functional (G. Turgeon, personal communication). In an alternative approach, global transcriptional regulators that affect multiple PKSs have been identified and mutated and are being tested for virulence (G. Adams, personal communication). In addition, the mycotoxin butenolide is also dispensable for virulence on wheat. A mutant disrupted for FGSG_08079 (putative cytochrome P450 monooxygenase) is defective for butenolide biosynthesis, but shows wild-type virulence on wheat heads (Harris *et al.*, 2007).

The second MIPS category, 'cellular communication/signal transduction', is composed of a number of genes encoding proteins belonging to signalling cascades required for specific adaptation to the environment. Core elements of these cascades include three mitogen-activated protein kinases (MAPKs) that are conserved in a wide range of plant and animal infecting pathogens as well as in numerous free living species (Rispail *et al.*, 2009). In *F. graminearum* the three MAPKs are called *GPMK1/MAP1 (S. cerevisiae FUS3/KSS1), MGV1 (S. cerevisiae SLT2),* and *HOG1/OS2 (S. cerevisiae HOG1)* (Hou *et al.*, 2002; Jenczmionka and Schäfer, 2003; Urban *et al.*, 2003, Ochiai *et al.*, 2007; Oide *et al.*, 2010). All three MAPKs are important for virulence. *GPMK1/MAP1* also regulates the secretion of various extracellular enzymes, including the

Table 4.2 Virulence genes characterized in *Fusarium graminearum*

FGSG ID*	Gene name (alias)	Protein function	Method of gene identification[†]	Cereal hosts [‡]	Point of pathogen arrest [††]	*In vitro* defects[‡‡]	DON production[§] (*in vivo*/*in vitro*)	Tri6 elements in promoter[§§]	Reference/PHI-base ID[§†]
Cellular communication/signal transduction mechanism									
FGSG_01665	*FSR1*	Putative signalling scaffold protein	R	B, M	RV (B, M)	h, ss			PHI:731
FGSG_00332	*FTL1* (*TBL1*)	*S. cerevisiae SIF2* (transducin beta-subunit)	F	W	A	h, sa,	ND/WT		PHI:446, Ding *et al*. (2009)
FGSG_09614	*GPA2* (*GzGPA2*)	Guanine nucleotide-binding protein alpha-3 subunit	R	B	RV (B)	h		2	PHI:1013, Yu *et al*. (2008)
FGSG_04104	*GPB1* (*GzGPB1*)	Guanine nucleotide-binding protein beta subunit	R	B	RV (B)	h	ND/enhanced	2	PHI:1015, Yu *et al*. (2008)
FGSG_09612	*HOG1* (*OS2*)	*S. cerevisiae HOG1* (osmotic stress MAPK)	R	W	E	h, sa, ss	ND/reduced		PHI:2327, Ochiai *et al*. (2007), Oide *et al*. (2010)
FGSG_06385	*MAP1* (*GPMK1*)	*S. cerevisiae KSS1*/*FUS3* (mating/filamentation MAPK)	R	W	B	h, sa, ss			PHI:309
FGSG_10313	*MGV1*	*S. cerevisiae SLT2* (cell integrity MAPK)	R	W	B	sa, ss		1	PHI:266
FGSG_09908	*PKAR*	Protein kinase A regulatory subunit	R	W	C	h, sa, ss	Reduced/ND	1	Beacham (2011)
FGSG_10114	*RAS2*	Ras GTPase	R	W, M	C, RV (M)	h, sa			PHI:861
FGSG_09897	*SNF1*	Sucrose non-fermenting protein kinase	R	B, W	E	h, sa, ss	ND/reduced		Lee *et al*. (2009b), M. Urban and K.E. Hammond-Kosack (unpublished)
FGSG_08948	*SSK1* (*RRG1*)	Histidine kinase two-component response regulator	R	W	E				PHI:1164, Jiang *et al*. (2011b), Oide *et al*. (2010)

FGSG_05484	*STE11*	MAPKKK; hypersensitive to MsDEF1	F	W	B			PHI:1016
FGSG_09903	*STE7*	MAPKK; hypersensitive to MsDEF1	R	W	B		2	PHI:1004
FGSG_04982	*TEP1*	Tensin-like phosphatase 1 Phosphatidylinositol-3 kinase signalling	R	WC	RV (WC)	H, sa	1	PHI:2325, Zhang *et al*. (2010)
Metabolism								
FGSG_02506	*ADE5*	Phosphoribosylamine-glycine ligase	F	B	RV(B)	h, sa, ss		PHI:744
FGSG_01939	*ARG2*	Acetylglutamate synthase	F	B	RV(B)	h, sa, ss		PHI:743
FGSG_01932	*CBL1*	Cystathionine beta-lyase	F	W, M	F, RV (M)		1	PHI:443
FGSG_05906	*FGL1*	Secreted lipase	R	W, M	F, RV (M)		1	PHI:432
FGSG_05955	*GCS1*	Glycosylceramide synthase (sphingolipid biosynthesis)	R	W, M	F	h, sa		PHI:1002
FGSG_05658	*GzmetE*	Homoserine O-acetyltransferase	F	B, M	RV(B, M)	SA, ss	1	PHI:355
FGSG_09197	*HMR1*	3-Hydroxy-3-methylglutaryl-coenzyme A reductase involved in isoprenoid biosynthesis	F	W	B			PHI:1006
FGSG_09896/ FGSG_00176	*ICL1 MCL1*	Double mutant of isocitrate and methylisocitrate lyase	R	W, B	RV(B, M)	h, ss		Lee *et al*. (2009a)
FGSG_10825	*MSY1*	Methionine synthase	R	W, M	C, RV (M)	h, SA		PHI:442
FGSG_05593	*MT2* (*FgMT2*)	Sphingolipid C-9-methyltransferase	R	W	C	h, SA		PHI:2409, Ramamoorthy *et al*. (2009)

Table 4.2 Continued

FGSG ID*	Gene name (alias)	Protein function	Method of gene identification†	Cereal hosts ‡	Point of pathogen arrest †‡	*In vitro* defects‡‡	DON production§ (*in vivo*/*in vitro*)	Tri6 elements in promoter§§	Reference/PHI-base ID§†
FGSG_09254	*NPC1*	Sterol transporter	R	W	G				PHI:1167, Breakspear *et al.* (2011)
FGSG_09895	*NTH1*	Neutral trehalase	R	W	F	h, SS	WT/ND	1	Beacham (2011)
FGSG_05371	*SID1*	Siderophore biosynthetic gene	R	W	C	h		1	PHI:1010
FGSG_02549	Transposon mutant	Putative phosphoglycerate mutase family	T	W	A	H, SS			PHI:1089, Dufresne *et al.* (2008)
Energy									
FGSG_06039	*ACL1*	ATP citrate lyase	R	W	A	h, sa,ss	ND/reduced	2	PHI:2386, Son *et al.* (2011a)
FGSG_12857	*ACL2*	ATP citrate lyase	R	W	A	h, sa,ss	ND/reduced		PHI:2387, Son *et al.* (2011a)
FGSG_00376	*NOS1*	NADH:Ubiquinone oxidoreductase	F	W, M	C, RV (M)	H			PHI:445
Protein fate									
FGSG_02519	*ATG15*	Autophagic death protein	R	W	F	H, sa		1	PHI:1166, Nguyen *et al.* (2011)
FGSG_02095	*FBP1*	F-box protein involved in ubiquitin-mediated degradation	F	B	RV(B)	ss			PHI:733
Biogenesis of cellular components									
FGSG_01964	*CHS5*	Myosin-motor like chitinase	F	B	RV(B)	h, sa, ss		1	Kim *et al.* (2009)
FGSG_12039	*CHS7*	Myosin-motor like chitinase	R	B	RV(B)	h, sa, ss		1	Kim *et al.* (2009)
Interaction with the environment									
FGSG_12970	*PAC1*[c]	Cys_2His_2 zinc finger transcription repressor	R	W	B	H, SA		1	PHI:1431, Merhej *et al.* (2011), J. Merhej and C. Barreau (personal communication)

FGSG_01974	none designated	Similar to HET-C2 glycolipid transfer protein	T	W	B	H, ss			PHI:1097, Dufresne *et al*. (2008)
FGSG_11955	*VE1* (*VEA*)	Light-responding activator *velvet1*	R	W	A	H, sa	Reduced/none		PHI:1431, Merhej *et al*. (2012)
Cellular transport, transport facilities and transport routes									
FGSG_00950	*SYN1* (*GzGSYN1*)	SNARE protein (transport docking and vesicle fusion)	F	B	RV(B)	h, SA, SS			PHI:2396, Hong *et al*. (2010)
FGSG_09928	*SYN2* (*GzGSYN2*)	SNARE protein (transport docking and vesicle fusion)	R	B	RV(B)	h, SA, ss			PHI:2396, Hong *et al*. (2010)
FGSG_00416	None designated	Putative major facilitator superfamily	T	W	B	H, SS		1	PHI:1086, Dufresne *et al*. (2008)
Cell cycle									
FGSG_04355	*CID1*	Cyclin-C-like gene required for infection and DON production	R	W, M	C, RV (M)	h, sa, ss	Reduced/ND		PHI:2419, Zhou *et al*., (2010)
Transcription and DNA modification									
FGSG_01353	*HDF1*	*S. cerevisiae HOS2* (Class II histone deacetylase)	R	W, M	C, RV (M)	H, SA, ss	Reduced/ND		PHI:1168, Li *et al*. (2011)
FGSG_04324	*HDF2*	*S. cerevisiae HDA1* (Class II histone deacetylase)	R	W, M	F, RV (M)	H, sa, nd	WT/ND		PHI:1169, Li *et al*. (2011)
FGSG_10129	*STUA*	APSES transcription factor	F	W	A	h, sa, ss	Reduced/reduced		PHI:1290, Lysoe *et al*. (2011a), Son *et al*. (2011b)
FGSG_06874	*TOP1*	Topoisomerase 1	F	W	C	h. sa, ss	ND/reduced	2	PHI:1291, Baldwin *et al*. (2010b)
FGSG_10057	*ERB1*	Putative transcription factor $(Zn(II)_2Cys_6$ domain)	T	W	C	h, SS			PHI:1163, Dufresne *et al*. (2008), Zhao *et al*. (2011)
FGSG_01555	*ZIF1*	B-ZIP transcription factor	F	W, M	C, RV (M)	H, SA, ss	Reduced/ND		PHI:444 (Son *et al*., 2011b; Wang *et al*., 2011b)

Table 4.2 Continued

FGSG ID*	Gene name (alias)	Protein function	Method of gene identification[†]	Cereal hosts [‡]	Point of pathogen arrest [†‡]	*In vitro* defects[‡‡]	DON production[§] (*in vivo*/*in vitro*)	Tri6 elements in promoter[§§]	Reference/PHI-base ID[§†]
Cell rescue, defence and virulence									
	daf10	146 genes deleted from end of chromosome 1 incl. FGSG_00071 *(tri1)*	F	W	C	SA, SS	None/ ND		Baldwin *et al*. (2010a)
FGSG_00007	*FGSG_00007*	Cytochrome P450 monooxygenase (DON biosynthesis)	R	W	Hyper-virulent		Increased DON *in vitro*	3	PHI:2393, Gardiner *et al*. (2009)
FGSG_10397	*FGSG_10397*	Unknown function	R	W	Hyper-virulent		Increased DON *in vitro*	3	PHI:2394, Gardiner *et al*. (2009)
FGSG_03747	*NPS6*	Non-ribosomal peptide synthetase for biosynthesis of extracellular siderophores	R	W	F				PHI:1007
FGSG_04111	*PTC1*	Type 2C protein phosphatase	R	WC	RV (WC)	SA		2	PHI:2326, Jiang *et al*. (2010)
FGSG_03538	*TRI10*	Regulatory protein	R	W	C		Reduced/ND		PHI:2328, Seong *et al*. (2009)
FGSG_03543	*TRI14*	Unknown	R	W, M	D, RV (M)	H, SA, SS	None/WT	1	PHI:525 (Dyer *et al*., 2005)
FGSG_03537	*TRI5*	Trichodiene synthase	R	W, M	D, RV (M)	H, SA, SS	None	3	PHI:44 (Proctor *et al*., 1995)
FGSG_03536	*TRI6*	Transcription factor	R	W	D		None		PHI:439 (Son *et al*., 2011b)

Unclassified protein									
FGSG_06631	*CPS1*	Adenylate-forming enzyme	R	W	C	ss			PHI:443
FGSG_09907	*FCV1*	Conserved hypothetical protein	R	W	C	h, sa, SS	WT/ ND		Beacham (2011)
FGSG_06680	*MES1*	Role in cell-surface organization	R	W	F	h, sa, ss			PHI:1078
FGSG_02077	None designated	Conserved hypothetical protein	T	W	A	H, SS			PHI:1093, Dufresne *et al.* (2008)
FGSG_12019	None designated	Hypothetical protein	T	W	B	H, SS		1	PHI:1098, Dufresne *et al.* (2008)
FGSG_12753	None designated	Hypothetical protein	T	W	B	H, SS			PHI:1092, Dufresne *et al.* (2008)

Abbreviations: ATP, adenosine triphosphate); APSES, 100 amino acid protein domain; GTP, guanosine triphosphate; MAPK, mitogen-activated protein kinase; MAPKKK, mitogen-activated protein kinase kinase kinase.

*FGSG ID (*F. graminearum* locus identifier) was taken from the *F. graminearum* genome version 3.0 (Broad and MIPs)

†Methods of gene identification are forward genetics (F), reverse genetics (R), transposon mutagenesis (T)

‡Cereal hosts tested are wheat spikes (W), barley spikes (B), wheat coleoptiles (WC), maize (M).

††Points of pathogen arrest are on wheat spikes A to F as indicated in Fig. 4.6. Reduced virulence on barley spikes, maize tissue and wheat coleoptiles is reported as RV(B), RV(M), RV(WC), respectively.

‡‡*In vitro* defects are changes in hyphal growth/morphology (h), asexual sporulation (sa) or sexual sporulation (ss). Wild-type phenotypes are indicated using capitals. The field is left blank, when no *in vitro* test is reported in the literature.

§DON production lists the ability to produce DON *in planta* or *in vitro*. The values are 'enhanced', 'reduced' and wild-type (WT) level or not determined (ND). The field is left blank, when no DON test is reported in the literature.

§§The number of Tri6 binding elements in the promoter sequence is given.

§†Reference/PHI-base ID: For literature reference also see PHI-base website (www.PHI-base.org) when PHI-base ID is available.

Table 4.3 Global studies by single gene deletion analysis of *F. graminearum* transcription factors and protein kinases (the kinome)

Protein type	Details*
Transcription factors‡	*Loss of pathogenicity†: n=11*
	04134, 05388, 06071, 06291, 06948, 08572, 08719, 09992, 10716, 12781, 12977
	Reduced virulence†‡: n=50
	00147, 00324, 00385, 00477, 00515, 00574, 00719, 00729, 01022, 01106, 01176, 01182, 01293, 01307, 01341, 01350, 01555, 02527, 04083, 04220, 05171, 05304, 06228, 06427, 06651, 06871, 06944, 07067, 07133, 07928, 08028, 08182, 08481, 08617, 08737, 08769, 09019, 09339, 09654, 09832, 09868, 10129, 10142, 10179, 10384, 10517, 11416, 13120, 13711, 13746
Protein kinases‡‡	*Loss of pathogenicity: n=8*
	00362, 00408, 01188, 02795, 04053, 06326, 07329, 10037
	Reduced virulence: n=27
	00472, 01312, 01641, 04382, 04484, 04770, 04947, 05547, 05734, 06793, 06878, 06939, 06957, 07251, 07295, 07344, 08468, 08635, 08691, 08701, 08906, 09274, 10066, 10095, 10228, 10381, 13318

*Listed are the phenotype, gene number per phenotype and the FGSG *F. graminearum* locus identifier from genome version 3.0 (Broad and MIPs).
†No obvious plant tissue necrosis observed (stage A in Table 4.4).
‡Data taken from Son *et al.* (2011b) (Figure S5).
†‡Tissue necrosis is observed, but disease development is reduced compared with wild type.
‡‡Data taken from Wang *et al.* (2011a). Loss-of-pathogenicity mutants with mutant disease index ≤0.5 (similar to FGSG_06385 MAP kinase mutant Gpmk1/map1) or lower than wild type for reduced virulence mutants.

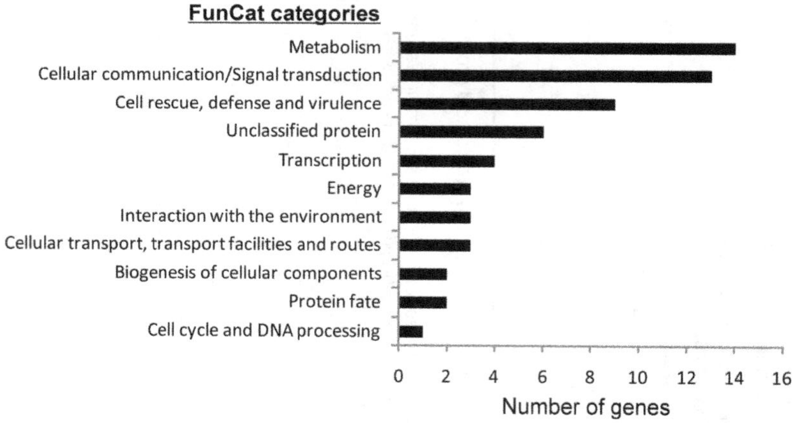

Figure 4.5 Functional classification of *F. graminearum* virulence genes listed in Table 4.2 using the MIPS FunCat functional annotation scheme for proteins.

secreted lipase *FGL1* required for virulence (Bluhm *et al.*, 2007). *MGV1* was shown to be essential for cell integrity and has a role in female fertility and heterokaryon formation. *HOG1/OS2* is required for mycotoxin induction in response to osmotic stress. Mutants in *GPMK1/MAP1* and *MGV1*, but not *HOG1*, were reported to be more sensitive to the seed defensin peptide MsDef1 from *Medicago sativa* (Ramamoorthy *et al.*, 2007). Therefore, the three MAP kinase proteins in *F. graminearum* possess overlapping as well as non-redundant functions.

Upstream components characterized for the *GPMK1/MAP1* MAPK cascade include the *RAS2*, *STE11* and *STE7* proteins. It is thought that upstream activation is mediated by G protein signalling. The RAS2-GTPase, the hetero-trimeric G protein subunits GPA2 (alpha subunit) and GPB1 (beta subunit) were also shown to regulate different aspects in vegetative growth, sexual development and toxin production (Bluhm *et al.*, 2007; Yu *et al.*, 2008). However, mutants in two other putative G protein beta subunits, *GPA1* and *GPA3*, showed wild-type virulence on barley (Yu *et al.*, 2008). The recent high-throughput study by Wang *et al* (2011a) identified 35 additional protein kinases contributing to virulence (Table 4.3). Single gene deletions of eight protein kinases caused loss of pathogenicity, similar to the phenotype observed for a *gpmk1/map1* mutant. It is likely that these kinases identify additional cell-signalling pathways required for fungal growth, development and plant infection processes.

Tetraspanins are membrane-associated proteins with four transmembrane domains that are proposed to be possible receptors for extracellular signalling molecules. In appressoria-forming pathogens, the *PLS1* tetraspanin is a known virulence gene, but was shown in *F. graminearum* to be dispensable for virulence on wheat (Rittenour and Harris, 2008).

The third MIPS classification group, 'cell rescue, defence and virulence', is represented by bona fide virulence genes such as the trichothecene biosynthetic genes essential for DON production. This group also includes two genes with a function in DON regulation (FGSG_00007 and FGSG_10397). Mutants defective for these genes produce increased amounts of DON *in vitro* and are hyper-virulent on wheat heads (Gardiner *et al.*, 2009). All known genes belonging to this category affect virulence on wheat heads.

The fourth MIPS category, 'transcription and DNA modification', includes conserved as well as *Fusarium* specific genes. The APSES type transcription factor encoding *STUA* gene regulates fungal development and cell cycle progression (Lysoe *et al.*, 2011a). The Pac-C transcription factor represses DON mycotoxin production *in vitro* under alkaline conditions. Under low-pH conditions, the Pac-C repressor becomes inactive. Overexpression of the *PAC-C* gene was recently shown to reduce virulence during wheat head infection (J. Merhej and C. Barreau, personal communication). In *Aspergillus nidulans* the LaeA methyltransferase is a global regulator of secondary metabolism and regulates sporulation in response to light. A putative *F. graminearum* homologue of this gene has recently been identified in a yeast two-hybrid assay using the interacting velvet protein as a bait (Jiang *et al.*, 2011a). While the molecular characterization of this gene has not been reported yet, gene deletion of the interacting velvet gene *VE1/VEA* was shown to have an effect on both virulence and DON production (Jiang *et al.*, 2011a; Merhej *et al.*, 2012).

Mycotoxin production and virulence is regulated by the *Fusarium* specific *TRI6* and *TRI10* transcription factors with a global effect on gene expression but also regulating the expression of the trichothecene biosynthetic genes (Seong *et al.*, 2009). The Tri6 consensus-binding site is relatively short (YNAGGCC) and is found in multiple copies in the promoters of trichothecene and isoprenoid biosynthetic genes in *F. graminearum*. For example, the *TRI5* gene has three Tri6 binding sites in its promoter sequence (Table 4.2). Other virulence genes with three Tri6 binding sites are FGSG_00007 and FGSG_10397, which are likely to be regulated by *TRI6* (Gardiner *et al.*, 2009).

The high-throughput study by Son *et al* (2011b) reported on the functional analysis of the whole transcription factor set for *F. graminearum* strain GZ3639. From the 709 genes annotated as transcription factors, 657 single gene deletion mutants were generated. Each mutant was tested for a range of phenotypes including virulence on wheat heads. This impressive study revealed an additional 11 transcription factors absolutely required for virulence and 50 transcription factors contributing to virulence (Table 4.3). No doubt further analysis of these genes and proteins will reveal additional functions important for virulence.

Certain enzymes involved in DNA modification can also have a specific function during virulence, while only having a minor effect on *in vitro* growth. This fact is revealed by gene mutations in the topoisomerase and histone deacetylase genes. A type 1 topoisomerase (*TOP1*) mutant severely affected virulence in both *F. culmorum* and *F. graminearum* and may control transcription of specific DNA regions during virulence (Baldwin *et al.*, 2010b). Deletion of two class II histone deacetylase genes demonstrated that *HDF1* (*S. cerevisiae SIF2*) has a greater effect on virulence than *HDF2* (*S. cerevisiae HDA1*). In yeast Sif2 is a component of the Set3 complex. Deletion of a third histone deacetylase gene *HDF3* had no effect on *F. graminearum* virulence. Histone deacetylases typically function as transcriptional corepressors.

Virulence genes belonging to other MIPS functional categories highlight the importance of cell wall synthesis during *in planta* growth, for example a need for two myosin-motor like chitinases (Kim *et al.*, 2009). Insertional mutagenesis also identified an F-box protein encoding gene *FBP1* believed to be involved in the regulation of ubiquitin-mediated degradation of proteins during disease development and sexual reproduction (Han *et al.*, 2007).

Our survey of published virulence genes for *F. graminearum* includes all publications accessed via ISI Web of Knowledge (Thomson Reuters, UK) on 30 May 2012. Combining the 63 virulence genes analysed in the highly focused gene deletion studies (Table 4.2) with the virulence genes identified in the two high-throughput studies surveying all transcription factors and protein kinases (Table 4.3), a total of 159 virulence genes are known for *F. graminearum.*

A seven-stage floral disease assessment key to assist in the intercomparison of mutant phenotypes

Precise molecular information is available for most *F. graminearum* genes found to play a role in virulence. In contrast, information on the detailed phenotypic analysis describing the mutant block is often patchy and makes comparisons of mutant behaviour during wheat head infection difficult (Table 4.2). A considerable challenge in pinpointing the exact block during disease development is the complexity of cereal heads with their diverse tissue types. We have therefore developed a *Fusarium* fate map for wheat head infection to enable comparative mutant analysis (Fig. 4.6). The qualitative key takes into account the reported phenotypes for various virulence genes and dissects the *F. graminearum* infection process into seven stages (A to G) during *in planta* growth. The key events are (A) germination and ectopic hyphal growth in inoculated floret tissue, (B) ovary abortion, (C) growth from floret to floret within the same spikelet, (D) invasion of the rachis node, (E) entry into the rachis node as indicated by internal rachis browning, (F) bleaching of adjacent uninoculated spikelets, (G) full disease development and invasive growth into the peduncle. The wild-type fungus can transition through all stages. A mutant may be blocked during any of these stages or may be delayed in reaching the peduncle in stage G (see Table 4.4 for details). The key also allows the recording of other important features: Normal hyphal growth/morphology (H), asexual sporulation (SA) and perithecial development leading to sexual sporulation (SS), DON mycotoxin production *in vitro* (TP) and *in vivo* (TIV). To use this key, wild type phenotypes are recorded in capital letters, while mutant phenotypes are recorded in lower-case letters similar to the symbols in use for genetic nomenclature. It is expected that the consequent use of such a phenotype key will allow a more detailed comparative analysis of the disease progression in wheat heads.

We applied the *Fusarium* fate map described above to annotate the virulence genes listed in Table 4.2 using the information and figures provided in the publications. For 38 genes (64%) a mutant block was described in sufficient detail in the literature to allow assigning one of the seven key stages, while for the remaining genes the phenotype for wheat head infection was not described, instead a reduction on maize and/or barley floral infections was noted. Twenty mutants were blocked in wheat head infection at stage C and did not transition the infection into the rachis node. This raises the possibility that the rachis node is a critical defence barrier during fungal invasion of attenuated strains. In a recent transcriptomics analysis of the wheat head infection process on a susceptible genotype, the greatest number of

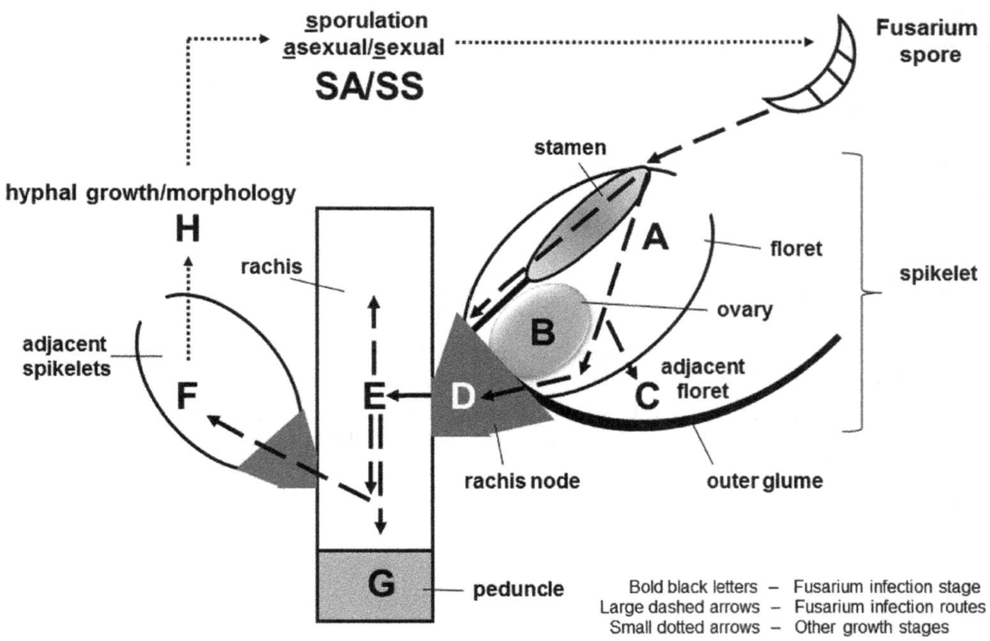

Figure 4.6 *Fusarium* fate map on wheat heads. (A) *Fusarium* spore germinates at the tip of a floret. Hyphae grow (dashed line) ectopically on stamen, filament and inner sides of palea and lemma (A), invade and abort ovary development (B). Hyphal growth then spreads into adjacent florets within the same spikelet (C). Infection of the rachis node (D) occurs. Hyphae transition into the rachis to infect the vascular system and grow upwards and downwards. Adjacent spikelets (F) become infected. The wild-type *F. graminearum* strain is able to grow into the peduncle (G). Hyphae grow on crop stubble or *in vitro* (H) and are able to produce asexual spores (SA) and sexual spores (SS) in perithecia.

differentially expressed *F. graminearum* genes was noted to occur at 5 days post spray inoculation. This coincides with the entry of hyphae into the rachis nodes (Lysoe *et al.*, 2011b). The virulence genes listed in Table 4.2 also reveal that very few mutants are arrested in pathogenicity at stage A, i.e. solely anther colonization. However, in view of the existence of the newly described symptomless colonization of rachis tissue, the phenotype of many of the reduced virulence mutants will need to be re-examined to pinpoint the exact cellular position of hyphal arrest. The use of strains that co-express a reporter protein help to simplify the *in planta* tracking of the hyphal growth patterns for each mutant. So far this has only been used for the *tri5* GFP mutant strain (Jansen *et al.*, 2005) and the histone deacetylase mutant *hdf1* GFP (Li *et al.*, 2011).

The highest proportion of mutants (50%) which give an infection type F is found in the MIPS functional category 'Metabolism' (Table 4.2). In these mutants, hyphae are able to advance through the rachis and cause visible disease on one or more neighbouring spikelets.

Two high-throughput gene deletion studies described the analysis of 709 predicted transcription factors (Son *et al.*, 2011b) in strain GZ3639 and 116 protein kinases (Wang *et al.*, 2011a) in strain PH-1. Both studies assayed mutants for changes in 17 phenotypes, including growth, conidiation, pathogenesis, stress responses, sexual reproduction and mycotoxin biosynthesis. In both studies wheat heads were droplet inoculated and disease development was assessed after 14 days. Taken together, ~13,000 phenotypic data were collected and an additional 96 virulence genes were identified (Table 4.3). However, information describing the *in planta* growth phenotypes is limited and possibly will be the topic of future publications.

For the transcription factor mutants, disease development was scored as 'wild type' or

Table 4.4 *Fusarium* – Wheat floral disease assessment key

Stage	Key[1]
A	Mutant only grows ectopically in floret and on stamens, but does not infect. No obvious plant tissue necrosis can be observed.
B	Mutant infects one floret (visible disease) and aborts ovary development. Visible infection does not spread to adjacent florets.
C	Mutant infects typically all adjacent florets, but does not grow via rachis node into the rachis. No rachis browning is observed.
D	Entry of hyphae into the rachis node tissue detected. Microscopy or testing for regrowth from sterilized tissue is required to observe this stage.
E	Mutant infects the rachis. Rachis browning is observed, but adjacent spikelets are not colonized. Mutant growth does not reach the peduncle (stem) tissue to cause macroscopically visible disease.
F	Mutant causes visible bleaching in adjacent spikelets. But disease development is reduced or delayed compared to wild type. Mutant growth does not reach the peduncle (stem) tissue to cause macroscopically visible disease.
G	Mutant causes visible bleaching on adjacent spikelets above and below the point-inoculated spikelet, but disease development is reduced or delayed compared to wild type. Mutant grows into the peduncle.
Other features[1]	
H	Hyphal morphology or growth pattern altered
SA	Sporulaton asexual (SA): Spores altered or not formed
SS	Sporulaton sexual (SS): Spores altered or not formed
TP	Trichothecene mycotoxin production *in planta*
TIV	Trichothecene mycotoxin production *in vitro*

[1]Please note that for the infection key, the letter when denoted in capitals indicates the last stage of infection completed by the mutant with a phenotype equivalent to that produced by the wild-type strain. For the hyphal morphology, asexual and sexual sporulation key, the capital letters (H, SA, SS) are used to indicate a wild-type phenotype. Small letters are used (h, sa, ss) to indicate an aberrant phenotype. Trichothecene mycotoxin production *in planta* (TP) and *in vitro* (TIV) can be quantified using the terms: increased, reduced and wild type.

'markedly reduced' and photographed (Fig. S5 in Son *et al.*, 2011b). Using both pieces of evidence, we classified 11 of these mutants as 'loss-of-pathogenicity' mutants. For these mutants, no obvious tissue necrosis was observed on wheat ears (see stage A in Fig. 4.6). The other 50 mutants were classified as 'reduced virulence' mutants. Here, the details provided, did not allow further separation using our *Fusarium*-wheat floral disease key (Table 4.4). Information on *in vitro* phenotypes and mycotoxin production *in planta* is available online at the website setup by the authors (http://kropbase.snu.ac.kr/cgi-bin/*Fusarium*/*Fusarium*_main.cgi).

Wang and colleagues (2011a) identified additional 35 protein kinases as virulence factors. The authors supplied information on the mean number of necrotic spikelets per wheat head for each mutant. We used available information for the MAP kinase gene *GPMK1/MAP1* to correlate the disease scores with our FUS disease stages. The GPMK1/MAP1 mutant was assigned a disease score of 0.5 (Table 4.2 in Wang *et al*, 2011a), corresponding to a 'loss-of-pathogenicity' phenotype described earlier (Jenczmionka and Schäfer, 2003; Urban *et al.*, 2003). Eight virulence protein kinases showed a comparable or lower score (loss-of-pathogenicity), while for the other 27 protein kinases a higher score, but less than for the wild type was reported (reduced virulence).

Interestingly, in both studies no hypervirulence mutants were identified. DON overproduction was reported for five transcription factor mutants.

Small secreted proteins

In many plant–pathogen interactions, there has been an intense search in recent years to identify the repertoire of small secreted effectors which contribute to the disease causing ability of the pathogen. These effectors typically operate by helping to suppress the plant defence responses triggered by the plant cells following recognition of essential components of the pathogen's surface. These components, such as chitin fragments, are collectively termed pathogen associated molecular patterns (Chisholm *et al.*, 2006; Jones and Dangl, 2006).

In certain host genotypes specific effectors are recognized. Recognition results in the activation of resistance gene mediated defences (Jones and Dangl, 2006). In the wheat–*F. graminearum* interaction, no gene-for-gene relationships have been identified even when extensive host germplasm and isolate collections have been screened. *F. graminearum* does not contain homologues of the *F. oxysporum* f sp. *lycopersici* (*Fol*) SIX genes (secreted in xylem) which activate defence conferred by different tomato disease resistance genes (see Chapter 3). In the list of experimentally verified *F. graminearum* virulence genes, the lipase encoded by *FLG1*, is predicted to be secreted into the wheat apoplast. Interestingly, in the absence of this protein, a massive defence response is induced in the neighbouring wheat cells early in the infection process (Voigt *et al.*, 2005). The second secreted virulence molecule is the siderophore triacetyl fusarinine C (TAFC) (Oide *et al.*, 2006). TAFC is required for iron import into the *Fusarium* cells and is also proposed to have a role in protecting hyphae against damage caused by hydrogen peroxide generated by wheat cell in response to infection. The third secreted virulence molecule is the DON mycotoxin, which inhibits protein synthesis (Kimura *et al.*, 1998). DON production is believed to suppress plant defence responses in advance of the *Fusarium* hyphae and several of the *TRI* genes are maximally expressed during the early symptomless phase (Brown *et al.*, 2011). In the genome of *F. graminearum* over 1369 (~10%) secreted proteins are predicted to be present of which 79% are shared with other *Fusarium* species and ~1.0% appear to be *F. graminearum* specific (Cuomo *et al.*, 2007; Brown *et al.*, 2012).

Bioinformatics approaches to analyse *F. graminearum* virulence

Prediction of virulence genes using a web-based database (PHI-base) that integrates phenotypic and molecular information

The number of genes confirmed by gene and/or transcript disruption experiments to be required for the disease causing ability of microbes has gradually increased over the past two decades. The availability of full genome sequence data for a large number of microbial pathogens will significantly impact the total number of genes and phenotypes described in the literature reporting on plant–microbe and animal–microbe interactions over the next decade. However, only a tiny fraction of this data deluge is currently being stored in formats readily accessible to computers. In addition, phenotypic information on virulence factors and their mechanisms of action, the data associating pathogens with their hosts, and environmental and experimental conditions is currently not consistently noted across pathosystems. Comparative analyses are at best cumbersome.

In an effort to reveal generic and novel themes of pathogenicity and to discover potentially generic intervention targets, we developed and curate the Pathogen-Host-Interactions database (PHI-base) which captures all information available on mutants described in the literature for many pathogenic species (Baldwin *et al.*, 2006; Winnenburg *et al.*, 2006, 2008). PHI-base was designed as part of a collaborative project between molecular plant pathologists and computational scientists at Rothamsted Research and was first launched in 2005. The database is available online at www.phibase.org. Currently the database contains information for 2420 genes from 107 pathogenic species of fungi, oomycetes and bacteria that infect plant, animal, insect and other host types. PHI-base also includes antifungal compounds and their target genes using information provided by the Fungicide Resistance Action Committee (www.frac.info).

A user-friendly search interface is provided on the entry portal for PHI-base (Fig. 4.7). PHI-base is accessed by scientists in over 80 countries across

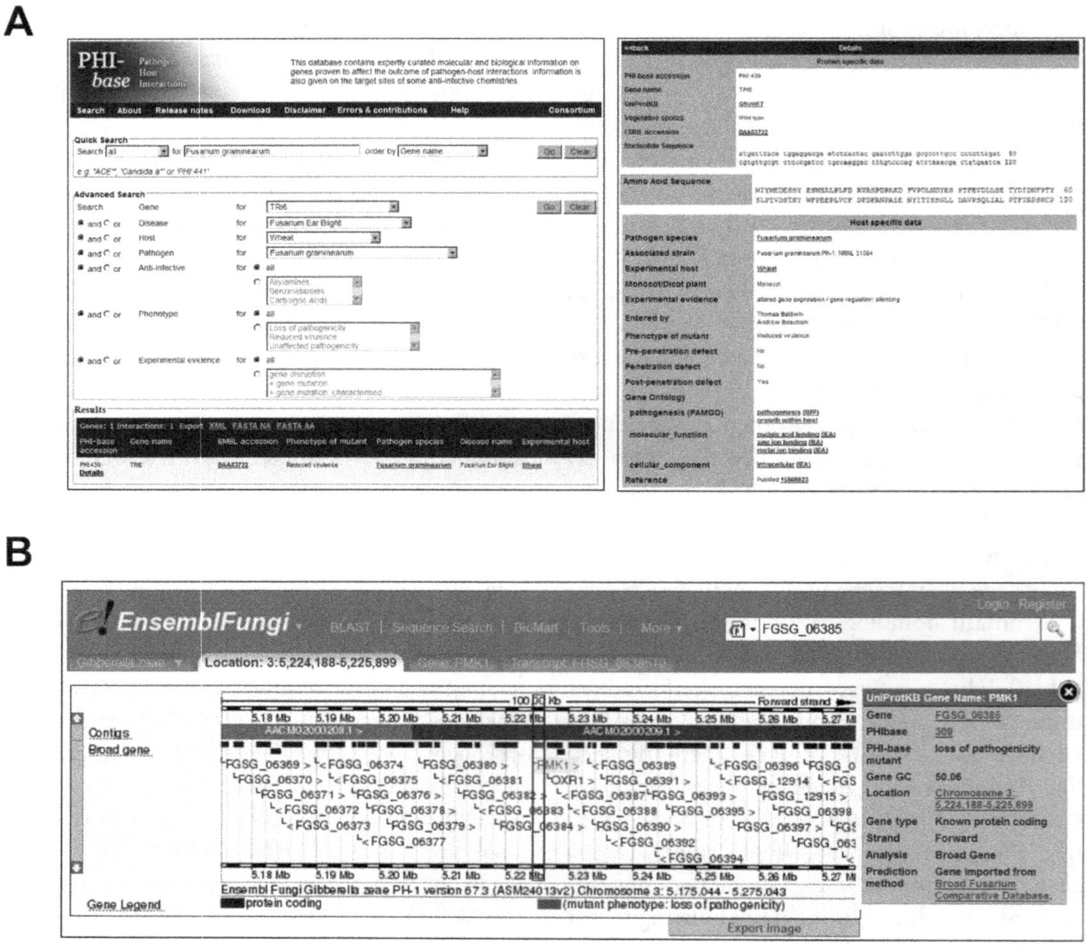

Figure 4.7 (A) PHI-base web portal. PHI-base database stores phenotypic and molecular information on host–pathogen interactions from 107 different pathogens. Left: Search interface for retrieving information on the *F. graminearum TRI6* gene. Right: Details page showing nucleotide and protein sequences automatically retrieved from EMBL database, Uniprot identifier, host links to Uniprot and EMBL database and phenotypic information including spore formation, virulence and experimental evidence retrieved from the publication by manual curation. (B) Ensembl Fungi *Gibberella zeae* genome browser displaying the FGSG_06385 *(PMK1)* gene (http://fungi.ensembl.org/Gibberella_zeae/Info/Index). The pop up window on the right displays gene and PHI-base identifier information including the phenotype of the gene deletion mutant (loss of pathogenicity).

six continents, and is being used for increasingly diverse types of studies. Since its inception, the database has included the full repertoire of virulence genes described for *F. graminearum* and the colonization of various plant species.

A prerequisite to identify computationally functional commonalities across pathogenic species and hosts is the use of a standardized language describing these interactions. An international effort has been made to employ and develop controlled vocabularies (ontologies). In ontologies each terms has a well-defined relationship and is often hierarchical. Examples for ontologies are the widely used Gene ontology (GO), PAMGO (Plant-Associated Microbe Gene Ontology) and more recently the PHI-base specific ontologies (Ashburner *et al.*, 2000; Torto-Alalibo *et al.*, 2009; Winnenburg *et al.*, 2006). An example of a species-specific controlled vocabulary for the *F. graminearum*-wheat head interaction to define the various *F. graminearum* virulence gene mutants is given in Table 4.4. The characterization of additional features of this interaction will lead to further refinements of the ontologies. It is important to note that the higher level generic ontologies permit inter-species comparisons across the greatest taxonomic distances.

The PHI-base database can be downloaded as a flat file for computational studies. Researchers recently employed downloaded copies of PHI-base to set up BLAST databases to identify known virulence genes in several newly sequenced pathogenic species (Bouzidi *et al.*, 2007; Jeon *et al.*, 2007; Kleemann *et al.*, 2008; Oeser *et al.*, 2009). The database has also been used as a source for experimental data in evolutionary studies of pathogenicity in Ascomycetes, and in comparative genomics analysis for pesticide target site identification (Sanchez-Rodriguez *et al.*, 2010; Sexton and Howlett, 2006).

In a recent development, the PhytoPath project integrates genome-scale data including *F. graminearum* with phenotypes of host infection from PHI-base within the Ensembl Fungi genome browsers (http://fungi.ensembl.org) (Fig. 4.7). The growing list of agriculturally important pathogenic species available in Ensembl Fungi can be accessed via the PhytoPath web site (www.phytopathdb.org).

Network analysis to predict virulence factors

InParanoid cluster analysis

We have used an InParanoid cluster analysis approach based on sequence similarity (Remm *et al.*, 2001) to predict the phenotypic effect of gene deletions in *F. graminearum* using PHI-base. In this approach, the whole *F. graminearum* genomic sequence (FG3 gene call) was surveyed. The generated protein clusters were displayed using the Ondex data integration and visualization software (Kohler *et al.*, 2006) and analysed for the presence of known virulence and non-virulence gene homologues from the 97 pathogens stored in PHI-base (Vers 3.3). Fig. 4.8 depicts three of the 209 clusters generated. Using the 'guilt by association' assertion (Petsko, 2009) phylogenetically related proteins are more likely to have a similar phenotype in gene deletion experiments. The cluster analysis shows that adenylate cyclase is a virulence factor in four plant- and two animal-pathogenic fungi (Fig. 4.8A). Thus, by association, the *F. graminearum* protein FGSG_01234 is predicted to be a virulence factor. In Fig. 4.8B showing a protein kinase regulatory subunit cluster, the associated phenotypes are diverse and include wild-type virulence (*M. oryzae SUM1*), reduced virulence (*Ustilago maydis* UBC1 and *Colletotrichum lagenarium* RPK1) and increased virulence (*Cryptococcus neoformans PKR1*). In this case, clustering of FGSG_09908 cannot suggest a phenotype and a role in virulence of wheat had to be determined through experimentation (Fig. 4.2) (Beacham, 2011). In Fig. 4.8C (neutral trehalase gene cluster) only two phenotypes occur: wild-type virulence (*Botrytis cinerea TRE1* and *Leptosphaeria maculans NTH1*) and reduced virulence (*M. oryzae PTH9* and MGG_09471). *PTH9* and MGG_09471 indicate the same gene was characterized in two independent studies. The phenotype of the *F. graminearum* neutral trehalase gene (FGSG_09895) has yet to be determined experimentally.

Protein–protein interaction networks

The disease formation process involves many proteins and other molecules, including enzymes, secreted effectors and toxins. Some of these proteins are likely to interact directly as part of a network; others may be kept apart through internal compartmentalization. Recently a protein–protein interaction network analysis approach was taken in *F. graminearum* to identify virulence gene candidates (Liu *et al.*, 2010). A total of 49 potential virulence proteins were identified of which nine have been previously connected to virulence (Fig. 4.9). This presumed intracellularly located network will provide guidance to future wet-biology based experimental analyses (Liu *et al.*, 2010). The network analysis involved determining the set of the interacting proteins to known virulence proteins, which are up regulated during Fusarium head blight disease and are either considered to be the interaction orthologues (interologues) of proteins shown to interact in other species or share interacting Pfam domains. The authors speculate that these subgroups of interacting proteins are also likely to be virulence factors. The analysis computationally

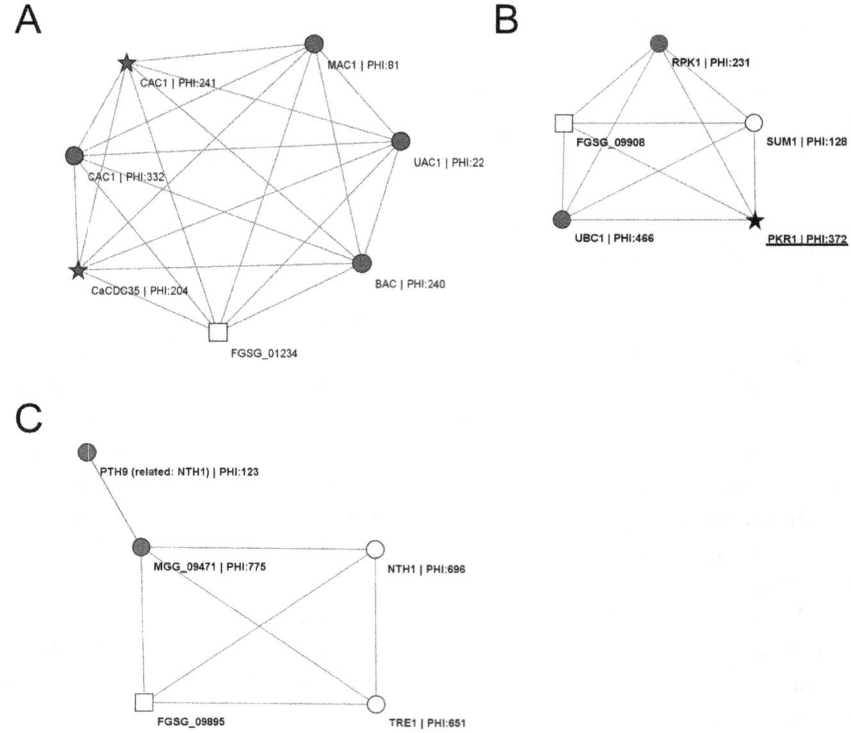

Figure 4.8 Identification of *F. graminearum* virulence gene homologues by cluster and network analysis. The clusters are (A) cAMP forming enzyme adenylate cyclases, (B) protein kinase A regulatory subunits and (C) neutral trehalases. In each of these clusters a proven role in pathogenicity or virulence has been shown in one or more pathogen species and the predicted *F. graminearum* member of each cluster is given. The predicted protein FGSG_09908 (panel B) is now known to have a role in virulence towards wheat spikes (Beacham, 2011). Symbols: Circles and stars indicate plant and animal pathogens, respectively. Phenotypes are indicated by shading: wild type (white), reduced virulence (grey) and increased virulence (black). Increased virulence was only observed for the *pkr1* mutant (underlined). White rectangles indicate *F. graminearum* gene homologues. Gene names and PHI-base accession numbers are given.

integrated the following three data sources: (1) the predicted protein-interactome from *F. graminearum* of 3745 proteins with 27,102 interactions using data generated in diverse species including *Drosophila,* nematode, human, mouse, baker's and fission yeast and *E. coli* (Zhao *et al.,* 2009); (2) the *F. graminearum* gene expression datasets stored in PLEXdb (Table 4.1) (Wise *et al.,* 2007) and (3) the multispecies virulence genes stored in PHI-base to seed the network (Winnenburg *et al.,* 2006).

The recent availability of a significant higher number of characterized *F. graminearum* virulence genes, arising from the two high-throughput studies, prompted Bennett *et al.* (2012) to refine this network analysis. An integrated network contains 9521 nodes (proteins) and 80,997 links was generated. Then by using various algorithms, Bennett *et al.* explored the disjoint and overlapping community structure of this network, the locations of the experimentally verified virulence proteins and other proteins annotated with Gene Ontology terms. This lead to the identification of a set of overlapping communities which contained 75 of the 98 verified virulence protein seeds which had mapped to the integrated network. At the intersection of these communities, proteins were identified that are highly connected, multifunctional and are predicted to have a crucial function in virulence. Interestingly, only four of the verified virulence proteins used to seed the networks were found to map to these intersections, i.e. belong to multiple communities, namely, FGSG_ 04104,

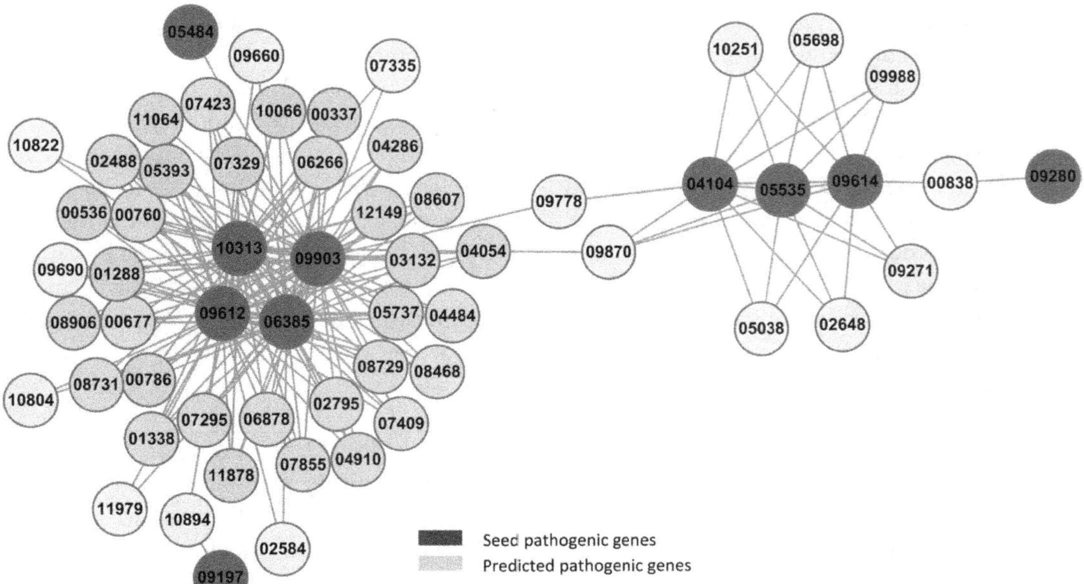

Figure 4.9 Systems biology network approach to predict virulence genes in *F. graminearum.* Liu and colleagues (2010) proposed that interacting proteins share similar functions and are involved in the same pathways. The depicted high confidence protein–protein interaction network (Sij≥4) consisting of 10 seed genes and 49 virulence gene candidates was created by identifying the subset of genes which show differential expression during *in planta* growth compared to *in vitro* growth of the fungus and displaying only the genes that interact with at least two seed genes. The seed pathogenicity genes are indicated as dark grey circles: *MGV1* (10313), *HOG1* (09612), *GPMK1/MAP1* (06385), *STE7* (09903), *STE11* (05484), *GPB1* (04104), *GPA1* (05535), *GPA2* (09614), *HMR1* (09197) and FGSG_09280. Most seed genes have a known role in MAPK signalling. The seed gene FGSG_09280 is not included in PHI-base v3.1 and was mistakenly included. Genes that interact with at least two seed genes (medium grey) encode potential virulence factors. Numbers are 'FGSG' gene identifiers.

FGSG_08028. FGSG_10142 and FGSG_03747 (Bennett *et al.*, 2012). The other proteins are hypothetical proteins with no annotation. It is proposed, that network analyses and similar systems biology strategies offers the potential to predict the function of the ~40% of proteins in *F. graminearum* that do not have annotation and will provide a rational basis for reverse genetics experiments of these 'hypothetical' proteins.

Both the InParanoid cluster analysis and the protein–protein interaction network analyses confirmed the importance of cell signalling processes for virulence. Various proteins involved in G-protein and MAP kinase mediated signalling pathways were identified as candidate virulence genes. The InParanoid cluster analysis, which does not rely on protein-protein interologue data from yeast and other non-fungal species, additionally identified several metabolic enzymes specific to filamentous fungi as putative virulence factors in *F. graminearum* (data to be reported elsewhere).

Spatial clustering of virulence gene homologs on chromosome 1

An increasing number of whole-genome sequences are available for many fungal species with a wide range of lifestyles (www.genomesonline.org). It is now possible to investigate the distribution of specific gene families and other gene groups within and across taxonomically related pathogenic and free living species. For *F. oxysporum* f. sp. *lycospersici* certain genes required to alter the host species range were shown to be located on the smallest of a total of 15 chromosomes (Ma *et al.*, 2010) (Chapter 3). These small chromosomes (< 3 Mb) were shown to be transmittable between the vegetative hyphae of individual isolates belonging to different *forma specialis*. However, for *F. graminearum* and several other cereal infecting *Fusarium* species, a similar scenario does not appear to exist because of the apparent lack of small chromosomes.

In *F. graminearum* very small gene clusters (typically 8–15 genes) encode proteins required for biosynthesis of various metabolites including trichothecenes, butenolide and aurofusarin (Frandsen *et al.*, 2006) (see Chapter 3). These clusters typically include genes, which code for transcription factors regulating other cluster gene members, cytochrome P450 proteins, polyketide synthases and/or non-ribosomal peptide synthetases (Brown *et al.*, 2004; Brown *et al.*, 2001; Gaffoor *et al.*, 2005; Harris *et al.*, 2007; Varga *et al.*, 2005). Many of these clusters reside in genomic regions with moderate to high recombination rates (Fig. 4.4).

In a recent study, our laboratory used a statistical bioinformatics approach to identify a novel type of gene cluster for virulence determinants. From the PHI-base database (www.phibase.org) the complete list of known virulence gene homologues was obtained (Winnenburg *et al.*, 2006). Using BLAST analysis (at a cut-off value of e^{-100}) all virulence gene orthologues were identified in the *F. graminearum* genome and their distribution across the four chromosomes was statistically analysed (Fig. 4.4A). A total of 211 hits were identified over the four chromosomes. The statistical analysis revealed the presence of a close grouping of five virulence gene homologues in a 37.6 kb region (15 genes) on chromosome 1 (Beacham, 2011).

Interestingly, the recombination rates obtained for this region in a cross between the sequenced strain PH-1 and strain MN00-676 from the same genetic lineage (Gale *et al.*, 2005) reveals that the identified micro-region resides within a region with a very low recombination frequency. It is possible that core virulence determinants are located in chromosomal regions with a reduced recombination rate to protect their integrity. Most of the proven *F. graminearum* virulence genes listed in Tables 4.2 and 4.3 reside within genomic regions with little or no recombination (Fig. 4.4B). Somewhat in contrast to this finding is the fact that genes specifically expressed during *in planta* growth are enriched in regions with a higher recombination rate and are frequently located in subtelomeric regions (Cuomo *et al.*, 2007; Guldener *et al.*, 2006a). We therefore propose that two classes of virulence factors exist: first, a core class of virulence factors with a more general role in virulence, which is conserved between a range of different pathogenic species; and secondly, a class of virulence factors, as suggested by Cuomo *et al.* (2007), that resides in regions with higher recombination rates to allow for adaptation in the arms race between host and pathogen.

Comparison of the evolution of pathogenicity and virulence genes versus the *TRI* genes in *Fusarium* species and isolates

Since the mid-1990s there has been a global re-emergence of *Fusarium* disease on the floral tissues of wheat as well as other small grain cereals. This increased disease occurrence and severity appears to be driven by global changes to our climate and altered agronomic practices (Leonard and Bushnell, 2003; McMullen *et al.*, 1997). Numerous *Fusarium* species have now been reported to cause FHB. However, these species are not uniformly distributed across the key cereal production regions (Chapter 1). Early studies indicated that *F. graminearum* was one of the most prevalent species globally. This apparently pandemic species was classified into several distinct genetic lineages based on marker analyses (O'Donnell *et al.*, 2000). Following further molecular investigations, some of the genetic lineages described have since been given distinct species names. It was also recognized that in specific parts of the world, where many *Fusarium* species once prevailed, one *Fusarium* species now pre-dominates, whereas in other regions *Fusarium* species diversity has been retained. Fortunately, vast ex-situ *Fusarium* species and isolate collections exist which contain isolates recovered from wheat and other cereal species over this entire period. The largest collections are maintained by three fungal stock centres, the USDA Northern Regional Research Lab Culture Collection (NRRL; Peoria, IL, USA), the Fungal Genetics Stock Center (FGSC; Kansas City, MO, USA) and the Centraalbureau voor Schimmelcultures

(CBS; Utrecht, The Netherlands). Therefore, dynamic changes to the global cereal infecting *Fusarium* population have been captured and can be analysed in the future.

Potential changes in *Fusarium* species distributions over time (166 years) have also been captured from a single field in the UK (Broadbalk). In this case, grain, stem/leaf and soil samples have been collected and stored from the field growing successive crops of wheat since 1846. From the Broadbalk long-term experiment situated at Rothamsted Research in the UK, linearly amplified DNA samples have now been successfully prepared for most years (Bearchell *et al.*, 2005). Using *Fusarium* species-diagnostic primers, it has been possible to determine that some species such as *F. culmorum* have been present since 1846, whereas others such as *F. graminearum* have only recently arrived, i.e. within the last 12 years (L. Fernandez-Ortuno and B. Fraaije, personal communication).

In this chapter, we have focused on the characterization of the pathogenicity and virulence genes of *F. graminearum*. Most molecular genetics investigations were done using the sequenced strain PH-1. This strain is a member of the pandemic genetic lineage 7 and was isolated in the USA from an infected wheat grain taken from a harvesting elevator in 1996. With the availability of different *F. graminearum* species and genetic lineages in the stock collections, it is possible to determine whether the virulence genes (Tables 4.2 and 4.3) exhibit different mutation rates. Using samples from the Broadbalk long term experiment, temporal changes in mutation rates for an entire *Fusarium* population grown within a single field under continuous wheat production could be explored. Neither type of analyses has yet been done but both could be highly informative.

Initial global studies of the *F. graminearum* population explored the evolution of specific genes. Various *TRI* genes important for the regulation and the production of the type B trichothecene mycotoxin appeared to have evolved independently from 12 other nuclear encoded non-*TRI* genes (O'Donnell *et al.*, 2000). Although the non-*TRI* genes analysed in this study do not appear to encode virulence proteins, one, encoding the β-tubulin protein, is a known fungicide target (www.frac.info, FRAC code list 2011).

Within a single *Fusarium* species, isolates that synthesize a specific trichothecene (e.g. DON, 3A-DON, 15A-DON or nivalenol (NIV)) are referred to as chemotypes and may exist in the same geographic regions. Over time, the ratio of chemotypes in a region can change. For example in the UK, 3A-DON, 15A-DON and NIV *F. graminearum* chemotypes have readily been isolated (Polley and Turner, 1995). Over the past 15 years the incidence of the 3A-DON chemotype has decreased whilst 15A-DON types have increased (HGCA Crop Monitor, http://www.cropmonitor.co.uk). In contrast, in Canada, over the same period of time the largely dominant 15A-DON chemotype is being replaced by 3A-DON chemotypes (Gilbert *et al.*, 2010).

Recently, the *TRI8* gene which codes for a trichothecene C-3 esterase has been shown to convert the diacetylated 3A,15A-DON into either 3A-DON or 15A-DON (Alexander *et al.*, 2011). The activity of this enzyme therefore defines the chemotype of the strain. With this genetic ability to define the chemotype of *Fusarium* species, it would also be interesting to explore whether chemotype differences correlate with changes in the sequence of other virulence genes. Also, by taking a molecular genetics approach, where new chemotypes of a species have recently emerged in a region, the other *TRI8* sequence could be inserted into these new strains to restore the originally prevalent chemotype. Then through comparative *in planta* testing it would be possible to ascertain if the heightened aggressiveness towards wheat floral tissue was due either to the chemotype change *per se* or due to other changes in the genome. If the later scenario was encountered, then by using a next generation re-sequencing approach, the underlying genomic differences and similarities between several isolates from the two chemotype groups could be revealed. This approach may provide a new way to defining additional virulence determinants and virulence protein–protein networks, which may be particularly relevant in specific geographical regions/production systems.

Inspection of the list of phenotypes associated with the deletion of single virulence genes described in Tables 4.2 and 4.3 indicates that many of the required *F. graminearum* proteins also have a role in asexual spore formation or sexual spore formation or both processes. The comprehensive transcription factor and kinome studies also reveal this possibility (Son *et al.*, 2011b, Wang *et al.*, 2011a. Therefore, these three very different biological processes appear to be mechanistically linked. Potentially, this observation, suggests this subset of gene sequences could be under severe evolutionary constraint and therefore show only minimal sequence variation.

This chapter has focused primarily on determining the function of specific genes and their gene products during infection of wheat floral tissue on living plants. However, a major practical consideration is the extent to which *Fusarium* can regrow on stored wheat grain which can lead to post-harvest spoilage on the farm, at the granary and/or during transportation (Magan *et al.*, 2003). So far none of the pathogenicity and virulence gene mutants generated have been tested for alterations in this aspect of the *Fusarium* life cycle. Gene modifications that altered a strain's ability to survive within dried grains would clearly influence patterns of *Fusarium* species dispersal and/or prevalence either locally or globally.

Concluding remarks

Species in the genus *Fusaria* can infect a single plant species or multiple plant species or may cause post-harvest disease problems. Often disease formation on certain cereal and non-cereal hosts is associated with multiple and often related fungal species (referred to as a species complex). The composition of a species complex can be influenced by local conditions. The production of specific toxins during infection has been shown to contribute directly to fungal virulence towards certain hosts. For example, in wheat the toxins deoxynivalenol (DON) and nivalenol (NIV) are essential virulence factors, whereas in maize only NIV was shown to contribute to virulence (Maier *et al.*, 2006). In contrast, both DON and NIV are dispensable for virulence of barley (Maier *et al.*, 2006). For other *in planta* interactions, whether the production of specific toxins is causally involved in disease formation or for another purpose(s) is not yet known. In the field, some *Fusarium* species which cause visible disease on potato tubers or the symptom- or symptomless infection of sugar beet roots can also cause considerable levels of disease and DON mycotoxin accumulation when inoculated onto wheat heads (Burlakoti *et al.*, 2007). The underlying genes, pathways and networks which define host species range and the degree of symptomless to visible infection in each host species have yet to be determined. In the taxonomically related species *Epichloe festucae* (an endophyte), different single gene alterations often resulted in an increase in intercellular fungal biomass and the appearance of visible disease symptoms (Tanaka *et al.*, 2006).

F. graminearum, the primary focus of this chapter, is currently one of the most important *Fusarium* species globally causing pre- and post-harvest disease on wheat and other small grain cereals. The now well-annotated four chromosome genome has revealed the existence of a 'two-speed' genome which accommodates the non-random distribution of different predicted gene types. The full evolutionary advantages and constraints of this spatial pattern, which is predicted to have emerged through the fusion of several smaller chromosomes, are unclear. However, the prevalence of the sexual cycle and the lack of transposon activity means that the basic blueprint of the *F. graminearum* genome is likely to be perpetuated for the foreseeable future. One of the most interesting observations to have arisen from the initial transcriptomics studies of the floral infection process, albeit on barley heads, was the over-representation of the *in planta*-expressed genes in the genomic regions with the highest levels of recombination. Could changes in these 'hot' genomic regions be the driver of host species range? So far, the various forward and reverse genetics experiments (and now the first *in silico* predictions) indicate that the genes coding for the various virulence factors reside predominantly within the genomic regions with the lowest levels of recombination. The only exception being those genes directly concerned with trichothecene toxin production or genes, which are strongly up-regulated under toxin inducing conditions. Whether this pattern will continue as more protein function results accrue will be informative. For a number of plant and animal pathogen species these genome wide analyses are increasingly used to test specific hypotheses concerning the evolution of pathogenicity, host defence suppression and host range.

The effective control of Fusarium head blight/head scab disease is now a high priority in all the major wheat-growing regions of the world to ensure the grain harvested is safe for food and feed consumption and for various fermentation purposes. So far, the wheat infecting fusaria have shown considerable intrinsic resistance to all the major classes of fungicides. In addition, adequate coverage of the various floral tissues with a single low volume fungicide application has proven to be technically challenging. Similarly, improvements in resistance of the wheat plants through genotype selection and marker-assisted selection have only achieved modest success. For example, the various major resistance QTLs identified to date in global wheat germplasm collections have not been very effective in high yielding elite commercial wheats. The new biological and molecular information described here, gained through the combined efforts of the international *Fusarium* community, and through comparative analyses with other pathosystems, provides a considerable deeper insight into the *Fusarium* – wheat interaction, than was available just a few years ago. Similarly, with the increasing use of next generation sequencing technologies, it will be feasible to acquire full genome information for numerous additional strains and species, which cause more or less disease on wheat. This comparative information will ensure that new *Fusarium* targets for intervention being considered are widespread, relatively monomorphic and reside in the most stable regions of the genome. With the increasing use of comparative species analyses within and beyond the *Fusaria* and *in silico* predictions to identify putative pathogenicity/virulence genes, the rate of gene function discovery for *F. graminearum* and the other wheat infecting species will greatly accelerate. By developing electronic models to describe the wheat head infection process, coupled to fully interactive *Fusarium* genome browsers, this will permit researchers to organize and interrogate the growing volumes of gene function and 'omics' data, to define pathways and networks of importance to disease formation and/or mycotoxin formation and to discern where key information is still missing. To realize effective and sustainable disease control several different intervention targets in both *Fusarium* and wheat will need to be found and pursued.

Acknowledgements

At Rothamsted Research all experiments involving transgenic strains of *F. culmorum* and *F. graminearum* use biological containment facilities under FERA licence number PHL 174G/6222 and with appropriate approval from the UK Health and Safety Executive. Dr John Antoniw provided invaluable input into the analysis of the *Fusarium graminearum* genome and in developing the OmniMapFree software used to generate Fig. 4.4. Drs Jacek Grzebyta, Mansoor Saqi, Andrea Splendiani, Michael Defoin-Platel and Chris Rawlings are thanked for continuing to assist in the maintenance and further development of the functionality of the Pathogen-Host Interactions database (PHI-base). We thank Drs Neil Brown and Andrew Beacham and Catherine Canevet for help in the preparation of Fig. 4.1, 4.3 and 4.9, respectively. Many members of the *Fusarium* community provided additional phenotyping information for the completion of Table 4.2 ahead of publication. This was highly appreciated. We also thank Dr Neil Brown and Professor John Lucas for critical reading earlier versions of this chapter. Rothamsted Research receives grant-aided support from the Biotechnology and Biological Sciences Research Council (BBSRC) of the UK. PHI-base receives additional support from the BBSRC Phytopath grant (Grant No BB/I001077/1) held jointly with the European Bioinformatics Institute, Cambridge, UK, and from BBSRC National Capability funds (Grant No. BB/J/004383/1).

References

Alexander, N.J., McCormick, S.P., Waalwijk, C., van der Lee, T., and Proctor, R.H. (2011). The genetic basis for 3-ADON and 15-ADON trichothecene chemotypes in *Fusarium*. Fungal Genet. Biol. *48*, 485–495.

Ashburner, M., Ball, C.A., Blake, J.A., Botstein, D., Butler, H., Cherry, J.M., Davis, A.P., Dolinski, K., Dwight, S.S., Eppig, J.T., *et al.* (2000). Gene ontology: tool for the unification of biology. The Gene Ontology Consortium. Nat. Genet. *25*, 25–29.

Audenaert, K., Callewaert, E., Hofte, M., De Saeger, S., and Haesaert, G. (2010). Hydrogen peroxide induced by the fungicide prothioconazole triggers deoxynivalenol (DON) production by *Fusarium graminearum*. BMC Microbiol. *10*, 112.

Bai, G.H., Desjardins, A.E., and Plattner, R.D. (2002). Deoxynivalenol-nonproducing *Fusarium graminearum* causes initial infection, but does not cause disease spread in wheat spikes. Mycopathologia *153*, 91–98.

Baldwin, T.K., Gaffoor, I., Antoniw, J., Andries, C., Guenther, J., Urban, M., Hallen-Adams, H.E., Pitkin, J., Hammond-Kosack, K.E., and Trail, F. (2010a). A partial chromosomal deletion caused by random plasmid integration resulted in a reduced virulence phenotype in *Fusarium graminearum*. Mol. Plant Microbe Interact. *23*, 1083–1096.

Baldwin, T.K., Urban, M., Brown, N., and Hammond-Kosack, K.E. (2010b). A role for topoisomerase I in *Fusarium graminearum* and *F. culmorum* pathogenesis and sporulation. Mol. Plant Microbe Interact. *23*, 566–577.

Baldwin, T.K., Winnenburg, R., Urban, M., Rawlings, C., Koehler, J., and Hammond-Kosack, K.E. (2006). The pathogen–host interactions database (PHI-base) provides insights into generic and novel themes of pathogenicity. Mol. Plant Microbe Interact. *19*, 1451–1462.

Beacham, A. (2011). Pathogenicity of *Fusarium graminearum* and *Fusarium culmorum* on wheat ears. PhD thesis in Plant Pathogen and Microbiology (Exeter University, Exeter).

Beacham, A.M., Antoniw, J., and Hammond-Kosack, K.E. (2009). A genomic fungal foray. Biologist *56*, 98–105.

Bearchell, S.J., Fraaije, B.A., Shaw, M.W., and Fitt, B.D. (2005). Wheat archive links long-term fungal pathogen population dynamics to air pollution. Proc. Natl. Acad. Sci. U S A. *102*, 5438–5442.

Bennett L., Lysenko, A., Papageorgiou, L., Urban, M., Hammond-Kosack K.E., Rawlings, C., Saqi, M., and Tsoka, S. (2012). Detection of multi-clustered genes and community structure for the plant pathogenic fungus *Fusarium graminearum*. CMSB Lecture Notes Computer Sciences *7605*, 69–86.

Bluhm, B.H., Zhao, X., Flaherty, J.E., Xu, J.R., and Dunkle, L.D. (2007). *RAS2* regulates growth and pathogenesis in *Fusarium graminearum*. Mol. Plant Microbe Interact. *20*, 627–636.

Boddu, J., Cho, S., Kruger, W.M., and Muehlbauer, G.J. (2006). Transcriptome analysis of the barley–*Fusarium graminearum* interaction. Mol. Plant Microbe Interact. *19*, 407–417.

Boenisch, M.J., and Schäfer, W. (2011). *Fusarium graminearum* forms mycotoxin producing infection structures on wheat. BMC Plant Biol. *11*, 110.

Bouzidi, M.F., Parlange, F., Nicolas, P., and Mouzeyar, S. (2007). Expressed sequence tags from the oomycete *Plasmopara halstedii*, an obligate parasite of the sunflower. BMC Microbiol. 7, 110.

Breakspear, A., Pasquali, M., Broz, K., Dong, Y., and Kistler, H.C. (2011). Npc1 is involved in sterol trafficking in the filamentous fungus *Fusarium graminearum*. Fungal Genet. Biol. *48*, 725–730.

Brown, D.W., Dyer, R.B., McCormick, S.P., Kendra, D.F., and Plattner, R.D. (2004). Functional demarcation of the *Fusarium* core trichothecene gene cluster. Fungal Genet. Biol. *41*, 454–462.

Brown, D.W., McCormick, S.P., Alexander, N.J., Proctor, R.H., and Desjardins, A.E. (2001). A genetic and biochemical approach to study trichothecene diversity in *Fusarium sporotrichioides* and *Fusarium graminearum*. Fungal Genet. Biol. 32, 121–133.

Brown, D.W., McCormick, S.P., Alexander, N.J., Proctor, R.H., and Desjardins, A.E. (2002). Inactivation of a cytochrome P-450 is a determinant of trichothecene diversity in *Fusarium* species. Fungal Genet. Biol. *36*, 224–233.

Brown, N.A., Antoniw, J., and Hammond-Kosack, K.E. (2012). The predicted secretome of the plant pathogenic fungus *Fusarium graminearum*: a refined comparative analysis. PLoS One *7*, e33731.

Brown, N.A., Bass, C., Baldwin, T.K., Chen, H., Massot, F., Carion, P., Urban, M., van de Meene, A., and Hammond-Kosack, K. (2011). Characterisation of the *Fusarium graminearum*-wheat floral interaction. J. Pathog. *2011*, Article ID 626345.

Brown, N.A., Urban, M., Van de Meene, A.M.L., and Hammond-Kosack, K.E. (2010). The infection biology of *Fusarium graminearum:* Defining the pathways of spikelet to spikelet colonisation in wheat ears. Fungal Biol. *114*, 555–571.

Buerstmayr, H., Ban, T., and Anderson, J.A. (2009). QTL mapping and marker-assisted selection for Fusarium head blight resistance in wheat: a review. Plant Breeding *128*, 1–26.

Burlakoti, R.R., Estrada, R., Rivera, V.V., Boddeda, A., Secor, G.A., and Adhikari, T.B. (2007). Real-time PCR quantification and mycotoxin production of *Fusarium graminearum* in wheat inoculated with isolates collected from potato, sugar beet, and wheat. Phytopathology *97*, 835–841.

Cappellini, R.A., and Peterson, J.L. (1965). Macroconidium formation in submerged cultures by a non-sporulating strain of *Gibberella zeae*. Mycologia *57*, 962–966.

Catlett, N.L., Lee, B.N., Yoder, O.C., and Turgeon, B.G. (2003). Split-marker recombination for efficient targeted deletion of fungal genes. Fungal Genet. Newslett. *50*, 9–11.

Chisholm, S.T., Coaker, G., Day, B., and Staskawicz, B.J. (2006). Host–microbe interactions: shaping the evolution of the plant immune response. Cell *124*, 803–814.

Cundliffe, E., Cannon, M., and Davies, J. (1974). Mechanism of inhibition of eukaryotic protein synthesis by trichothecene fungal toxins. Proc. Natl. Acad. Sci. U S A. *71*, 30–34.

Cuomo, C.A., Guldener, U., Xu, J.R., Trail, F., Turgeon, B.G., Di Pietro, A., Walton, J.D., Ma, L.J., Baker, S.E., Rep, M., *et al.* (2007). The *Fusarium graminearum* genome reveals a link between localized polymorphism and pathogen specialization. Science *317*, 1400–1402.

Cuzick, A., Urban, M., and Hammond-Kosack, K. (2008). *Fusarium graminearum* gene deletion mutants *map1* and *tri5* reveal similarities and differences in the pathogenicity requirements to cause disease on Arabidopsis and wheat floral tissue. New Phytol. *177*, 990–1000.

Desjardins, A.E. (2006). *Fusarium* Mycotoxins – Chemistry, Genetics and Biology (The American Phytopathological Society, St. Paul, MN).

Desjardins, A.E., Proctor, R.H., Bai, G.H., McCormick, S.P., Shaner, G., Buechley, G., and Hohn, T.M. (1996). Reduced virulence of trichothecene-nonproducing mutants of *Gibberella zeae* in wheat field tests. Mol. Plant Microbe Interact. *9*, 775–781.

Ding, S.L., Mehrabi, R., Koten, C., Kang, Z.S., Wei, Y.D., Seong, K.Y., Kistler, H.C., and Xu, J.R. (2009). Transducin beta-like gene *FTL1* is essential for pathogenesis in *Fusarium graminearum*. Eukaryot. Cell *8*, 867–876.

Dufresne, M., van der Lee, T., Ben M'barek, S., Xu, X., Zhang, X., Liu, T., Waalwijk, C., Zhang, W., Kema, G.H., and Daboussi, M.J. (2008). Transposon-tagging identifies novel pathogenicity genes in *Fusarium graminearum*. Fungal Genet. Biol. *45*, 1552–1561.

Dyer, R.B., Plattner, R.D., Kendra, D.F., and Brown, D.W. (2005). *Fusarium graminearum* TRI14 is required for high virulence and DON production on wheat but not for DON sythesis *in vitro*. J. Agric. Food. Chem. *53*, 9281–9287.

Feng, J., Liu, G.S., Selvaraj, G., Hughes, G.R., and Wei, Y.D. (2005). A secreted lipase encoded by *LIP1* is necessary for efficient use of saturated triglyceride lipids in *Fusarium graminearum*. Microbiology *151*, 3911–3921.

Frandsen, R., Nielsen, N., Maolanon, N., Sorensen, J., Olsson, S., Nielsen, J., and Giese, H. (2006). The biosynthetic pathway for aurofusarin in *Fusarium graminearum* reveals a close link between the naphthoquinones and naphthopyrones. Mol. Microbiol. *61*,1069 – 1080.

Frandsen, R.J.N., Andersson, J.A., Kristensen, M.B., and Giese, H. (2008). Efficient four fragment cloning for the construction of vectors for targeted gene replacement in filamentous fungi. BMC Mol. Biol. *9*, 70.

Gaffoor, I., Brown, D.W., Plattner, R., Proctor, R.H., Qi, W., and Trail, F. (2005). Functional analysis of the polyketide synthase genes in the filamentous fungus *Gibberella zeae* (anamorph *Fusarium graminearum*). Eukaryot. Cell *4*, 1926–1933.

Gaffoor, I., and Trail, F. (2006). Characterization of two polyketide synthase genes involved in zearalenone biosynthesis in *Gibberella zeae*. Appl. Environ. Microbiol. *72*, 1793–1799.

Gale, L.R., Bryant, J.D., Calvo, S., Giese, H., Katan, T., O'Donnell, K., Suga, H., Taga, M., Usgaard, T.R., Ward, T.J., *et al.* (2005). Chromosome complement of the fungal plant pathogen *Fusarium graminearum* based on genetic and physical mapping and cytological observations. Genetics *171*, 985–1001.

Gardiner, D.M., Kazan, K., and Manners, J.M. (2009). Novel genes of *Fusarium graminearum* that negatively regulate deoxynivalenol production and virulence. Mol. Plant Microbe Interact. *22*, 1588–1600.

Gilbert, J., Clear, R.M., Ward, T.J., Gaba, D., Tekauz, A., Turkington, T.K., Woods, S.M., Nowicki, T., and O'Donnell, K. (2010). Relative aggressiveness and production of 3-or 15-acetyl deoxynivalenol and deoxynivalenol by *Fusarium graminearum* in spring wheat. Can. J. Plant Pathol.32, 146–152.

Guenther, J.C., Hallen-Adams, H.E., Bucking, H., Shachar-Hill, Y., and Trail, F. (2009). Triacylglyceride metabolism by *Fusarium graminearum* during colonization and sexual development on wheat. Mol. Plant Microbe Interact. *22*, 1492–1503.

Guldener, U., Mannhaupt, G., Munsterkotter, M., Haase, D., Oesterheld, M., Stumpflen, V., Mewes, H.W., and Adam, G. (2006a). FGDB: a comprehensive fungal genome resource on the plant pathogen *Fusarium graminearum*. Nucleic Acids Res. *34*, D456–458.

Guldener, U., Seong, K.Y., Boddu, J., Cho, S., Trail, F., Xu, J.R., Adam, G., Mewes, H.W., Muehlbauer, G.J., and Kistler, H.C. (2006b). Development of a *Fusarium graminearum* Affymetrix GeneChip for profiling fungal gene expression *in vitro* and in *planta*. Fungal Genet. Biol. *43*, 316–325.

Hallen, H.E., Huebner, M., Shiu, S.H., Guldener, U., and Trail, F. (2007). Gene expression shifts during perithecium development in *Gibberella zeae* (anamorph *Fusarium graminearum*), with particular emphasis on ion transport proteins. Fungal Genet. Biol. *44*, 1146–1156.

Hallen, H.E., and Trail, F. (2008). The L-type calcium ion channel *cch1* affects ascospore discharge and mycelial growth in the filamentous fungus *Gibberella zeae* (anamorph *Fusarium graminearum*). Eukaryot. Cell 7, 415–424.

Hammond, T.M., and Keller, N.P. (2005). RNA silencing in *Aspergillus nidulans* is independent of RNA-dependent RNA polymerases. Genetics *169*, 607–617.

Han, Y.K., Kim, M.D., Lee, S.H., Yun, S.H., and Lee, Y.W. (2007). A novel F-box protein involved in sexual development and pathogenesis in *Gibberella zeae*. Mol. Microbiol. 63, 768–779.

Han, Y.K., Lee, T., Han, K.H., Yun, S.H., and Lee, Y.W. (2004). Functional analysis of the homoserine O-acetyltransferase gene and its identification as a selectable marker in *Gibberella zeae*. Curr. Genet. *46*, 205–212.

Harris, L.J., Alexander, N.J., Saparno, A., Blackwell, B., McCormick, S.P., Desjardins, A.E., Robert, L.S., Tinker, N., Hattori, J., Piche, C., *et al.* (2007). A novel gene cluster in *Fusarium graminearum* contains a gene that contributes to butenolide synthesis. Fungal Genet. Biol. *44*, 293–306.

Hohn, T.M., McCormick, S.P., Alexander, N.J., Desjardins, A.E., and Proctor, R.H. (1998). Function and biosynthesis of trichothecenes produced by *Fusarium* species. Mol. Genet. Host-Specific Toxins Plant Dis. *13*, 17–24.

Hong, S.Y., So, J., Lee, J., Min, K., Son, H., Park, C., Yun, S.H., and Lee, Y.W. (2010). Functional analyses of two syntaxin-like SNARE genes, *GzSYN1* and *GzSYN2*, in the ascomycete *Gibberella zeae*. Fungal Genet. Biol. 47, 364–372.

Hou, Z.M., Xue, C.Y., Peng, Y.L., Katan, T., Kistler, H.C., and Xu, J.R. (2002). A mitogen-activated protein kinase gene *(MGV1)* in *Fusarium graminearum* is required for female fertility, heterokaryon formation, and plant infection. Mol. Plant Microbe Interact. *15*, 1119–1127.

Jansen, C., von Wettstein, D., Schäfer, W., Kogel, K.H., Felk, A., and Maier, F.J. (2005). Infection patterns in barley and wheat spikes inoculated with wild-type and trichodiene synthase gene disrupted *Fusarium graminearum*. Proc. Natl. Acad. Sci. U S A. *102*, 16892–16897.

Jenczmionka, N.J., and Schäfer, W. (2005). The Gpmk1 MAP kinase of *Fusarium graminearum* regulates the induction of specific secreted enzymes. Curr. Genet. *47*, 29–36.

Jenczmionka, N.J., Maier, F.J., Losch, A.P., and Schäfer, W. (2003). Mating conidiation and pathogenicity of *Fusarium graminearum*, the main causal agent of the head-blight disease of wheat, are regulated by the MAP kinase gpmk1. Curr. Genet. *43*, 87–95.

Jeon, J., Park, S.Y., Chi, M.H., Choi, J., Park, J., Rho, H.S., Kim, S., Goh, J., Yoo, S., Choi, J., *et al.* (2007). Genome-wide functional analysis of pathogenicity genes in the rice blast fungus. Nat. Genet. *39*, 561–565.

Jiang, J., Liu, X., Yin, Y., and Ma, Z. (2011a). Involvement of a velvet protein FgVeA in the regulation of asexual development, lipid and secondary metabolisms and virulence in *Fusarium graminearum*. PLoS One *6*, e28291.

Jiang, J., Yun, Y., Fu, J., Shim, W.-B., and Ma, Z. (2011b). Involvement of a putative response regulator FgRrg-1 in osmotic stress response, fungicide resistance and virulence in *Fusarium graminearum*. Mol. Plant Pathol. *12*, 425–436.

Jiang, L.H., Yang, J.R., Fan, F.Y., Zhang, D.J., and Wang, X.L. (2010). The Type 2C protein phosphatase FgPtc1p of the plant fungal pathogen *Fusarium graminearum* is involved in lithium toxicity and virulence. Mol. Plant Pathol. *11*, 277–282.

Jones, J.D.G., and Dangl, J.L. (2006). The plant immune system. Nature *444*, 323–329.

Kadotani, N., Nakayashiki, H., Tosa, Y., and Mayama, S. (2003). RNA silencing in the phytopathogenic fungus *Magnaporthe oryzae*. Mol. Plant Microbe Interact. *16*, 769–776.

Kang, Z., and Buchenauer, H. (1999). Immunocytochemical localization of *Fusarium* toxins in infected wheat spikes by *Fusarium culmorum*. Physiol. Mol. Biol. Plants 55, 275–288.

Kang, Z., and Buchenauer, H. (2000). Cytology and ultrastructure of the infection of wheat spikes by *Fusarium culmorum*. Mycol. Res. *104*, 1083–1093.

Kim, J.E., Lee, H.J., Lee, J., Kim, K.W., Yun, S.H., Shim, W.B., and Lee, Y.W. (2009). *Gibberella zeae* chitin synthase genes, *GzCHS5* and *GzCHS7*, are required for hyphal growth, perithecia formation, and pathogenicity. Curr. Genet. 55, 449–459.

Kim, J.E., Myong, K., Shim, W.B., Yun, S.H., and Lee, Y.W. (2007). Functional characterization of acetylglutamate synthase and phosphoribosylamine-glycine ligase genes in *Gibberella zeae*. Curr. Genet. *51*, 99–108.

Kimura, M., Kaneko, I., Komiyama, M., Takatsuki, A., Koshino, H., Yoneyama, K., and Yamaguchi, I. (1998). Trichothecene 3-O-acetyltransferase protects both the producing organism and transformed yeast from related mycotoxins – Cloning and characterization of Tri101. J. Biol. Chem. 273, 1654–1661.

Kimura, M., Tokai, T., O'Donnell, K., Ward, T.J., Fujimura, M., Hamamoto, H., Shibata, T., and Yamaguchi, I. (2003). The trichothecene biosynthesis gene cluster of *Fusarium graminearum* F15 contains a limited number of essential pathway genes and expressed non-essential genes. FEBS Lett. *539*, 105–110.

Kleemann, J., Takahara, H., Stuber, K., and O'Connell, R. (2008). Identification of soluble secreted proteins from appressoria of *Colletotrichum higginsianum* by analysis of expressed sequence tags. Microbiology *154*, 1204–1217.

Kohler, J., Baumbach, J., Taubert, J., Specht, M., Skusa, A., Ruegg, A., Rawlings, C., Verrier, P., and Philippi, S. (2006). Graph-based analysis and visualization of experimental results with ONDEX. Bioinformatics *22*, 1383–1390.

Korves, T., and Colosimo, M.E. (2009). Controlled vocabularies for microbial virulence factors. Trends Microbiol. *17*, 279–285.

Lee, S.H., Han, Y.K., Yun, S.H., and Lee, Y.W. (2009a). Roles of the glyoxylate and methylcitrate cycles in sexual development and virulence in the cereal pathogen *Gibberella zeae*. Eukaryot. Cell *8*, 1155–1164.

Lee, S.H., Lee, J., Lee, S., Park, E.H., Kim, K.W., Kim, M.D., Yun, S.H., and Lee, Y.W. (2009b). *GzSNF1* is required for normal sexual and asexual development in the ascomycete *Gibberella zeae*. Eukaryot. Cell *8*, 116–127.

Leonard, K.J., and Bushnell, W.R. (2003). Fusarium head blight of wheat and barley (The American Phytopathological Society, Minnesota).

Leslie, J.F., and Summerell, B.A. (2006). The *Fusarium* Laboratory Manual, First edition (Blackwell Publishing Ltd., Ames, Iowa, USA).

Li, Y., Wang, C., Liu, W., Wang, G., Kang, Z., Kistler, H.C., and Xu, J.R. (2011). The *HDF1* histone deacetylase gene is important for conidiation, sexual reproduction, and pathogenesis in *Fusarium graminearum*. Mol. Plant Microbe Interact. *24*, 487–496.

Liu, X., Tang, W.H., Zhao, X.M., and Chen, L. (2010). A network approach to predict pathogenic genes for *Fusarium graminearum*. PLoS One *5*, e13021.

Lowe, R.G., Jubault, M., Canning, G., Urban, M., and Hammond-Kosack, K. (2011). The induction of trichothecene mycotoxins by cereal-infecting *Fusarium* species. In Plant Fungal Pathogens: Methods and Protocols, Bolton, M.D., and Thomma, B.P., eds. (Springer, London), pp. 439–455.

Lysoe, E., Pasquali, M., Breakspear, A., and Kistler, H.C. (2011a). The transcription factor FgStuAp influences spore development, pathogenicity, and secondary metabolism in *Fusarium graminearum*. Mol. Plant Microbe Interact. *24*, 54–67.

Lysoe, E., Seong, K.Y., and Kistler, H.C. (2011b). The transcriptome of *Fusarium graminearum* during the infection of wheat. Mol. Plant Microbe Interact. *24*, 995–1000.

Ma, L.J., van der Does, H.C., Borkovich, K.A., Coleman, J.J., Daboussi, M.J., Di Pietro, A., Dufresne, M., Freitag, M., Grabherr, M., Henrissat, B., *et al.* (2010). Comparative genomics reveals mobile pathogenicity chromosomes in *Fusarium*. Nature *464*, 367–373.

Magan, N., Hope, R., Cairns, V., and Aldred, D. (2003). Host-harvest fungal ecology: Impact of fungal growth and mycotoxin accumulation in stored grain. Eur. J. Plant Pathol. *109*, 723–730.

Maier, F.J., Maiz, S., Losch, A.P., Lacour, T., and Schäfer, W. (2005). Development of a highly efficient gene targeting system for *Fusarium graminearum* using the disruption of a polyketide synthase gene as a visible marker. FEMS Yeast Res *5*, 653–662.

Maier, F.J., Miedaner, T., Hadeler, B., Felk, A., Salomon, S., Lemmens, M., Kassner, H., and Schäfer, W. (2006). Involvement of trichothecenes in fusarioses of wheat, barley and maize evaluated by gene disruption of the trichodiene synthase *(Tri5)* gene in three field isolates of different chemotype and virulence. Mol. Plant Pathol. *7*, 449–461.

Malz, S., Grell, M.N., Thrane, C., Maier, F.J., Rosager, P., Felk, A., Albertsen, K.S., Salomon, S., Bohn, L., Schäfer, W., *et al.* (2005). Identification of a gene cluster responsible for the biosynthesis of aurofusarin in the *Fusarium graminearum* species complex. Fungal Genet. Biol. *42*, 420–433.

McDonald, T., Brown, D., Keller, N.P., and Hammond, T.M. (2005). RNA silencing of mycotoxin production in *Aspergillus* and *Fusarium* species. Mol. Plant Microbe Interact. *18*, 539–545.

McMullen, M., Jones, R., and Gallenberg, D. (1997). Scab of wheat and barley: a re-emerging disease of devasting impact. Plant Dis. *81*, 1340–1348.

Merhej, J., Urban, M., Dufresne, M., Hammond-Kosack, K.E., Richard-Forget, F., and Barreau, C. (2012). The *velvet* gene, *FgVe1*, affects fungal development and positively regulates trichothecene biosynthesis and pathogenicity in *Fusarium graminearum*. Mol. Plant Pathol. *13*, 363–374.

Meyer, V. (2008). Genetic engineering of filamentous fungi–progress, obstacles and future trends. Biotechnol Adv *26*, 177–185.

Mudge, A.M., Dill-Macky, R., Dong, Y.H., Gardiner, D.M., White, R.G., and Manners, J.M. (2006). A role for the mycotoxin deoxynivalenol in stem colonisation during crown rot disease of wheat caused by *Fusarium graminearum* and *Fusarium pseudograminearum*. Physiol. Mol. Biol. Plants *69*, 73–85.

Nganje, W.E., Kaitibie, S., Wilson, W.W., Leistriz, F.L., and Bangsund, D.A. (2004). Economic impacts of Fusarium head blight in wheat and barley: 1993–2001. In Agribusiness and Applied Economics Report (Fargo, ND 58105-5636, Department of Agribusiness and Applied Economics, Agricultural Experiment Station, North Dakota State University).

Nguyen, L.N., Bormann, J., Le, G.T., Starkel, C., Olsson, S., Nosanchuk, J.D., Giese, H., and Schäfer, W. (2011). Autophagy-related lipase Fg*ATG15* of *Fusarium graminearum* is important for lipid turnover and plant infection. Fungal Genet. Biol. *48*, 217–224.

Nicholson, P., Chandler, E., Draeger, R.C., Gosman, N.E., Simpson, D.R., Thomsett, M., and Wilson, A.H. (2003). Molecular tools to study epidemiology and toxicology of Fusarium head blight of cereals. Eur. J. Plant Pathol. *109*, 691–703.

O'Donnell, K., Kistler, C.H., Tacke, B.K., and Casper, H.H. (2000). Gene genealogies reveal global phylogeographic structure and reproductive isolation among lineages of *Fusarium graminearum*, the fungus causing wheat scab. Proc. Natl. Acad. Sci. U.S.A. *97*, 7905–7910.

O'Donnell, K., Ward, T.J., Geiser, D.M., Kistler, H.C., and Aoki, T. (2004). Genealogical concordance between the mating type locus and seven other nuclear genes supports formal recognition of nine phylogenetically distinct species within the *Fusarium graminearum* clade. Fungal Genet. Biol. *41*, 600–623.

Ochiai, N., Tokai, T., Nishiuchi, T., Takahashi-Ando, N., Fujimura, M., and Kimura, M. (2007). Involvement of the osmosensor histidine kinase and osmotic stress-activated protein kinases in the regulation of secondary metabolism in *Fusarium graminearum*. Biochem. Biophys. Res. Commun. *363*, 639–644.

Oeser, B., Beaussart, F., Haarmann, T., Lorenz, N., Nathues, E., Rolke, Y., Scheffer, J., Weiner, J., and Tudzynski, P. (2009). Expressed sequence tags from the flower pathogen *Claviceps purpurea*. Mol. Plant Pathol. *10*, 665–684.

Oide, S., Liu, J.Y., Yun, S.H., Wu, D.L., Michev, A., Choi, M.Y., Horwitz, B.A., and Turgeon, B.G. (2010). Histidine kinase two-component response regulator proteins regulate reproductive development, virulence, and stress responses of the fungal cereal pathogens *Cochliobolus heterostrophus* and *Gibberella zeae*. Eukaryot. Cell *9*, 1867–1880.

Oide, S., Moeder, W., Krasnoff, S., Gibson, D., Haas, H., Yoshioka, K., and Turgeon, B.G. (2006). *NPS6*, encoding a nonribosomal peptide synthetase involved in siderophore-mediated iron metabolism, is a conserved virulence determinant of plant pathogenic ascomycetes. Plant Cell *18*, 2836–2853.

Parry, D.W., Jenkinson, P., and McLeod, L. (1995). Fusarium ear blight (scab) in small grain cereals – a review. Plant Pathol. *44*, 207–238.

Paterson, R., M., R., and Lima, N. (2010). How will climate change affect mycotoxins in food?. Food Res. Int.*43*, 1902–1914.

Petsko, G.A. (2009). Guilt by association. Genome Biol. *10*, 104.

Polley, R.W., and Turner, J.A. (1995). Surveys of stem base diseases and *Fusarium* ear diseases in winter-wheat in England Wales and Scotland 1989–1990. Ann. Appl. Biol.*126*, 45–59.

Pritsch, C., Muehlbauer, G.J., Bushnell, W.R., Somers, D.A., and Vance, C.P. (2000). Fungal development and induction of defense response genes during early infection of wheat spikes by *Fusarium graminearum*. Mol. Plant Microbe Interact. *13*, 159–169.

Proctor, R.H., Hohn, T.M., and McCormick, S.P. (1995). Reduced virulence of *Gibberella zeae* caused by disruption of a trichothecene toxin biosynthetic gene. Mol. Plant Microbe Interact. *8*, 593–601.

Proctor, R.H., Hohn, T.M., and McCormick, S.P. (1997). Restoration of wild-type virulence to *Tri5* disruption mutants of *Gibberella zeae* via gene reversion and mutant complementation. Microbiology *143*, 2583–2591.

Proctor, R.H., Hohn, T.M., McCormick, S.P., and Desjardins, A.E. (1995b). *Tri6* encodes an unusual zinc-finger protein involved in regulation of trichothecene biosynthesis in *Fusarium sporotrichioides*. Appl. Environ. Microbiol. *61*, 1923–1930.

Ramamoorthy, V., Cahoon, E.B., Thokala, M., Kaur, J., Li, J., and Shah, D.M. (2009). Sphingolipid C-9 methyltransferases are important for growth and virulence but not for sensitivity to antifungal plant defensins in *Fusarium graminearum*. Eukaryot. Cell *8*, 217–229.

Ramamoorthy, V., Zhao, X., Snyder, A.K., Xu, J.R., and Shah, D.M. (2007). Two mitogen-activated protein kinase signalling cascades mediate basal resistance to antifungal plant defensins in *Fusarium graminearum*. Cell Microbiol. *9*, 1491–1506.

Remm, M., Storm, C.E., and Sonnhammer, E.L. (2001). Automatic clustering of orthologs and in-paralogs from pairwise species comparisons. J. Mol. Biol. *314*, 1041–1052.

Rispail, N., Soanes, D.M., Ant, C., Czajkowski, R., Grunler, A., Huguet, R., Perez-Nadales, E., Poli, A., Sartorel, E., Valiante, V., *et al.* (2009). Comparative genomics of MAP kinase and calcium-calcineurin signalling components in plant and human pathogenic fungi. Fungal Genet. Biol. *46*, 287–298.

Rittenour, W.R., and Harris, S.D. (2008). Characterization of *Fusarium graminearum* Mes1 reveals roles in cell-surface organization and virulence. Fungal Genet. Biol. *45*, 933–946.

Rittenour, W.R., and Harris, S.D. (2010). An *in vitro* method for the analysis of infection-related morphogenesis in *Fusarium graminearum*. Mol. Plant Pathol. *11*, 361–369.

Ruepp, A., Zollner, A., Maier, D., Albermann, K., Hani, J., Mokrejs, M., Tetko, I., Guldener, U., Mannhaupt, G., Munsterkotter, M., *et al.* (2004). The FunCat, a functional annotation scheme for systematic classification of proteins from whole genomes. Nucleic Acids Res. *32*, 5539–5545.

Salameh, A., Almaghrabi, H., and Buerstmayr, H. (2010). 60. Jahrestagung der Vereinigung der Pflanzenzuchter und Saatgutkaufleute Osterreichs, 24–26 November 2009, Raumberg-Gumpenstein., 191.

Sanchez-Rodriguez, A., Martens, C., Engelen, K., Van de Peer, Y., and Marchal, K. (2010). The potential for pathogenicity was present in the ancestor of the Ascomycete subphylum *Pezizomycotina*. BMC Evol. Biol. *10*, 318.

Savard, M.E., Sinha, R.C., Seaman, W.L., and Fedak, G. (2000). Sequential distribution of the mycotoxin deoxynivalenol in wheat spikes after inoculation with *Fusarium graminearum*. Can. J. Plant Pathol.*22*, 280–285.

Seong, K., Hou, Z.M., Tracy, M., Kistler, H.C., and Xu, J.R. (2005). Random insertional mutagenesis identifies genes associated with virulence in the wheat scab fungus *Fusarium graminearum*. Phytopathology *95*, 744–750.

Seong, K., Zhao, X., Xu, J., Guldener, U., and Kistler, H. (2008). Conidial germination in the filamentous fungus *Fusarium graminearum*. Fungal Genet. Biol. *45*, 389 – 399.

Seong, K.Y., Pasquali, M., Zhou, X., Song, J., Hilburn, K., McCormick, S., Dong, Y., Xu, J.R., and Kistler, H.C. (2009). Global gene regulation by *Fusarium* transcription factors Tri6 and Tri10 reveals adaptations for toxin biosynthesis. Mol. Microbiol. 72, 354–367.

Sexton, A.C., and Howlett, B.J. (2006). Parallels in fungal pathogenesis on plant and animal hosts. Eukaryot. Cell *5*, 1941–1949.

Soanes, D.M., Skinner, W., Keon, J., Hargreaves, J., and Talbot, N.J. (2002). Genomics of phytopathogenic fungi and the development of bioinformatic resources. Mol. Plant Microbe Interact. *15*, 421–427.

Son, H., Lee, J., Park, A.R., and Lee, Y.W. (2011a). ATP citrate lyase is required for normal sexual and asxexual development in *Gibberella zeae*. Fungal Genet. Biol. *48*, 408–417.

Son, H., Seo, Y.S., Min, K., Park, A.R., Lee, J., Jin, J.M., Lin, Y., Cao, P., Hong, S.Y., Kim, E.K., *et al.* (2011b). A phenome-based functional analysis of transcription factors in the cereal head blight fungus, *Fusarium graminearum*. PLoS Pathog. *7*, e1002310.

Strange, R.N., and Smith, H. (1978). Specificity of choline and betaine as stimulants of *Fusarium graminearum*. Trans. Br. Mycol. Soc. *70*, 187–192.

Tag, A., Hicks, J., Garifullina, G., Ake, C., Jr., Phillips, T.D., Beremand, M., and Keller, N. (2000). G-protein signalling mediates differential production of toxic secondary metabolites. Mol. Microbiol. *38*, 658–665.

Tanaka, A., Christensen, M.J., Takemoto, D., Park, P., and Scott, B. (2006). Reactive oxygen species play a role in regulating a fungus-perennial ryegrass mutualistic interaction. Plant Cell *18*, 1052–1066.

Torto-Alalibo, T., Collmer, C.W., and Gwinn-Giglio, M. (2009). The plant-associated microbe gene ontology (PAMGO) consortium: community development of new gene ontology terms describing biological processes involved in microbe–host interactions. BMC Microbiol. *9 Suppl 1*, 1–5.

Trail, F., and Common, R. (2000). Perithecial development by *Gibberella zeae:* a light microscopy study. Mycologia *92*, 130–138.

Turgeon, B.G., Condon, B., Liu, J., and Zhang, N. (2010). Protoplast transformation of filamentous fungi. Methods Mol. Biol. *638*, 3–19.

Urban, M., Mott, E., Farley, T., and Hammond-Kosack, K. (2003). The *Fusarium graminearum MAP1* gene is essential for pathogenicity and development of perithecia. Mol. Plant Pathol. *4*, 347–359.

Varga, J., Kocsube, S., Toth, B., and Mesterhazy, A. (2005). Nonribosomal peptide synthetase genes in the genome of *Fusarium graminearum*, causative agent of wheat head blight. Acta. Biol. Hung. *56*, 375–388.

Voigt, C.A. (2011). Abstract, p. 26. Asilomar Fungal Genetics Conference.

Voigt, C.A., Schäfer, W., and Salomon, S. (2005). A secreted lipase of *Fusarium graminearum* is a virulence factor required for infection of cereals. Plant J. *42*, 364–375.

Wang, C., Zhang, S., Hou, R., Zhao, Z., Zheng, Q., Xu, Q., Zheng, D., Wang, G., Liu, H., Gao, X., *et al.* (2011a). Functional analysis of the kinome of the wheat scab fungus *Fusarium graminearum*. PLoS Pathog. *7*, e1002460.

Wang, Y., Liu, W., Hou, Z., Wang, C., Zhou, X., Jonkers, W., Ding, S., Kistler, H.C., and Xu, J.R. (2011b). A novel transcriptional factor important for pathogenesis and ascosporogenesis in *Fusarium graminearum*. Mol. Plant Microbe Interact. *24*, 118–128.

Watson, R.J., Burchat, S., and Bosley, J. (2008). A model for integration of DNA into the genome during transformation of *Fusarium graminearum*. Fungal Genet. Biol. *45*, 1348–1363.

Weld, R.J., Plummer, K.M., Carpenter, M.A., and Ridgway, H.J. (2006). Approaches to functional genomics in filamentous fungi. Cell Res. *16*, 31–44.

Winnenburg, R., Baldwin, T.K., Urban, M., Rawlings, C., Kohler, J., and Hammond-Kosack, K.E. (2006). PHI-base: a new database for pathogen host interactions. Nucleic Acids Res. *34*, D459–464.

Winnenburg, R., Urban, M., Beacham, A., Baldwin, T.K., Holland, S., Lindeberg, M., Hansen, H., Rawlings, C., Hammond-Kosack, K.E., and Kohler, J. (2008). PHI-base update: additions to the pathogen host interaction database. Nucleic Acids Res. *36*, D572–576.

Wise, R., Caldo, R., Hong, L., Shen, L., Cannon, E., and Dickerson, J. (2007). BarleyBase/PLEXdb: A unified expression profiling database for plants and plant pathogens. In Methods Mol. Biol., D. Edwards, ed. (Totowa, New Jersey, Humana Press), pp. 347–363.

Yu, H.Y., Seo, J.A., Kim, J.E., Han, K.H., Shim, W.B., Yun, S.H., and Lee, Y.W. (2008). Functional analyses of heterotrimeric G protein G alpha and G beta subunits in *Gibberella zeae*. Microbiology *154*, 392–401.

Yu, J.H., Hamari, Z., Han, K.H., Seo, J.A., Reyes-Dominguez, Y., and Scazzocchio, C. (2004). Double-joint PCR: a PCR-based molecular tool for gene manipulations in filamentous fungi. Fungal Genet. Biol. *41*, 973–981.

Zhang, D., Fan, F., Yang, J., Wang, X., Qiu, D., and Jiang, L. (2010). FgTep1p is linked to the phosphatidylinositol-3 kinase signalling pathway and plays a role in the virulence of *Fusarium graminearum* on wheat. Mol. Plant Pathol. *11*, 495–502.

Zhao, C., Waalwijk, C., de Wit, P.J., van der Lee, T., and Tang, D. (2011). *EBR1*, a novel Zn(2)Cys(6) transcription factor, affects virulence and apical dominance of the hyphal tip in *Fusarium graminearum*. Mol. Plant Microbe Interact. *24*, 1407–1418.

Zhao, X.M., Zhang, X.W., Tang, W.H., and Chen, L. (2009). FPPI: *Fusarium graminearum* protein–protein interaction database. J. Proteome Res. *8*, 4714–4721.

Zhou, X., Heyer, C., Choi, Y.E., Mehrabi, R., and Xu, J.R. (2010). The *CID1* cyclin C-like gene is important for plant infection in *Fusarium graminearum*. Fungal Genet. Biol. *47*, 143–151.

Applying Proteomics to Investigate the Interactions between Pathogenic *Fusarium* Species and their Hosts

5

Linda J. Harris, Thérèse Ouellet and Rajagopal Subramaniam

Abstract

The recent technical advances in mass spectrometry (MS) such as tandem MS (MS/MS) and gel and non-gel-based methods of protein separation and quantification have galvanized research efforts to study the proteomes of pathogenic fungi. The acquisition of high quality genomic sequences has provided the species-specific protein databases essential for successful assignment of sequenced peptides to fungal proteins. Proteomics can be defined as describing the complete set of protein species present in a living organism. While transcriptomics describes the transcripts available for translation, proteomics tells us which proteins have actually been translated, modified, and are available for facilitating cellular processes. This review touches on how proteomics has been exploited to further explore biological questions involving *Fusarium* species.

Introduction

Plants depend on their innate immunity to defend themselves against pathogen attack (Jones and Dangl, 2006). Potential microbial pathogens are initially recognized through pathogen-associated molecular patterns (PAMPs) by pattern recognition receptors, leading to the induction of defence responses. PAMPs are usually conserved microbial molecules that the pathogen cannot readily change or do without, including components of the fungal cell wall (glycosylated fungal proteins, chitin and glucan) (Nurnberger and Kemmerling, 2009). PAMP-triggered immunity (PTI) is an important element of non-host resistance. The other type of plant immune response, effector-triggered immunity (ETI), is the basis for pathogen race/host cultivar-specific plant resistance. Pathogens secrete effectors that can interfere with PTI. The recognition of these effectors by plant resistance proteins unique to individual plant cultivars allows the plant immune response to successfully overcome the pathogen. An understanding of PTI and ETI with respect to *Fusarium* species and their plant hosts is in its infancy. Potential candidates for PAMPs or effectors may be found in the proteins or metabolites secreted by *Fusarium* or found exposed on the cell surface.

Diverse protein classes can play a key role in pathogenicity or fungal fitness. *Fusarium* species employ a wide range of cell wall-degrading enzymes to not only facilitate host penetration and spread but also to access nutrients (Kikot *et al.*, 2009). For example, the *F. solani* f. sp. *pisi* genome contains two cutinase genes. The *cut2* gene is constitutively expressed, causing the initial discharge of free cutin monomers from the host cell wall, which in turn induce up-regulation of a second cutinase gene, *cut1* (Woloshuk and Kolattukudy, 1986; Li *et al.*, 2002). Targeted disruption of *cut1* significantly reduces virulence on non-wounded pea tissue (Rogers *et al.*, 1994). Thus, these secreted enzymes are important in cuticle penetration. A major component of the extracellular proteins secreted by *F. oxysporum* cultures, Nep1 (necrosis and ethylene-inducing peptide 1), has been shown to induce ethylene production and necrosis in leaves of coca (Bailey, 1995). Subsequently, additional Nep1-like proteins (NLPs) secreted by plant pathogenic fungi, oomycetes, and bacteria have been shown to induce plant defence responses and cell death in dicots (Qutob *et al.*, 2006). More recently, characterization of the tomato–*F. oxysporum* pathosystem has revealed that cultivar-specific forms of *F. oxysporum* exhibited a gene-for-gene interaction with their host plant. Three avirulence proteins have been identified in the fungus, which are recognized by three host resistance proteins (for review, see Takken and Rep, 2010). The identification of avirulence proteins and of Nep1 was facilitated by proteomic analysis of *F. oxysporum*-infected tomato xylem sap and fungal cultures, respectively. Hence, the majority of proteomic studies to date in *Fusarium* species have focused on describing the secreted proteome (secretome), the cell wall proteome, and the proteome associated with mycotoxin production. We will summarize these and additional proteomics studies on *Fusarium* species in the following sections.

Discussion

Secretome

The initial plant/pathogen interaction depends on the fungus sensing signals from the host to induce a successful pathogenic response; this can be mediated by proteins secreted by the fungi. As many *Fusaria* are broad host pathogens, they need to be capable of producing a diverse set of hydrolytic enzymes to enable infection and growth on different carbon and nitrogen sources. Thus, describing the secretome is a logical approach to characterize the initial infection mechanisms of pathogenic fungi. Although the secretome has been defined as containing both the proteins of the protein secretion machinery and the native secreted proteins (Tjalsma *et al.*, 2000), most studies in this section concentrate on the secreted proteins alone.

The first avirulence protein identified in *F. oxysporum* was SIX1 (Secreted In Xylem; renamed Avr3) (Rep *et al.*, 2004; Houterman *et al.*, 2008). Using 1-D SDS-PAGE gels, a fungal-derived 12 kDa protein was isolated from xylem sap only during compatible interactions between tomato plants and *F. oxysporum* f. sp. *lycopersici* (Rep *et al.*, 2002). Three peptide sequences identified from this protein by MS were used to design degenerate primers to clone the *F. oxysporum Six1* gene (Rep *et al.*, 2004). This cysteine-rich protein is required for *I-3*-mediated resistance of tomato against *F. oxysporum* f. sp. *lycopersici*, as demonstrated by gene disruption and complementation studies.

The discovery of SIX1 protein in xylem vessels colonized by *F. oxysporum* inspired more in-depth proteomics analyses of intracellular spaces. Houterman and co-workers (Houterman *et al.*, 2007) used 2-DE to separate proteins extracted from the xylem sap of *F. oxysporum*-infected tomato plants. These researchers concentrated 40 ml of sap collected from 80 infected plants for a single 2-DE gel, pH range 3 to 10 and then a second sample was run in a second gel with a pH range of 6 to 11, revealing more basic proteins. A comparison with non-infected sap revealed thirteen plant proteins and seven fungal proteins only present in the infected sap, which were identified using MALDI-TOF MS and LC-QTOF MS/MS. The seven fungal proteins included SIX1 and two other avirulence proteins, Six4/Avr1 and Six3/Avr2, which are recognized by tomato resistance genes *I* and *I-2*, respectively. In all, 11 candidate effector proteins have been identified in *F. oxysporum* (Takken and Rep, 2010). The *SIX* genes encoding these proteins can be used in PCR assays to rapidly distinguish between *F. oxysporum* isolates (Lievens *et al.*, 2009).

A broad spectrum of the *F. graminearum* secretome was sampled using a diverse range of carbon sources to supplement standard liquid media (Paper *et al.*, 2007). The carbon sources included sucrose, pectin, xylan, collagen, maize bran, maize and carrot cell walls and AFEX-extracted dried distillers' grains. Proteins were extracted from culture filtrates, separated on 1DE gels and the gels cut into eighteen slices. Each slice was in-gel alkylated and digested prior to MS/MS analysis of the digested peptides. This permitted the bulk analysis of proteins, with the exception of proteins less than 5 kDa (cut-off MW of desalting column). They identified from 37 to 154 *F. graminearum* proteins in the thirteen *in vitro* conditions, for a combined total of 228 proteins. Of these, 46 proteins were generally constitutively expressed while 56 were only detected under a single *in vitro* condition. Monocot and dicot cell walls induced 96 common proteins, of 126 and

120 total proteins, respectively. Surprisingly, the plant cell wall media each induced a shared set of ~50 proteins with the growth media containing sucrose.

Phalip *et al.* (2005) also compared the secretome on simple sugars versus plant cell walls by growing *F. graminearum* in M3 minimal media without shaking with either 1% glucose or 0.8% hop cell wall preparation as the sole carbon source. Proteins were extracted from culture filtrate when maximal growth had been achieved (6 days in glucose and 9 days in hop cell walls). As only ~1 µg/ml protein was collected from glucose cultures, the researchers separated proteins on a 1DE gel and identified 23 unique proteins. The hop cell wall culture yielded much higher protein concentrations (~25 µg/ml) and a more complex protein mixture. Thus, protein spots were excised from 2-DE and 84 *F. graminearum* proteins identified using nano-LC MS/MS. Protein identification and classification was revised in a subsequent publication (Phalip *et al.*, 2009). All but 10 of these proteins were also identified in the secretome by Paper *et al.* (2007). As an alternative approach, *F. graminearum* was grown *in vitro* with either finely ground wheat or barley flour and 69 fungal proteins identified in the supernatant by 2DE and MALDI-TOF-TOF MS, including 24 proteins not identified in the previous secretome studies (Yang *et al.*, 2012). The proteins identified in the presence of either glucose or hop cell walls included 24 categories of hydrolytic enzymes, and the data showed that a whole battery of enzymes were being secreted *de novo* that had the predicted capability to digest the many components of the plant cell wall. Complementary transcription analysis with a focus on xylan-manipulating enzymes has reinforced these results and indicated that the diversity and apparent redundancy of xylan-related genes observed on the cell wall substrate can be broken down into simpler patterns when more specific substrates were used (Hatsch *et al.*, 2006). Differential expression would confer flexibility to adapt to variations in the environment. Of the many cell wall-degrading enzymes secreted by *F. graminearum* when infecting plants, a lipase, some proteinases and pectic enzymes have been shown so far to directly contribute to pathogenicity (Kikot *et al.*, 2009).

The secretome of *F. verticillioides* was analysed for cellulosic biofuel applications by growing in media containing maize bran fraction and using 1-DE followed by LC-MS/MS (Ravalason *et al.*, 2012). Of 166 proteins identified, 57 were predicted to be cell-wall degrading enzymes through homology to known proteins. When added to commercial β-glucosidase and *T. reesei* cellulases, the *F. verticillioides* secretome fraction enhanced the hydrolysis of wheat straw, possibly due to the wider range of hemicellulases found in *Fusarium*.

Brown and colleagues (2012) recently examined the database of *F. graminearum* predicted proteins (MIPS version 3.2: http://mips.helmholtz-muenchen.de/genre/proj/FGDB/) using a combination of predictive software programs to identify signal peptides, membrane spanning regions, and cellular location and produce a refined secretome of 574 proteins. They further predicted 19 *F. graminearum*-specific and 31 *Fusarium*-specific secreted proteins ($P < 10^{-5}$), based on BlastP analysis of 57 fungal and oomycete predicted proteomes.

Secretome analysis is often complicated by the low concentration of secreted proteins in culture media. Using liquid phase IEF followed by dialysis and lyophilization to prepare proteins can dramatically increase the number of resolved spots in subsequent narrow pH-range 2-DE gels (Vincent *et al.*, 2009). This procedure avoids an extraction precipitation step during which a significant proportion of secreted proteins are lost.

Cell wall proteome

The fungal cell wall is important for maintaining cell shape, nutrient uptake, protection against environmental stress, and interaction with the host. *Fusarium* cell walls are composed of an electron-dense outer layer enriched in glycoproteins that is attached to an inner chitin/glucan layer (Barran *et al.*, 1975; Barbosa and Kemmelmeier, 1993; Schoffelmeer *et al.*, 1999). A similar structure has been observed in many other ascomycete species (Klis *et al.*, 2010). Detailed characterization of the highly glycosylated fungal cell wall proteins has presented technical difficulties due to the variable molecular mass and number of negative charges associated with the carbohydrate side chains (Yin *et al.*, 2008). Thus, researchers often need to modify their proteomic analysis to identify glycoproteins.

Prados-Rosales and associates used both direct tryptic digestion as well as sequential extractions (designed to be specific to the type of protein-cell wall bonds) of isolated cell walls to characterize *F. oxysporum* cell wall proteins (Prados-Rosales *et al.*, 2009). The first method was applied to wild-type strains grown under adhesion-inducing conditions; LC-MS/MS analysis of tryptic digests of purified cell walls led to the identification of 174 *F. oxysporum* proteins. Nineteen proteins had a predicted signal peptide and ten of these lacked transmembrane domains while still containing a predicted motif for glycosylphosphatidylinositol modification of cell wall proteins, the major type of modification of cell wall proteins in ascomycete fungi. The number of predicted classical cell wall proteins detected in this proteomics study was much lower that the number of cell wall proteins predicted *in silico* using the genomic sequence of *F. oxysporum*. However, detection of a low number of predicted cell wall proteins has also been reported by other fungal researchers (Castillo *et al.*, 2008). The second method of consecutive cell wall extractions was used when comparing wild-type and non-pathogenic mutant strains of *F. oxysporum*. The MAPK *fmk1* gene has been shown to play a role in pathogenicity and one of the phenotypes of *fmk1* gene disruption is loss of adhesion to tomato roots. To explore the basis for the adhesion loss, they compared proteins extracted from cell walls of the wild-type and *fmk1* mutant by 2-D DIGE (2-D fluorescence difference gel electrophoresis) (Prados-Rosales *et al.*, 2009). In 2D-DIGE, fluorescence labelling (cy3 and cy5) of two protein samples permits qualitative and quantitative comparisons. Proteins deemed to be differentially expressed (> 1.3-fold change; $P < 0.05$) between wild type and mutant were identified through MALDI-TOF-MS. They identified eight proteins over-represented and six under-represented in the *fmk1* deletion strain relative to wild type, however only two of these proteins contained a putative signal peptide. In light of their results, the authors recommend the direct tryptic digestion of cell walls coupled with a direct MS analysis method to characterize the *Fusarium* cell wall proteome (Prados-Rosales *et al.*, 2009).

Proteome profiling during mycotoxin biosynthesis

Fusarium verticillioides can produce fumonisin mycotoxins which have been associated with cancer and neural tube defects in mammals, while *F. graminearum* can produce trichothecene mycotoxins, associated with feed refusal, weight loss, and abdominal distress in animals (Council for Agricultural Science and Technology, 2003). In order to propose novel methods to reduce mycotoxin contamination of food and feed, researchers have been actively studying the biosynthesis and regulation of mycotoxin production (see Chapter 8).

F. verticillioides FCC1 protein is a global regulator of gene expression that influences both secondary metabolism and fungal development; fumonisin production and conidiation is impaired in *fcc1* gene disruption mutants (Shim and Woloshuk, 2001). Choi and Shim (2008) used 2-DE to separate and compare proteins extracted from wild-type and *fcc1* mutant strains grown in liquid media conducive to fumonisin biosynthesis. Proteins from two biological replicates (three technical replicates each) of each strain were extracted and 1150 proteins were resolved (pH 4–7). Forty-eight differential protein spots were chosen for MS analysis (MALDI-TOF MS, QTOF-MS) that exhibited > 5-fold expression differences between strains, but only 37 provided sufficient MS data leading to identification. Many of the proteins present in higher levels in the wild-type strain were associated with protein metabolism (e.g. peptidases), leading the authors to speculate that post-translational modification of proteins may be required for proper regulation of secondary metabolism. Surprisingly, none of the proteins identified in the wild type were associated with fumonisin biosynthesis. EST-based microarrays were used previously to compare gene expression profiles of wild-type and *fcc1*-disrupted strains grown under similar conditions (Prittila *et al.*, 2004). The increased abundance of stress response proteins detected in the *fcc1*$^-$ mutant strain correlated with the previous gene expression data. The corresponding genes of 18 proteins were chosen for qRT-PCR analysis; 10 genes did not exhibit expression profiles that correlated with the protein expression observed.

When *F. graminearum* is grown in nitrogen-depleted but carbon-rich liquid media, the fungus produces trichothecene mycotoxins and other secondary metabolites. The crude fungal filtrate extracted from these cultures is much more toxic to cells than the trichothecenes alone (S. Gleddie, personal communication). As trichothecenes are also induced during plant infection, we chose to determine protein profiles during trichothecene-inducing liquid culture to partially mimic plant infection conditions (Taylor *et al.*, 2008). Thus, we may identify protein changes during the switch to mycotoxin biosynthesis as well as other proteins potentially important for pathogenicity.

A non-gel-based quantitative shotgun proteomics method was chosen to maximize the number of proteins identified. iTRAQ (isobaric tags for relative and absolute quantification) allows relative quantification of peptides and proteins between multiple samples (Aggarwal *et al.*, 2006). Three biological replicates of *F. graminearum* proteins were extracted from mycelia at four time points (time 0, 4 days, 9 days, 12 days) in the nitrogen-depleted liquid cultures (Taylor *et al.*, 2008). After TCA precipitation, protein samples underwent reducing and denaturing, cysteine blocking, trypsin digestion and labelling with one of four isobaric tags. Peptide separation by strong cation exchange (SCX) chromatography was followed by MS/MS analysis and data analysis using Applied Biosystems ProteinPilot software. Five proteins directly involved in trichothecene biosynthesis were identified in at least one replicate and were increased in abundance over the time course; however, only FGSG_07986 (trichothecene 3-O-acetyltransferase) was observed in all three replicates. A core set of 435 *F. graminearum* proteins was identified in at least two of the three biological replicates and statistical analysis revealed 130 proteins that exhibited significant differential expression between time points. The most down-regulated proteins over the time course mainly fell into the primary metabolism, protein translation and protein chaperone categories. Up-regulated proteins also included FGSG_08077 (part of a co-regulated gene cluster predicted to be required for butenolide biosynthesis), predicted secreted proteins and homologues of other fungal virulence proteins. 2-DE was also performed with proteins extracted from time 0 and 4d samples grown under the same trichothecene-inducing liquid culture conditions to validate results of the iTRAQ experiments. Fourteen protein spots that appeared more abundant in the 4 day gel than in the 0 day gel were excised and subjected to LC-MS/MS. Ten proteins showed up-regulation in both iTRAQ and 2-DE, three proteins were not identified by iTRAQ, and the abundance of one protein did not agree between the two methods. Of the proteins found in greater abundance over time in the trichothecene-producing media, 99% are encoded by genes known to be expressed *in planta*.

It is often difficult to reproduce proteomics experiments between laboratories as protein extraction methods can vary and are often briefly described in publications. Pasquali and associates have produced a video demonstrating a protocol for growing *F. graminearum* cultures under toxin-inducing conditions and subsequent protein extraction, based on the protein extraction method of Taylor *et al.* (2008), with some modifications (Pasquali *et al.*, 2010). This provides a visual guide for researchers new to protein extraction and adds value to a written protocol.

Protein phosphorylation events are involved in the signalling pathway regulating trichothecene biosynthesis in *F. graminearum*. Experiments monitoring changes in the phosphoproteome under mycotoxin-inducing conditions will be reviewed in the subsequent section on post-translational modification events.

In planta proteomics

The majority of the *Fusarium* proteomics experiments are conducted *in vitro* as the complexity of the proteome is greatly increased when extracting both fungal and plant proteins. Often, the high ratio of plant to fungal biomass means that it is hard to detect all but the most abundantly expressed fungal proteins. However, the availability of full genomic sequence for three *Fusarium* species has proven invaluable to identify fungal proteins in infected plant tissue samples.

Using protein extracts from *F. graminearum*-infected wheat heads sampled 1 to 3 days after inoculation, Zhou and coworkers (2006) detected about 1380 protein spots on 2-DE gels stained with SyproRuby. Of 41 protein spots that were differentially regulated and analysed further with LC-MS/MS, eight were identified as *Fusarium* proteins. These abundant fungal proteins included proteins with predicted function in primary metabolism (oxidoreductase, fructose aldolase, glyceraldehyde 3-phosphate dehydrogenase), translation initiation (IF5A), protein folding (peptidyl prolyl *cis–trans* isomerase), signal transduction (small GTPase) and defence response (superoxide dismutase). Yang *et al.* (2010) obtained similar results when using a gel-based strategy to look at changes in protein levels in a susceptible barley cultivar infected with *F. graminearum*. Of 80 protein spots that showed an increased level in the water-soluble albumin fraction in response to *F. graminearum*, nine were fungal proteins and included two superoxide dismutases, an L-xylose reductase, a peptidyl prolyl *cis–trans* isomerase, a triosephosphate isomerase, and proteins of unknown function.

Paper *et al.* (2007) identified about 120 fungal proteins (based on two or more peptides each) in the apoplastic fluid of *F. graminearum*-infected wheat heads sampled at 4–14 days after inoculation using vacuum filtration. The two largest classes were secreted cell wall-degrading enzymes and proteases. All of the 26 cell wall-degrading enzymes, including twenty-one glycosyl hydrolases, one lyase and two esterases, had signal peptides. Eleven of the 15 proteases contained a signal peptide. Of the 120 fungal proteins identified, about 44% had no predicted signal peptide, including at least thirteen proteins predicted to be involved in primary metabolism. To check if the vacuum filtration treatment used to extract the *in planta* secretome was responsible for the higher percentage of proteins lacking canonical signal peptides, these researchers extracted proteins from mycelial mats using vacuum filtration. The mycelial mats yielded forty-seven identified proteins, 94% of which had signal peptides, suggesting that vacuum filtration was not causing an inordinate amount of tissue damage and non-secreted protein extraction. Whether or not the proteins without signal peptides are actively secreted by the fungus remains a point of debate.

Although we have conducted shotgun proteomics through iTRAQ analysis of maize and wheat infected with *F. graminearum*, fungal proteins were not identified (L. Harris and T. Ouellet, unpublished results). The Paragon algorithm used to analyse iTRAQ data is designed to identify related peaks present in all samples. Since we were mainly interested in relative protein quantification between uninoculated and inoculated susceptible and tolerant cereal lines, our experiments compared mock- and fungal-inoculated samples. Consequently, the software did not recognize fungal proteins as they were present in only half of the samples.

Proteomic studies of sexual reproduction

The MAT locus is a master regulator for sexual reproduction in many ascomycetes, controlling sexual initiation to ascospore production. To complement microarray transcriptional profiling studies investigating which genes are controlled by the MAT locus or associated signalling pathway, a proteomics approach was used to compare wild-type and MAT gene deletion strains (Lee *et al.*, 2008). A self-fertile wild-type progenitor strain as well as a self-sterile transgenic strain deleted for the *MAT1-1* mating-type gene were grown on carrot agar plates under perithecia-inducing conditions. Six days after perithecial production in the wild-type strain, proteins were extracted from perithecia (wild type) or mycelia (*mat1-1* deletion) and 380 protein spots were resolved on 2D gels. Three technical replicates of two biological replicates of each strain were compared and 13 proteins were significantly differentially expressed ($P < 0.05$; ≥ 3-fold change between strains) and identified using either MALDI-TOF or ESI-MS/MS. Many of these proteins were cell wall-localized or cell wall biogenesis proteins, suggesting that the MAT locus may directly or indirectly regulate formation of the perithecial cell wall. A different set of cell wall-associated genes were also identified during the previous transcriptomic study (Lee *et al.*, 2006), demonstrating the value of conducting both proteomics and transcriptomics analysis.

Proteomics of *Fusarium*/viral and bacterial interactions

Infection of *F. graminearum* with double-stranded RNA mycoviruses is associated with slower growth rate, hypovirulence, increased pigmentation and decreased trichothecene production (Chu *et al.*, 2002). How mycoviruses affect the fungal cell is unknown; however, Kwon and associates (2009) chose to use protein profiling to investigate what effect *F. graminearum* virus-DK21 (FgV-DK21) has on its host cell. Mycelial proteins were extracted from virus-free and FgV-DK21-infected strains after 5 days' growth in complete media (CM) broth and their protein profiles compared in 2-DE (three biological replicates, three technical replicates of each). They were able to reproducibly visualize 470 spots and followed up with in-gel digestion and ESI-MS/MS of thirty-three protein spots that consistently showed a ≥3-fold difference in intensity between virus-free and virus-infected strains. Ten protein spots could not be identified due to poor MS data or the presence of contaminating keratin while one up-regulated spot was putatively identified as an FgV-DK21 viral protein. Six up-regulated proteins were identified, implicated in differentiation, protein synthesis, glycolysis, and purine metabolism. The 16 identified proteins down-regulated in virus-infected cells are mainly involved in glycolysis and antioxidant activities. The authors will focus on the proteins affected by mycovirus infection to determine whether they play a role in hypovirulence or other viral-related symptoms.

The non-pathogenic *F. oxysporum* MSA 35 describes a fungus associated with a suite of rhizobacteria and has been used as a biocontrol agent to suppress *Fusarium* wilt. In the absence of the associated rhizobacteria, the *F. oxysporum* isolate is virulent on lettuce. To investigate this fungal/bacterial interaction, Moretti *et al.* (2010) extracted protein from *F. oxysporum* MSA 35 mycelia grown on potato dextrose agar media. They performed both 2-DE followed by MALDI-TOF/TOF MS and 1-DE with LC-ESI-MS/MS to maximize protein identification. A total of 197 fungal and 162 bacterial proteins were identified with ~20% overlap between methods. To reveal which proteins were differentially expressed between the *F. oxysporum*/bacterial consortium and the same *F. oxysporum* strain cured of the associated rhizobacteria, 2-DE and MALDI_TOF MS/MS was applied. This proteomic approach provided a rich database of candidate proteins contributing to understanding the interactions of a microbial community with biocontrol applications.

Proteomics of specific post-translational modifications

Proteins can undergo more than 300 different kinds of post-translational modifications (PTMs), including acetylation, methylation, phosphorylation, and simple and complex glycosylation, providing a higher level of complexity to the cellular proteome. Proteomic techniques have been developed to enrich for and study PTMs, to allow studies into how PTMs regulate cellular activities (Zhao *et al.*, 2009b).

Treating *Fusarium* cultures with either the phosphatase inhibitor okadaic acid or the kinase inhibitor staurosporine demonstrated that protein phosphorylation impacts trichothecene biosynthesis in *F. graminearum* (Rampitsch *et al.*, 2010). Spiking liquid cultures with the phosphatase inhibitor led to increased DON production while the kinase inhibitor had the opposite effect. Consequently, Rampitsch and co-workers (2010) used three different strategies to describe the phosphoproteome of *F. graminearum* under mycotoxin-inducing conditions. Using ProQ Diamond to stain phosphorylated proteins, 2-DE was used to resolve proteins from time 0, which represents the switch from growth in a nutrient-rich medium to a nitrogen-poor medium (trichothecene-inducing conditions) and then every two days until day 8. Since only twenty phosphoproteins were identified with this approach although many more phosphoproteins were visualized in the gels, the researchers moved on to higher throughput GeLC-MS and gel-free methods. They chose to sample earlier after induction to optimize the opportunity to observe signalling events and examined proteins from time 0, 6 and 12 hours after induction. In all, 241 phosphoproteins were identified, containing 348 putative phosphorylation sites. Many of the phosphopeptides identified contain target phosphorylation sequences recognized by yeast or mammalian kinases. Proteins which were only detected after induction included six transcription factors and six protein kinases that exhibited differential phosphorylation over the time course; these proteins and others identified in this study are candidates for future studies into trichothecene regulation.

In a subsequent project (Rampitsch *et al.*, 2012), a high-throughput proteomic analysis was conducted of phosphopeptides from *F. graminearum* strain DAOM233423 (derived from strain GZ3639) grown *in vitro* without nutritional limitation to obtain a base profile of the phosphoproteome. Using a combination of strong-cation exchange (SCX) and immobilized metal affinity chromatography (IMAC) followed by LC-MS/MS, 2902 putative phosphopeptides were identified with homologous matches to 1496 different proteins. Functional classification of the annotated protein set revealed that phosphopeptides from nuclear proteins with ATP-binding function were the most abundant. There are indications that phosphorylation sites from well-characterized phosphoproteins representing diverse biological processes are conserved in *F. graminearum*.

The generation of reactive oxygen species by fungal NADPH oxidases (NOX) is required for rice blast disease development (Egan *et al.*, 2007). NOX catalyses the oxidation of NADPH to yield superoxide, and Egan and associates (2007) proposed that this reactive oxygen species promotes protein cross-linking to strengthen cell walls, essential for appressorial development. As the disruption of NADPH oxidase genes in *F. graminearum* has a significant effect on virulence in wheat (R. Subramaniam, unpublished data), FHB disease development may also be dependent on redox signalling in *F. graminearum*. NOX activity can lead to the oxidative PTM of cysteine thiol residues to create disulfide bridges. Subramaniam and Rampitsch (unpublished data) are using two complementary strategies (gel-based and non-gel-based) to define the redox peptidome of *F. graminearum* under nitrogen-starved conditions by comparing proteins extracted from wild-type and *nox* mutants. In both approaches, free cysteine thiol groups are blocked, followed by breakdown of disulphide bridges. As the thiol-specific fluorophore monobromobimane (mBBr) fluoresces after conjugation with free thiol groups, it was added to extracted proteins to distinguish between reduced and oxidized cysteine residues prior to 2-DE separation. In the non-gel-based (or shotgun) strategy (method described by McDonagh *et al.*, 2009), the peptides whose cysteine thiol groups are tagged with biotin are enriched on a streptavidin-Sepharose resin column before undergoing LC-MS/MS to identify cross-linked proteins. These approaches have led to the identification of ~ 15 bimane-labelled proteins using 2-DE and ~ 50 proteins using the shot-gun approach, with some overlap between datasets. Further optimization of these techniques will allow us to expand our knowledge of how PTMs influence the regulation of mycotoxin biosynthesis and other processes impacting pathogenicity.

Proteomics vs. transcriptomics

The emphasis during the genomics era has been on analysing global gene expression data to tackle biological questions. However, we know that there are numerous post-transcriptional regulation strategies that can affect protein abundance and transcript expression profiles can not always be depended upon to consistently predict protein abundance (reviewed in Nie *et al.*, 2007; Maier *et al.*, 2009). Experiments with yeast (*Saccharomyces cerevisiae*) have suggested that protein and gene transcript turn-over, translational regulation, as well as experimental error are major contributors to the lack of correlation. Protein turn-over is considered to be the most important feature influencing the correlation between transcriptomic and proteomic datasets and is dependent on protein stability, post-translational processes such as ubiquitination and phosphorylation, and cellular localization (Wu *et al.*, 2008). Stronger correlations of gene and protein abundance are seen, however, when the transcript exhibited high ribosomal occupancy (high percentage of ribosome-bound mRNAs). Hence, the data and results obtained from a proteomic approach can certainly complement any transcriptomic data collected from the same samples.

Web-based resources for protein identification

The successful identification of proteins depends on the accuracy of the protein database. *Fusarium graminearum* strain PH-1 was sequenced to 10-fold coverage in 2003 and annotation efforts led to an initial set of 11,640 predicted genes (Cuomo *et al.*, 2007). The genome assembly and comparative genomics of *F. verticillioides* and *F. oxysporum* have improved *Fusarium* gene annotation (Ma *et al.*, 2010). Assembly version FG3.1 of *F. graminearum* at the Broad Institute is in 31 supercontigs and a combination of automated and manual annotation has predicted 13,332 genes. In addition, MIPS has annotated 13,826 protein coding genes including greater than 1900 genes that either have a different exon/intron structure or are not present in the Broad gene set (FGDB v3.2, http://mips.helmholtz-muenchen.de/genre/proj/FGDB/; Wong *et al.*, 2011). The first genome sequenced from the '*Fusarium solani* species complex' is *Nectria haematococca* mating population VI (asexual stage *Fusarium solani*) and this genome has 15,707 predicted genes (annotation release version 2.0; Coleman *et al.*, 2009). The gene calling will improve with the genome sequencing of additional *Fusarium* genomes and the use of RNASeq expression data to amend exon/intron predictions. *F. graminearum* has a remarkably low number of high identity paralogous proteins; only two pairs of proteins share greater than 80% identity (Cuomo *et al.*, 2007). This lack of high identity paralogues greatly increases the probability of unambiguous protein identification.

To optimize protein identification of MS data, it is important to construct a highly relevant and inclusive protein database based on the most recent gene calls. The searched protein database should also contain common contaminating proteins (e.g. keratin). An estimation of false positives should always be conducted to ensure parameters are sufficient to yield high quality results. We have found it useful to randomize or reverse all the protein sequences contained in the search database and add these randomized sequences to the file of forward sequences so that both types of sequences are searched simultaneously. This provides a quick estimation of the rate of false positives so that search parameters can be modified if necessary.

Researchers in the *Fusarium* community and beyond have developed web-based tools and databases useful for *Fusarium* proteomics studies.

1. *Fusarium* predicted protein databases can be found at the following websites.
   a. MIPS *Fusarium graminearum* genome database (Güldener *et al.*, 2006; Wong *et al.*, 2011): http://mips.helmholtz-muenchen.de/genre/proj/FGDB/.
   b. *Fusarium* comparative database (Ma *et al.*, 2010): http://www.broadinstitute.org/annotation/genome/fusarium_group/MultiHome.html
   c. JGI *Nectria haematococca* Mating Population VI (MPVI) (*Fusarium solani*) genome database (Coleman *et al.*, 2009): http://genome.jgi-psf.org/Necha2/Necha2.home.html
2. Blastx annotations and a non-redundant set of fungal proteins with subcellular localizations assigned were used to train a classifier program (FGsub) to predict where 12,786 *F. graminearum* proteins may be localized (Sun *et al.*, 2010). Although none of the predictions were verified in the lab, localization predictions for 12,786 proteins can be found at http://csb.shu.edu.cn/fgsub/.
3. Choi and associates have constructed a fungal secretome database (http://fsd.snu.ac.kr/), which uses a three-layer classification rule (based on nine prediction programs) to categorize putative secretory proteins in fungal and oomycete genomes (Choi *et al.*, 2010).
4. Protein–protein interactions (PPI) have been predicted by identifying *F. graminearum* protein orthologues in seven organisms (*S. cerevisiae*, *S. pombe*, *C. elegans*, *D. melanogaster*, *H. sapiens*, *M. musculus* and *E. coli*) and integrating this information with public interactome datasets for the seven organisms as well as data based on predicted domain–domain interactions (Zhao *et al.*, 2009a). A core high-confidence PPI set was compiled which contained 27,102 interactions between 3745 proteins. To verify their predictions, the researchers chose twelve PPI (including three high-confidence PPI) to test using the yeast two-hybrid assay and confirmed the interaction of one high-confidence pair (FGSG_06103 and FGSG_07404). Thus, this dataset can provide predicted PPI data but will undoubtedly contain many

false positives and any PPI will need to be independently verified. The *F. graminearum* protein–protein interaction (FPPI) database is available on an interactive web server (http://csb.shu.edu.cn/fppi) and also contains homology information on *F. verticillioides* and *F. oxysporum*.

Conclusions

Coupled with structural genomics, transcriptomics, metabolomics, and mathematical modelling, proteomics is an essential facet of systems biology, a multidisciplinary approach to tackling complex biological questions. The intricacy of plant–pathogen interactions will require the integrative analysis of systems biology to lead to a deeper understanding of fungal infection and plant resistance mechanisms. At this time, researchers have captured brief snapshots of *Fusarium* proteomes at different development stages and environmental conditions and of different genotypes. The increased adoption of non-gel-based high-throughput proteomic technologies will broaden the depth and resolution of these proteomes. Many more proteomics experiments will be required to gain a broad understanding of *Fusarium* proteomes. As more fungal researchers incorporate proteomics into their studies, we can all learn from their improvements in subcellular extraction methods and experimental design. With the increased sensitivity and developing applications of MS and accelerated genome sequencing, *Fusarium* researchers should certainly consider including proteomics in their repertoire of tools to dissect the plant–*Fusarium* interaction.

References

Aggarwal, K., Choe, L.H., and Lee, K.H. (2006). Shotgun proteomics using the iTRAQ isobaric tags. Brief. Funct. Genom. *5*, 112–120.

Bailey, B. (1995). Purification of a protein from culture filtrates of *Fusarium oxysporum* that induces ethylene and necrosis in leaves of *Erythroxylum coca*. Phytopathology *85*, 1250–1255.

Barbosa, L.P., and Kemmelmeier, C. (1993). Chemical composition of the hyphal wall from *Fusarium graminearum*. Exp. Mycol. *17*, 274–283.

Barran, L.R., Schneider, E.F., Wood, P.J., Madhosingh, C., and Miller, R.W. (1975). Cell wall of *Fusarium sulphureum*; I. Chemical composition of the hyphal wall. Biochim. Biophys. Acta. *392*, 148–158.

Castillo, L., Calvo, E., Martinez, A.I., Ruiz-Herrera, J., Valentin, E., Lopez, J.A., and Sentandreu, R. (2008). A study of the *Candida albicans* cell wall proteome. Proteomics *8*, 3871–3881.

Choi, J., Park, J., Kim, D., Jung, K., Kang, S., and Lee, Y.-H. (2010). Fungal secretome database: Integrated platform for annotation of fungal secretomes. BMC Genomics *11*, 105.

Choi, Y.-E., and Shim, W.-B. (2008). Identification of genes associated with fumonisin biosynthesis in *Fusarium verticillioides* via proteomics and quantitative real-time PCR. J. Microbiol. Biotechnol. *18*, 648–657.

Chu, Y.-M., Jeon, J.-J., Yea, S.-J., Kim, Y.-H., Yun, S.-H., Lee, Y.-W., and Kim, K.-H. (2002). Double-stranded RNA mycovirus from *Fusarium graminearum*. Appl. Env. Microbiol. *68*, 2529–2534.

Coleman, J.J., Rounsley, S.D., Rodriguez-Carres, M., Kuo, A., Wasmann, C.C., Grimwood, J., Schmutz, J., Taga, M., White, G.J., Zhou, S., *et al.* (2009). The genome of *Nectria haematococca*: Contribution of supernumerary chromosomes to gene expansion. PLoS Genet. *5*, e1000618.

Council for Agricultural Science and Technology. (2003). Mycotoxins: Risks in Plant, Animal and Human Systems. Task Force Report No. 139. (Council for Agricultural Science and Technology, Ames, IA).

Cuomo, C.A., Guldener, U., Xu, J.-R., Trail, F., Turgeon, B.G., Di Pietro, A., Walton, J.D., Ma, L.-J., Baker, S.E., Rep, M., *et al.* (2007). The *Fusarium graminearum* genome reveals a link between localized polymorphism and pathogen specialization. Science *317*, 1400–1402.

Egan, M.J., Wang, Z.-Y., Jones, M.A., Smirnoff, N., and Talbot, N.J. (2007). Generation of reactive oxygen species by fungal NADPH oxidases is required for rice blast disease. Proc. Natl. Acad. Sci. U.S.A.*104*, 11772–11777.

Güldener, U., Mannhaupt, G., Münsterkötter, M., haase, D., Oesterheld, M., Stümpflen, V., Mewes, H.W., and Adam, G. (2006). FGDB: a comprehensive fungal genome resource on the plant pathogen *Fusarium graminearum*. Nucleic Acids Res. *34*, D456-D458.

Hatsch, D., Phalip, V., Petkovski, E., and Jeltsch, J.-M. (2006). *Fusarium graminearum* on plant cell wall: no fewer than 30 xylanase genes transcribed. Biochem. Biophys. Res. Commun. *345*, 959–966.

Houterman, P.M., Speijer, D., Dekker, H.L., De Koster, C.G., Cornelissen, B.J.C., and Rep, M. (2007). The mixed xylem sap proteome of *Fusarium oxysporum*-infected tomato plants. Mol. Plant Pathol. *8*, 215–221.

Houterman, P.M., Cornelissen, B.J.C., and Rep, M. (2008). Suppression of plant resistance gene-based immunity by a fungal effector. PLoS Pathog. *4*, e1000061.

Jones, J.D.G., and Dangl, J.L. (2006). The plant immune system. Nature *444*, 323–329.

Kikot, G.E., Hours, R.A., and Alconada, T.M. (2009). Contribution of cell wall degrading enzymes to pathogenicity of *Fusarium graminearum*: a review. J. Basic Microbiol. *49*, 231–241.

Klis, F.M., Brul, S., and De Groot, P.W.J. (2010). Covalently linked wall proteins in ascomycetous fungi. Yeast *27*, 489–493.

Kwon, S.-J., Cho, S.-Y., Lee, K.-M., Yu, J., Son, M., and Kim, K.-H. (2009). Proteomic analysis of fungal host factors differentially expressed by *Fusarium graminearum* infected with *Fusarium graminearum virus-DK-21*.Virus Res. *144*, 96–106.

Lee, S.H., Lee, S., Choi, D., Lee, Y.W., and Yun, S.H. (2006). Identification of the down-regulated genes in a *mat1-1*-deleted strain of *Gibberella zeae*, using cDNA subtraction and microarray analysis. Fungal Genet. Biol. *43*, 295–310.

Lee, S.-H., Kim, Y.-K., Yun, S.-H., and Lee, Y.-W. (2008). Identification of differentially expressed proteins in a *mat1-2*-deleted strain of *Gibberella zeae*, using a comparative proteomics analysis. Curr. Genet. *53*, 175–184.

Li, D., Sirakova, T., Rogers, L., Ettinger, W.F., and Kolattukudy, P.E. (2002). Regulation of constitutively expressed and induced cutinase genes by different zinc finger transcription factors in *Fusarium solani* f. sp. *pisi* (*Nectria hematococca*). J. Biol. Chem. *277*, 7905–7912.

Lievens, B., Houterman, P.M., and Rep, M. (2009). Effector gene screening allows unambiguous identification of *Fusarium oxysporum* f. sp. *lycopersici* races and discrimination from other *formae speciales*. FEMS Microbiol. Lett. *300*, 201–215.

Ma, L.-J., van der Does, H.C., Borkovich, K.A., Coleman, J.J., Daboussi, M.-J., Di Pietro, A., Dufrense, M., Freitag, M., Grabherr, M., Henrissat, B., *et al.* (2010). Comparative genomics reveals mobile pathogenicity chromosomes in *Fusarium*. Nature *464*, 367–373.

Maier, T., Guell, M., and Serrano, L. (2009). Correlation of mRNA and protein in complex biological samples. FEBS Lett. *583*, 3966–3973.

McDonagh, B., Ogueta, S., Lasarte, G., Padilla, C.A., and Barcena, J.A. (2009). Shotgun redox proteomics identifies specifically modified cysteines in key metabolic enzymes under oxidative stress in *Saccharomyces cerevisiae*. J. Proteom. *72*, 677–689.

Moretti, M., Grunau, A., Minerdi, D., Gehrig, P., Roschitzki, B., Eberl, L., Garibaldi, A., Gullino, M.L., and Reidel, K. (2010). A proteomics approach to study synergistic and antagonistic interactions of the fungal-bacterial consortium *Fusarium oxysporum* wild-type MSA 35. Proteomics *10*, 3292–3320.

Nie, L., Wu, G., Culley, D.E., Scholten, J.C.M., and Zhang, W. (2007). Integrative analysis of transcriptomic and proteomic data: Challenges, solutions and applications. Crit. Rev. Biotechnol. *27*, 63–75.

Nurnberger, T., and Kemmerling, B. (2009). PAMP-triggered basal immunity in plants. Adv. Botanical Res. *51*, 1–38.

Paper, J.M., Scott-Craig, J.S., Adhikari, N.D., Cuomo, C.A., and Walton, J.D. (2007). Comparative proteomics of extracellular proteins *in vitro* and *in planta* from the pathogenic fungus *Fusarium graminearum*. Proteomics *7*, 3171–3183.

Pasquali, M., Giraud, F., Lasserre, J.P., Planchon, S., Hoffman, L., Bohn, T., and Renaut, J. (2010). Toxin induction and protein extraction from *Fusarium* spp. cultures for proteomic studies. J. Vis. Exp. *36*, e1690. http://www.jove.com/index/Details.stp?ID=1690.

Phalip, V., Delalande, F., Carapito, C., Goubet, F., Hatsch, D., Leize-Wagner, E., Dupree, P., Van Dorsselaer, A., and Jeltsch, J.-M. (2005). Diversity of the exoproteome of *Fusarium graminearum* grown on plant cell wall. Curr. Genet. *48*, 366–379.

Phalip, V., Forster, A., Carapito, R., Habrylo, O., and Jeltsch, J.-M. (2009). Tackling *Humulus lupulus* fungal diseases by studying a hop cell wall/*Fusarium graminearum* model. Acta Hort. *848*, 103–113.

Prados-Rosales, R., Luque-Garcia, J.L., Martinez-Lopez, R., Gil, C., and Di Pietro, A. (2009). The *Fusarium oxysporum* cell wall proteome under adhesion-inducing conditions. Proteomics *9*, 4755–4769.

Prittila, A.M., McIntyre, L.M., Payne, G.A., and Woloshuk, C.P. (2004). Expression profile analysis of wild-type and *fcc1* mutant strains of *Fusarium verticillioides* during fumonisin biosynthesis. Fungal Genet. Biol. *41*, 647–656.

Qutob, D., Kemmerling, B., Brunner, F., Kufner, I., Engelhardt, S., Gust, A.A., Luberacki, B., Seitz, H.U., Stahl, D., Rauhut, T., *et al.* (2006). Phytotoxicity and innate immune responses induced by Nep1-like proteins. Plant Cell *18*, 3721–3744.

Rampitsch, C., Subramaniam, R., Djuric-Ciganovic, S., and Bykova, N.V. (2010). The phosphoproteome of *Fusarium graminearum* at the onset of nitrogen starvation. Proteomics *10*, 124–140.

Rampitsch, C., Tinker, N.A., Subramaniam, R., Barkow-Oesterreicher, S., and Laczko, E. (2012). Phosphoproteome profile of *Fusarium graminearum* grown *in vitro* under nonlimiting conditions. Proteomics *12*, 1002–1005.

Ravalason, H., Grisel, S., Chevret, D., Favel, A., Berrin, J.-G., Sigoillot, J.-C., and Herpoel-Gimbert, I. (2012). *Fusarium verticillioides* secretome as a source of auxiliary enzymes to enhance saccharification of wheat straw. Bioresour. Technol. *114*, 589–596.

Rep, M., Dekker, H.L., Vossen, J.H., de Boer, A.D., Houterman, P.M., Speijer, D., Back, J.W., de Koster, C.G., and Cornelissen, B.J.C. (2002). Mass spectrometric identification of isoforms of PR proteins in xylem sap of fungus-infected tomato. Plant Phys. *130*, 904–917.

Rep, M., van der Does, H.C., Meijer, M., van Wijk, R., Houterman, P.M., Dekker, H.L., de Koster, C.G., and Cornelissen, B.J.C. (2004). A small, cysteine-rich protein secreted by *Fusarium oxysporum* during colonization of xylem vessels is required for I-3-mediated resistance in tomato. Mol. Microbiol. *53*, 1373–1383.

Rogers, L.M., Flaishman, M.A., and Kolattukudy, P.E. (1994). Cutinase gene disruption in *Fusarium solani* f. sp. *pisi* decreases in virulence on pea. Plant Cell *6*, 935–945.

Schoffelmeer, E.A.M., Klis, F.M., Sietsma, J.H., and Cornelissen, B.J.C. (1999). The cell wall of *Fusarium oxysporum*. Fungal Genet. Biol. *27*, 275–282.

Shim, W.-B., and Woloshuk, C.P. (2001). Regulation of fumonsin B1 biosynthesis and conidiation in *Fusarium verticillioides* by a cyclin-like (C-type) gene, *FCC1*. Appl. Environ. Microbiol. *67*, 1607–1612.

Sun, C., Zhao, X.-M., Tang, W., and Chen, L. (2010). FGsub: *Fusarium graminearum* protein subcellular localizations predicted from primary structures. BMC Systems Biol. *4* (Suppl. 2), S12.

Takken, F., and Rep, M. (2010). The arms race between tomato and *Fusarium oxysporum*. Mol. Plant Pathol. *11*, 309–314.

Taylor, R.D., Saparno, A., Blackwell, B., Anoop, V., Gleddie, S., Tinker, N.A., and Harris, L.J. (2008). Proteomic analyses of *Fusarium graminearum* grown under mycotoxin-inducing conditions. Proteomics *8*, 2256–2265.

Tjalsma, H., Bolhuis, A., Jongbloed, J.D.H., Bron, S., and Van Dijl, J.M. (2000). Signal peptide-dependent protein transport in *Bacillus subtilis*: a genome-based survey of the secretome. Microbiol. Mol. Biol. Rev. *64*, 515–547.

Vincent, D., Balesdent, M.-H., Gibon, J., Claverol, S., Lapaillerie, D., Lomenech, A.-M., Blaise, F., Rouxel, T., Martin, F., Bonneu, M., *et al.* (2009). Hunting down fungal secretomes using liquid-phase IEF prior to high resolution 2-DE. Electrophoresis *30*, 4118–4136.

Woloshuk, C.P., and Kolattukudy, P.E. (1986). Mechanism by which contact with plant cuticle triggers cutinase gene expression in the spores of *Fusarium solani* f. sp. *pisi*. Proc. Natl. Acad. Sci. U.S.A.*83*, 1704–1708.

Wong, P., Walter, M., Lee, W., Mannhaupt, G., Munsterkotter, M., Mewes, H.-W., Adam, G., and Güldener, U. (2011). FGDB: revisiting the genome annotation of the plant pathogen *Fusarium graminearum*. Nucleic Acids Res. *39*, D637–639.

Wu, G., Nie, L., and Zhang, W. (2008). Integrative analyses of posttranscriptional regulation in the yeast *Saccharomyces cerevisiae* using transcriptomic and proteomic data. Curr. Microbiol. *57*, 18–22.

Yang, F., Jensen, J.D., Spliid, N.H., Svensson, B., Jacobsen, S., Jorgensen, L.N., Jorgensen, H.J.L., Collinge, D.B., and Finnie, C. (2010). Investigation of the effect of nitrogen on severity of Fusarium Head Blight in barley. J. Proteomics *73*, 743–752.

Yang, F., Jensen, J.D., Svensson, B., Jorgensen, H.J.L., Collinge, D.B., and Finnie, C. (2012). Secreteomics identifies *Fusarium graminearum* proteins involved in the interaction with barley and wheat. Mol. Plant Pathol. *13*, 445–453.

Yin, Q.Y., de Groot, P.W.J., de Koster, C.G., and Klis, F.M. (2008). Mass spectrometry-based proteomics of fungal cell wall glycoproteins. Trends Microbiol. *16*, 20–26.

Zhao, X.-M., Zhang, X.-W., Tang, W.-H., and Chen, L. (2009a). FPPI: *Fusarium graminearum* protein–protein interaction database. J. Proteom. Res. *8*, 4714–4721.

Zhao, Y., and Jensen, O.N. (2009b). Modification-specific proteomics: Strategies for characterization of post-translational modifications using enrichment techniques. Proteomics *9*, 4632–4641.

Zhou, W., Eudes, F., and Laroche, A. (2006). Identification of differentially regulated proteins in response to a compatible interaction between the pathogen *Fusarium graminearum* and its host, *Triticum aestivum*. Proteomics *6*, 4599–4609.

Repeat-induced Point Mutation, DNA Methylation and Heterochromatin in *Gibberella zeae* (Anamorph: *Fusarium graminearum*)

6

Kyle R. Pomraning, Lanelle R. Connolly, Joseph P. Whalen, Kristina M. Smith and Michael Freitag

Abstract

Multiple mechanisms control genome stability in filamentous fungi. To limit the expansion of repeated DNA, e.g. transposable elements (TEs), a group of filamentous ascomycetes make use of a duplication-dependent mutator system, called 'repeat-induced point mutation' (RIP). This phenomenon was the first eukaryotic genome defence system identified and characterized in the 1980s by Selker and colleagues in pioneering studies with *Neurospora crassa*. RIP detects gene-sized duplications and aligns homologous copies by an unknown homology search mechanism during pre-meiosis. Transition mutations (C:G to T:A) are introduced, typically into both or all copies of the DNA segments. In 2007, RIP was first described in *Gibberella zeae* (anamorph: *Fusarium graminearum*) by Kistler and colleagues. Here we review previous experiments and add our recent data, which confirm that RIP occurs at relatively high frequencies in this homothallic species. We show that the *G. zeae rid* homologue is required for RIP, as had been found in *N. crassa*. In contrast to *N. crassa*, DNA methylation does not seem to be a common consequence of RIP. Lastly, we discuss potential evolutionary consequences of RIP in heterothallic and homothallic fungi.

Mechanisms that maintain genome stability in filamentous fungi

Fungal genomes, like those of other eukaryotes, harbour repetitive sequences that can range from short simple sequence repeats (Karaoglu *et al.*, 2005; Kim *et al.*, 2008; Labbe *et al.*, 2010; Murat *et al.*, 2010), families of transposable elements (Aleksenko and Clutterbuck 1996; Daboussi, 1997; Hua-Van *et al.*, 2000; Daboussi and Capy, 2003; Clutterbuck, 2004; Galagan *et al.*, 2005; Wortman *et al.*, 2009), or functional gene duplications to partial or entire chromosome duplications (Miao *et al.*, 1991). Fungi also can acquire or evolve 'dispensable', 'lineage-specific', or 'B' chromosomes (Miao *et al.*, 1991; Ma *et al.*, 2010; Rep and Kistler, 2010). Repetitive sequences within genomes of a single species are plastic. They can move or be moved around, change in copy number and actively contribute to rearrangements of non-repeat sequence (Daviere *et al.*, 2001; Rep *et al.*, 2005; Bergemann *et al.*, 2008; Ogasawara *et al.*, 2009; Carr *et al.*, 2010).

Multiple mechanisms to control the expansion of repeats have been discovered in eukaryotes, the first in the ascomycete *Neurospora crassa* by Selker and colleagues (Selker *et al.*, 1987). This fungus utilizes a genome-wide, DNA-based homology search process that occurs only during the sexual cycle after fertilization but before karyogamy (Fig. 6.1A; reviewed in Selker 1990). The phenomenon was first spotted by Southern analyses as a mixture of gene-sized deletions and restriction fragment length polymorphism (RFLP) that revealed putative genome instability, and thus initially referred to as 'rearrangements induced premeiotically' (Selker *et al.*, 1987). Subsequently it was shown that premeiotic recombination events can be separated from the mechanism responsible for the observed RFLPs and that duplications of gene-sized regions were required (Selker and Garrett 1988). A DNA pairing-dependent protein complex was hypothesized to introduce the many

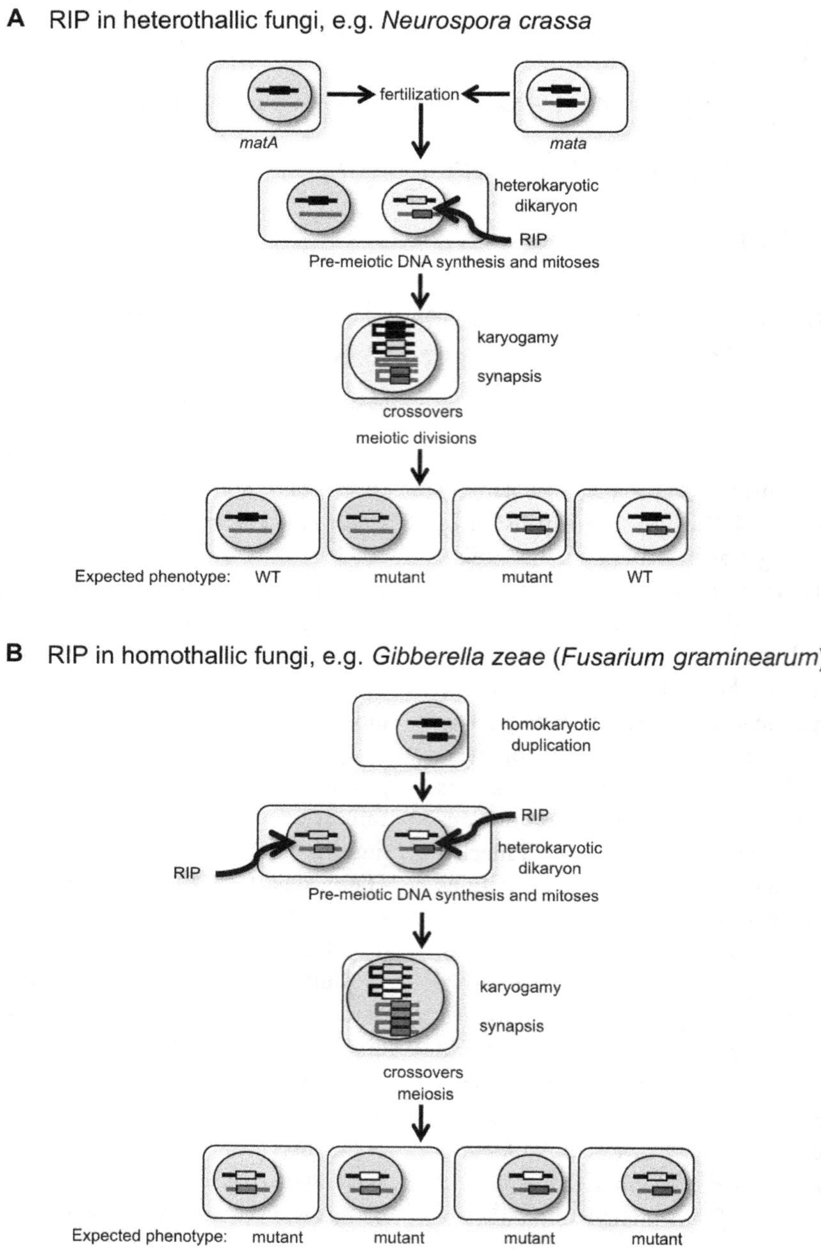

Figure 6.1 Expected differences in RIP in (A) heterothallic and (B) homothallic fungi. In heterothallic fungi (A), mating of two strains is required to generate a heterokaryon. RIP occurs between fertilization and karyogamy and affects only nuclei in which duplications occur. After meiosis, four meiotic products contain segregated copies of the duplications. The black boxes indicate wild-type alleles and shaded grey boxes indicate differentially mutated alleles. If the mutated gene is recessive and if mutation is severe enough to cause phenotypes, two phenotypically wild-type and two phenotypically mutant alleles are recovered after a cross (adapted from (Selker *et al.*, 1987; Cambareri *et al.*, 1989; Selker 1990; Singer *et al.*, 1995)). In homothallic fungi (B), no mating partner is required. After initiation of the sexual phase all nuclei are subject to RIP, but it remains unclear at precisely what stage of development the homology search and mutagenesis occurs. If mutation is severe enough, all four meiotic products are expected to express mutant phenotypes (various shades of grey and white boxes). In both schemes RIP events will escape detection if mutation is light and phenotypes are not easily discernible. DNA sequencing is the only method by which the extent of RIP can be truly quantified.

transition mutations, always C:G to T:A, which were observed, and this observation resulted in a new moniker, 'repeat-induced point mutation', or RIP (Cambareri *et al.*, 1989). Previously unmethylated DNA segments that underwent RIP were often found to also acquire methylated cytosines (Selker *et al.*, 1987; Selker and Garrett, 1988; Singer *et al.*, 1995).

One potential function for RIP is as a defence mechanism against the amplification of transposable elements (TE). So far, only one active TE has been identified in *Neurospora*, the LINE-like element Tad that exists in more than ten copies in a strain from the Ivory Coast, Adiopodoumé (Kinsey, 1989, 1993; Kinsey and Helber, 1989; Cambareri *et al.*, 1994). RIP of TE coding sequences is expected to inactivate the TE replicative machinery, as shown with studies on Tad (Kinsey *et al.*, 1994; Anderson *et al.*, 2001). Moreover, any region-specific mutator similar to RIP will reduce TE repeat homology, e.g. of terminal repeats, and thus may compromise their ability to align. This would be expected to reduce the frequency of homologous recombination between TEs that are dispersed throughout the genome in non-homologous regions. Thus, the overall effect of highly effective RIP in TEs should be a significant increase in overall genome stability, although this has not yet been addressed by careful laboratory evolution experiments. Below we will discuss the expected repercussions of this mutator phenomenon. While negative effects of RIP on genome evolution have been emphasized more recently (Galagan and Selker 2004), even the first paper on RIP (Selker *et al.*, 1987) suggested that potential benefits may be derived from accelerated sequence diversification.

In the ascomycete *Ascobolus immersus* and the basidiomycete *Coprinus cinereus, de novo* cytosine DNA methylation can occur within repeats that are longer than ~400 bp (Zolan and Pukkila 1986; Goyon and Faugeron 1989; Faugeron *et al.*, 1990; Rhounim *et al.*, 1992; Freedman and Pukkila 1993). The timing of the methylation events is similar to RIP, in that they occur after fertilization and development of specialized dikaryons but prior to meiosis, and has thus been referred to as 'methylation induced premeiotically', or MIP (Rhounim *et al.*, 1992). The newly acquired, mainly CpG methylation is subsequently maintained through many mitoses and has been directly shown to effectively block recombination between methylated repeats (Rhounim *et al.*, 1992; Colot *et al.*, 1996; Goyon *et al.*, 1996; Maloisel and Rossignol, 1998). Thus, in each sexual cycle repeat detection prior to karyogamy followed by cytosine methylation (MIP) or by a combination of mutagenesis and subsequent cytosine methylation (RIP) serves to reduce the potentially harmful effects of mobile TEs and other repetitive sequences that were introduced during vegetative growth.

Additional RNA-directed pathways that silence expression from repetitive DNA and may thus serve as genome defence mechanisms have been described in filamentous fungi and have been recently reviewed (Li *et al.*, 2010). Both RNA interference-like 'quelling' (Romano and Macino, 1992; Pickford *et al.*, 2002; Hammond *et al.*, 2008; Tinoco *et al.*, 2010), found in the vegetative stage of the life cycle, and 'meiotic silencing' (Aramayo and Metzenberg, 1996; Shiu *et al.*, 2001; Shiu and Metzenberg, 2002), a meiosis-specific phenomenon that relies on the presence of unpaired DNA, are able to silence transgenes. Both phenomena require a similar and minimal set of proteins that have been implicated in RNA-based regulatory phenomena in fungi and other organisms. These include Dicer, Argonaute and in many – but not all – cases RNA-dependent RNA polymerases (Pickford *et al.*, 2002; Shiu and Metzenberg, 2002; Galagan *et al.*, 2003; Hammond and Keller, 2005; Ma *et al.*, 2010). Additional factors have been identified by both genetic and biochemical analyses (Li *et al.*, 2010). Several additional pathways have now been identified by which (presumably regulatory) small RNA species can be generated, at least in *Neurospora* (Lee *et al.*, 2010).

Currently there is little – if any – evidence to suggest that small RNA generated during quelling or meiotic silencing have an impact on RIP, *de novo* or maintenance cytosine DNA methylation, or silenced chromatin states, at least in *Neurospora* (Freitag *et al.*, 2004). This topic has not yet been addressed in sufficient depth in other filamentous fungi, however.

Repeat-induced point mutation (RIP) in heterothallic and homothallic fungi

RIP is not restricted to *Neurospora*, as additional filamentous fungi, i.e. *Podospora anserina* (Graia *et al.*, 2001), *Magnaporthe grisea* (Ikeda *et al.*, 2002), *Leptosphaeria maculans* (Idnurm and Howlett 2003), *Gibberella zeae* (Cuomo *et al.*, 2007), and *Nectria haematococca* (Coleman *et al.*, 2009) have been experimentally shown to produce the hallmarks of RIP, i.e. reporter gene inactivation, generation of RFLPs and/or cytosine methylation after a cross (Table 6.1). RIP or MIP have not been discovered outside of the filamentous fungi, which correlates with the distribution of a family of DNA methyltransferase-like genes that appear to be restricted to the filamentous fungi (see below). At the same time, not all filamentous fungi use RIP or MIP, as studies in *Aspergillus nidulans* (Selker 1990; Lee *et al.*, 2008), *Sordaria macrospora* (Nowrousian *et al.*, 2010) and *Cochliobolus heterostrophus* (Selker 1990) showed no evidence for either process.

What made the discovery of RIP in *G. zeae* especially interesting was the fact that this species is homothallic, which has repercussions on the genetic diversity of mutated elements. In heterothallic species (e.g. *N. crassa*), fertilization between strains of two different mating types is required to form a heterokaryotic strain (Fig. 6.1A). RIP occurs after fertilization but before karyogamy and will generate transition mutations only in the nucleus that carries duplicated copies of gene-sized segments. If a certain element is present in only one of the mating partners, if RIP is severe, and if the gene in question is recessive, the four meiotic products will segregate as two mutant and two wild-type alleles (Fig 6.1A). Conversely, in a homothallic species that does not require a partner, one would expect RIP to act on all copies present in the original homokaryotic dikaryon to generate heterokaryotic states by differential mutagenesis of all copies (Fig. 6.1B). If again RIP is assumed to be severe and the gene is recessive, all recovered strains should be mutant. Thus, in the most extreme case, one would expect that all progeny issuing from a *G. zeae* cross with duplications in an essential or selectable gene should be mutated and non-viable. This is not observed, however, largely because RIP is hardly ever severe enough to affect all alleles to the point of complete inactivation, and segregation during meiosis will generate duplication strains with one active and one inactivated copy.

While we understand RIP as a phenomenon and as a tool for mutagenesis, we still lack insight into the mechanisms of the homology search and pairing mechanism, as well as the details of the actual mutagenesis. As biochemistry is difficult to carry out on the tissues in question, many attempts have been directed towards the isolation of RIP mutants in laboratory settings or by screening of a large library of wild-collected *Neurospora* strains. At least three wild-collected strains appear to be at least partially defective for RIP, one of those the Adiopodoumé strain (Noubissi *et al.*, 2000), which also carries the only known active *Neurospora* transposon, Tad. One of the difficulties inherent in a mutagenesis scheme in the heterothallic *N. crassa* is the fact that most induced RIP mutants will be recessive and thus complemented by the mating partner after heterokaryon formation subsequent to fertilization. This will result in the isolation of alleles that are

Table 6.1 Organisms in which RIP has been observed by experiment

Organism	Duplications tested	First reference
Neurospora crassa	Many	Selker *et al.* (1987)
Podospora anserina	*mat* cosmid	Graia *et al.* (2001)
Magnaporthe grisea	*hph*	Ikeda *et al.* (2002)
Leptosphaeria maculans	*hph*	Idnurm and Howlett (2003)
Gibberella zeae	*hph*	Cuomo *et al.* (2007)
Nectria haematococca	*hph*	Coleman *et al.* (2009)

for the most part dominant or semi-dominant. Homothallic fungi do not need mating partners, thus a second, more utilitarian, aspect of homothallism is worth noting here. In the homothallic *G. zeae*, mutagenesis should yield strains that, when selfed, may generate progeny in which RIP defects can be directly observed even when alleles are recessive.

Direct evidence for RIP in *G. zeae*

The homothallic *G. zeae* relies on disseminated ascospores to cause head blight of wheat, thus giving it ample opportunity to accumulate mutations caused by RIP in consecutive sexual cycles that are part of its natural annual life cycle. Sequencing of the *G. zeae* PH1 genome revealed a number of repetitive elements (e.g. Fot1fg) with many RIP-type transition mutations (Cuomo *et al.*, 2007). Strains with two copies of the *hph* gene, encoding hygromycin phosphotransferase that confers resistance to hygromycin (Hyg), had been generated for a different study (Seong *et al.*, 2005) and were selfed to experimentally confirm that RIP occurs (Cuomo *et al.*, 2007). A high percentage (42%) of progeny had no Hyg resistance (Hyg^R) when two copies of *hph* were present, while all progeny with a single copy of *hph* retained resistance. This suggested that duplicated *hph* genes were highly unstable. The size of the unlinked *hph* duplication was above the limits described for RIP in *Neurospora*, where ectopically inserted tandem and unlinked repeats of >400 and 1000 base pairs (bp), respectively, can be efficiently detected and mutated (Selker 1990; Watters *et al.*, 1999). The *hph* alleles from five Hyg-sensitive (Hyg^S) *G. zeae* progeny strains were cloned and sequenced to determine whether loss of resistance could be attributed to point mutations or some other cause. In the five alleles, a total of 348 mutations were identified, 99% of which were C:G to T:A transition mutations (Cuomo *et al.*, 2007). In addition, these mutations primarily occurred in the CpA dinucleotide context, similar to what had been observed in *N. crassa* (Selker 1990).

Deletion of *G. zeae rsp1*, a homologue of the *Roundspore* locus of *N. crassa* that has been widely used for studies on meiotic silencing, triggered abnormal ascospore development during selfing (Son *et al.*, 2011). Outcrosses between the mutant and wild-type strains also resulted in ascospores with mutant phenotype. When strains with one additional copy of *rsp1* were selfed, they mostly produced normal ascospores but very few ascospores (<1%) were abnormal. Both endogenous and ectopically integrated genes contained numerous RIP-type point mutations. This study showed that *G. zeae* has functional meiotic silencing, which is triggered by unpaired DNA as in *N. crassa*. It also showed that *G. zeae* RIP occurs at relatively low frequency.

We have since assayed for RIP in many additional *G. zeae* strains that were derived from multiple independent transformations of the PH1 wild type strain with a vector conferring hygromycin resistance (Fig. 6.2A). While many single-spore isolates harboured only one copy of the plasmid, a significant number showed tandem integrations or mixtures of linked and unlinked copies of *hph* by PCR or Southern analyses (data not shown). Hyg^R strains were purified by single spore isolation, selfed on carrot agar (Pasquali and Kistler 2006), and the ratio of Hyg^R to Hyg^S progeny determined. In summary, we found variable RIP frequencies, ranging from 10% to 39% (Table 6.2). In addition, we also examined many of our Hyg^S progeny and found evidence of RFLPs in *hph* alleles after digest of genomic DNA with the *RsaI* restriction endonuclease followed by Southern analyses (see example in Fig. 6.2B). Although RIP was clearly occurring at a significant frequency that was much higher than observed for *rsp1* (Son *et al.*, 2011), it was not as frequent as in *N. crassa* where linked duplications can undergo RIP at frequencies between 60–95% and unlinked duplications show RIP frequencies of 30–50% (Selker, 1990; M. Freitag, unpublished data).

To prove active RIP one needs to examine the sequence of presumably mutated alleles and show presence of C:G to T:A mutations. As an example we show here the results from sequencing of partial mixed *hph* alleles from 14 progeny that were obtained after selfing of FMF60 (7 Hyg^R and 7 Hyg^S strains) and compared them to the *hph* alleles obtained from FMF60. We found 0 to 33 C:G to T:A transition mutations per kilobase (Fig.

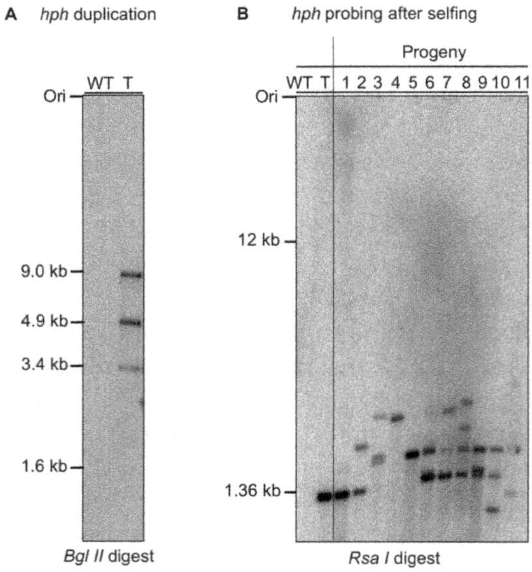

Figure 6.2 RIP of *hph* duplication in *G. zeae* PH1. (A) Generation of a duplication strain. A transformant (FMF69; T) was generated by ectopic transformation with an *hph* gene driven by the *A. nidulans trpC* promoter. Genomic DNA was prepared (Pomraning *et al.*, 2009), digested with *BglII*, separated in a 1% agarose gel and analysed by Southern blot (Miao *et al.*, 2000). Blots were probed with the *hph* cassette and three copies at most likely two insertion sites were detected in the transformant (T) compared to WT. (B) RFLPs suggest RIP. Genomic DNA of wild type PH1 (WT), the transformant FMF60 (T) and 11 progeny (1 to 11) were digested with *RsaI*, separated in a 1.0% agarose gel, blotted to nylon membrane and probed with the *hph* cassette. In T only a single band from three *hph* copies is detected. In the progeny, numerous RFLPs, presumably caused by RIP, were identified.

6.3A). We next examined the dinucleotide context of the C:G to T:A mutations and found that 48% occurred at CpAs and 47% at CpGs (Fig. 6.3B and Table 6.3), suggesting less CpA bias than in the previous study (78%; Cuomo *et al.*, 2007) or in *N. crassa* (74%; Selker 1990, and M. Freitag, unpublished data). The frequency of C to T versus G to A mutations is not random, suggesting that RIP acts processively on a single strand of each repeat copy at one time (Selker 1990; Watters *et al.*, 1999). In further support of this hypothesis, we found one *hph* allele (XJW7–25) with 26 mutations that were all C to T changes on one strand, while the other alleles discussed here had mixtures of C:G to T:A transitions (data not shown).

DNA methylation of sequences mutated by RIP in *G. zeae*

We next examined progeny from selfings for potential RIP-induced cytosine DNA methylation. In *N. crassa*, DNA methylation is often – but not always – a consequence of RIP (Selker *et al.*, 1987; Singer *et al.*, 1995). There is an almost perfect correlation between AT-rich regions of the genome that bear the signature of RIP (high TpA/ApT content and low CpA content) and the distribution of DNA methylation (Selker *et al.*, 2003; Lewis *et al.*, 2009; Smith *et al.*, 2011). DNA methylation in *Neurospora* is catalysed by a single cytosine DNA methyltransferase, DIM-2 (Kouzminova and Selker 2001). In pioneering studies, Selker and colleagues showed that cytosine methylation depends on histone H3 lysine 9 trimethylation (H3K9me3) catalysed by the SET domain histone methyltransferase DIM-5 (Tamaru and Selker, 2001; Tamaru *et al.*, 2003). The Heterochromatin Protein 1 (HP1) binds to trimethylated H3K9 (Freitag *et al.*, 2004) and serves as a platform for DIM-2 binding (Honda and Selker 2008). Just like DIM-2, DIM-5 and HP1 are essential for DNA methylation in *Neurospora* (Kouzminova and Selker 2001; Tamaru and Selker 2001; Freitag *et al.*, 2004). Homologues for the three proteins are present in all sequenced *Fusarium* species (Ma *et al.*, 2010), *F. fujikuroi* (B. Tudzynski and M. Freitag, unpublished data) and *Nectria haematococca* (Coleman *et al.*, 2009), lending support to the idea that the DNA methylation machinery is present in these species. Transcript and protein levels of DIM-2, DIM-5, and HP1 have not been studied in detail in any *Fusarium* species, however, and there are no recent reports on levels of DNA methylation in *G. zeae*.

We first investigated methylation of RIP inactivated *hph* duplications in progeny obtained from selfing of FMF60 by Southern analyses of genomic DNA that had been digested with 5-methylcytosine sensitive (*Sau3AI*) and insensitive (*DpnII*) isoschizomers (Fig. 6.4). We found no evidence for DNA methylation in any of the *hph* alleles that had undergone RIP. In *Neurospora* heavy RIP induces DNA methylation, while less severely mutated alleles remain unmethylated (Singer *et al.*, 1995). The degree of RIP in the

Table 6.2 Strains used in this study to assay RIP frequencies. In all cases the RIP frequency (RIP%) is based on tests for hygromycin resistance (Hyg^R) on rich YPD medium with 50–100 μg/ml Hyg, typically in 48-well plates. The RIP% is calculated as the number of Hyg sensitive (Hyg^S) colonies divided by the total number of strains examined. Failure to isolate Hyg^S strains is denoted as 'not detected'. Duplications are indicated as 'Dup', followed by the name of the duplicated gene (*hph*) and the number of ectopic copies. Single ectopic integrations are labelled as '*hph*^{ec}'

Strain	Genotype	Assay	% RIP (Hyg^S/total)	Notes
IS1	*Dup(hph^{2ec})*	Hyg^R	42% (108/256)	Cuomo *et al.* (2007)
IN10	*hph^{ec}*	Hyg^R	Not detected (0/256)	Cuomo *et al.* (2007)
FMF60	*Dup(hph^{3ec})*	Hyg^R	39% (38/98)	This study
KRP#7	*Dup(hph^{4ec})*	Hyg^R	32% (7/22)	This study
JWT1–7	*Dup(hph^{2ec})*	Hyg^R	37% (29/79)	This study
LRCX9–1	*Dup(hph^{2ec})*	Hyg^R	11% (8/72)	This study
KRP#8	*hph^{ec}, Δrid::hph$^+$*	Hyg^R	Not detected (0/81)	This study
KRP#9	*Dup(hph^{2ec}), Δrid::hph$^+$*	Hyg^R	Not detected (0/47)	This study
FMF96	*Dup(hph^{2ec}), Δrid::neo$^+$*	Hyg^R	Not detected (0/128)	This study
FMF97	*Dup(hph^{2ec}), Δrid::neo$^+$*	Hyg^R	Not detected (0/116)	This study
FMF98	*Dup(hph^{2ec}), Δrid::neo$^+$*	Hyg^R	Not detected (0/173)	This study

A

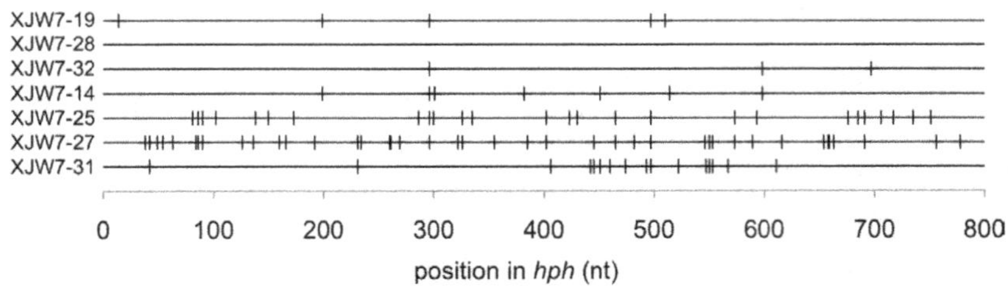

B Number and context of mutations

Strain	CpA	CpT	CpG	CpC	total
XJW7-19	3	0	2	0	5
XJW7-28	0	0	0	0	0
XJW7-32	1	1	1	0	3
XJW7-14	1	0	6	0	7
XJW7-25	13	1	12	0	26
XJW7-27	24	0	16	1	41
XJW7-31	10	0	6	0	16

Figure 6.3 Sequence analysis of *hph* in progeny from a selfing of an *hph* duplication strain. (A) Position of mutations along the sequenced region of the *hph* duplication. DNA sequencing of seven Hyg^S strains showed variable numbers of mutations. Only half of the *hph* duplications (~800bp of the ~1.6kb duplication) were analysed. In all but XJW7–25, the results are likely from mixed alleles, as often sequencing resulted in mixed nucleotide read-outs. The genomic location of the duplication was not determined and PCR amplification may have occurred on both alleles. In XJW7–25, all mutations were homozygous (only one nucleotide was detected by sequencing), suggesting that the sequence is derived from a single allele. (B) Tabulation of the number and dinucleotide context of the mutations found in the sequenced portion of *hph* alleles shown above. These results suggest that CpA bias is not always as strong as suggested by previous results with *G. zeae* (Cuomo *et al.*, 2007) or in *N. crassa* (Selker 1990).

Table 6.3 Dinucleotide context of RIP mutations in *G. zeae* (*F. graminearum*). The number of transition mutations in all sequenced alleles available was calculated and tabulated. Overall, CpA and CpG are by far the preferred contexts in *G. zeae*

Context	CpA	CpT	CpG	CpC	Total	Study
Mutations	109	10	106	3	228	This study
%	47.8	4.4	46.5	1.3		
Mutations	136	0	39	0	175	Cuomo *et al.* (2007)
%	77.7	0	22.3	0		

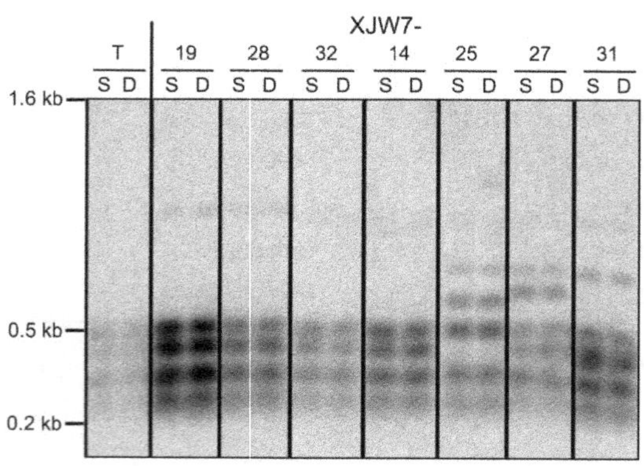

Figure 6.4 Mutations caused by RIP did not result in DNA methylation in *G. zeae*. Genomic DNA was digested with the 5-methylcytosine sensitive *Sau3AI* (S) and insensitive *DpnII* (D) isoschizomers, separated in a 1% agarose gel and analysed by Southern blot. Blots were probed with the *hph* cassette used for transformation. While RFLPs are discernible in most strains when DNA was digested with these enzymes (e.g. XJW7–19, -28, -14, -25, -27, -31), in no case did we observe differences in the *Sau3*AI- and *Dpn*II-digested DNA, which suggests that RIP mutagenesis was either not severe enough to generate a *de novo* DNA methylation signal, or that DNA methylation is not a consequence of RIP in *G. zeae*.

alleles we tested here was relatively light, even in the most heavily mutated allele investigated (33 transitions/kb). This level of RIP is typically associated with *de novo* DNA methylation in *Neurospora* (Singer *et al.*, 1995). Frequency of mutation is not the only determinant for *de novo* methylation in *Neurospora*, however. Studies on *Neurospora* DNA methylation signals suggest that AT content and enrichment for TpA dinucleotides, even in short, ~230 bp fragments increases the likelihood of *de novo* methylation (Miao *et al.*, 2000). The best signals for DNA methylation are regions enriched for TAAA or TTAA repeats (Tamaru and Selker 2003). In our study, the wild type *hph* gene was 60% GC, while the wild-type *Neurospora am* gene was 54% GC (Singer *et al.*, 1995). Even in our most heavily mutated *hph* allele, the GC content was only slightly reduced, from 60% to 56.5%. Based solely on their AT content and TpA enrichment, we would expect that our mutated *hph* alleles would also not be methylated in *Neurospora*. Further studies on the fate of DNA subjected to RIP in *G. zeae* are warranted.

Genome wide DNA methylation in *G. zeae*

The results obtained with *hph* alleles that were mutated by RIP encouraged us to examine the levels of genome-wide cytosine methylation in *G. zeae*. We used two methods to check for global cytosine methylation. First, we performed MeDIP-seq (Pomraning *et al.*, 2009) on the sequenced PH1 reference strain to isolate regions of the genome enriched for 5-methylcytosine. This method did not reveal regions of enrichment above background outside of the heavily repeated rDNA, suggesting little – if any – cytosine methylation in this species (data not shown). To validate this result, we probed Southern blots of genomic DNA digested with the 5-methylcytosine sensitive (*Sau3AI*) and insensitive (*Dpn*II) isoschizomers with a fragment of the rDNA untranscribed spacer, which is typically methylated in fungi that have cytosine methylation (e.g. *N. crassa*). We found very little DNA methylation in these regions (Fig. 6.5), and the ethidium bromide-stained DNA showed no global DNA methylation, which would have been noticed as a high-molecular weight fraction in the *Sau3*AI lanes (data not shown). The same was found for two different widely used strains, GZ3639, and a mating type deletion strain derived from GZ3639 (Lee *et al.*, 2003). Overall, our results suggest that there is little, or no DNA methylation in *G. zeae*.

We also examined the levels of two post-translational modifications of core histones that are markers for heterochromatic silencing, histone H3K9 di- and trimethylation, by ChIP in combination with high-throughput sequencing, following previously described protocols (Pomraning *et al.*, 2009, Smith *et al.*, 2010). To our surprise we found very little H3K9 methylation in the *G. zeae* genome, while a modification associated with transcription, H3K4me2, was found overlapping with promoters of active genes (L.R. Connolly, K.M. Smith and M. Freitag, in preparation). This suggests that silent 'heterochromatin' of the type previously studied in *Neurospora* appears restricted to few regions in the *G. zeae* genome. To further expore this, we constructed strains in which the *G zeae* homologues of *Neurospora* DIM-5 and HP1 have been deleted. In our analyses (L.R. Connolly, K.M. Smith and M. Freitag, in preparation), we did not detect the obvious growth and developmental phenotypes that were previously observed in *Neurospora dim-5* and *hpo* mutants (Tamaru and Selker, 2001, Freitag *et al.*, 2004), and that were reported recently for a *G. zeae* HP1 deletion strain (Reyes-Dominguez *et al.*, 2011).

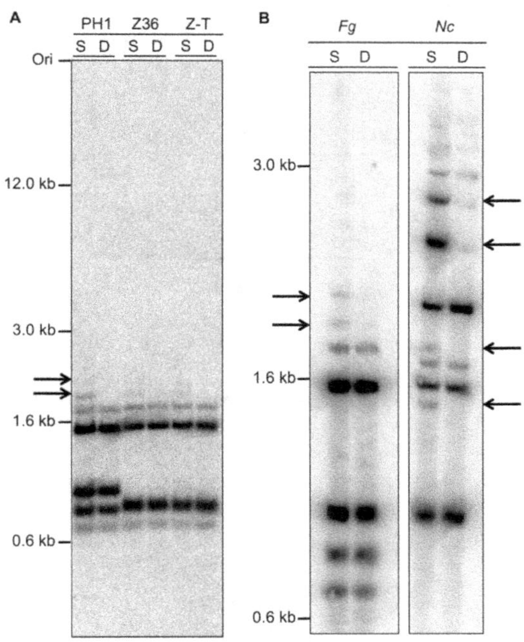

Figure 6.5 Little rDNA methylation in *G. zeae*. A. RFLPs in untranscribed rDNA spacer regions of PH1 and Z3639. Genomic DNA from three different *G. zeae* strains (PH1; Z3639, 'Z36'; Z3639-deribed transformant, 'Z-T') was digested with the 5-methylcytosine-sensitive *Sau3A*I (S) and insensitive *Dpn*II (D) isoschizomers, separated in a 1% agarose gel and blotted to nylon membrane. Blots were probed with a fragment of the rDNA untranscribed spacer derived by PCR of DNA between the large and small ribosomal subunits. Very little DNA methylation was detected (arrows). B. Comparison of rDNA methylation levels in *G. zeae* and *N. crassa*. Genomic DNA of *G. zeae* PH1 and *N. crassa* OR74A was treated as in (A). Gibberella (*Fg*) and *Neurospora* (*Nc*) samples were run on the same gel and blotted together, membranes were cut after transfer and hybridized with rDNA untranscribed spacer probes derived by PCR from Gibberella or *Neurospora* genomic DNA. Blots in (B) were stripped and re-probed with Gibberella and *Neurospora hH3* fragments to assess complete digestion (data not shown).

RID is essential for RIP in *G. zeae*

The molecular mechanism of RIP is still poorly understood, partly because it has been difficult to isolate RIP mutants and because biochemical studies on the limited amount of essentially non-synchronized fertilized tissue is difficult to carry out. Important insights were gained from studies on homologues of DNA methyltransferases of *A. immersus*, *N. crassa* and *A. nidulans*, however. In *Ascobolus*, Rossignol and colleagues discovered a DNA methyltransferase-like protein they called Masc1 (Malagnac *et al.*, 1997). Inactivation of the *masc1* gene resulted in a block of sexual development in homozygous crosses and absence of MIP in heterozygous crosses, which led to the conclusion that *masc1* may encode the DNA methyltransferase required for MIP (Malagnac *et al.*, 1997). Maintenance of pre-existing DNA methylation was not affected in *masc1*. The *Neurospora* Masc1 homologue, RID, is essential for RIP as shown by homozygous crosses with *rid* alleles that had been inactivated by RIP (Freitag *et al.*, 2002). In contrast to Masc1, RID is not involved in sexual development. It is also not involved in DNA methylation in the vegetative phase. Heterozygous *rid* crosses showed reproducibly diminished RIP frequencies, ~10–15% lower than those observed with wild-type alleles, which suggested that the *rid* alleles recovered were recessive (Freitag *et al.*, 2002; K. R. Pomraning and M. Freitag, unpublished results).

These results lend support to two hypotheses for how RIP may cause mutations. Mutations may be generated by either (1) methylation of cytosine to 5-methylcytosine by a DNA methyltransferase, which is followed by enzymatic deamination to thymine, or (2) direct deamination of cytosine to uracil (Selker 1990). Both hypotheses require that DNA is replicated from T:G to T:A, or U:G to U:A to T:A before any repair of presumed mismatches occurs. Support for the first hypothesis stems from the finding that DNA methyltransferase homologues are essential for both MIP and RIP. Nevertheless, biochemical activity for Masc1 and RID has not yet been shown with extracts or purified proteins. Curiously, a Masc1 homologue from *A. nidulans*, DmtA, is also defective in sexual development (Lee *et al.*, 2008), but this fungus does not use either MIP or RIP, and has no measurable DNA methylation in vegetative tissues. Support for the second hypothesis comes from work in bacteria where direct deamination of C to U has been detected as a result of DNA methyltransferase activity in the presence of low *S*-adenosylmethionine levels (Shen *et al.*, 1992; Yebra and Bhagwat, 1995; Macintyre *et al.*, 2001). In summary, the requirement of DNA methyltransferase-like homologues for MIP in *A. immersus* (Malagnac *et al.*, 1997) and RIP in *N. crassa* (Freitag *et al.*, 2002) support the notion that DNA methylation is somehow involved in RIP or that the substrate goes through an important DNA methylation intermediate during RIP.

The putative DNA methyltransferase RID was identified by reverse genetics (Freitag *et al.*, 2002). Homologues of *masc1* or *rid* genes are present in many ascomycete genomes and form a distinct group of eukaryotic DNA methyltransferases (Colot and Rossignol, 1999; Freitag *et al.*, 2002; Goll and Bestor, 2005). In our recent studies, we disrupted the *G. zeae rid* homologue (FGSG_08648.3) by two independent methods, and in two independent strains derived from wild-type PH1, to assess RID function. In the first approach, transformation of strain XJW7–19 replaced *rid* with a gene encoding neomycin phosphotransferase, which provides resistance to the antibiotic G418 (strain FMF96–98; Fig. 6.6). Strain XJW7–19 had been previously generated by transformation of PH1 with a vector conferring resistance to hygromycin and was found to contain two or three copies of the *hph* gene. In the second approach, transformation of PH1 replaced *rid* with the *hph* gene. PCR and Southern analyses

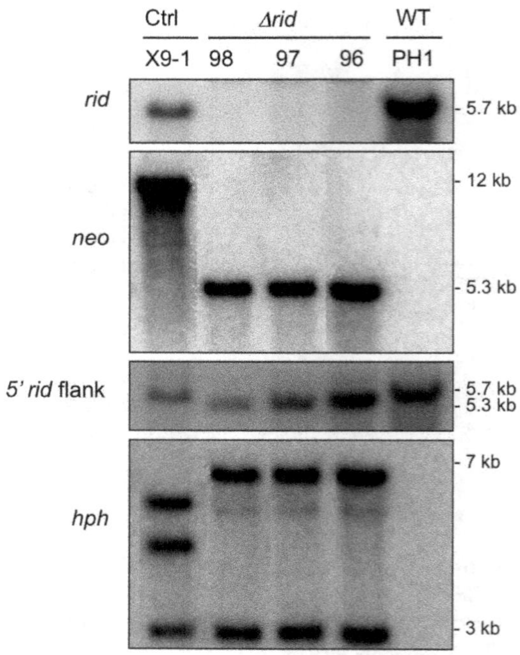

Figure 6.6 Deletion of the *G. zeae rid* homologue. Genomic DNA from *G. zeae* wild type PH1, three independently derived single spore isolates in which *rid* had been replaced with *neo* (FMF96, 97, 98) and an independent transformant with an ectopic *neo* integration (X9–1) was digested with *Bgl*II, separated in a 1% agarose gel and blotted to nylon membrane. Blots were successively probed with fragments containing the whole coding regions of *rid*, *neo* or *hph* and the 5′ flank of *rid*. Expected sizes for fragments after correct replacement of *rid* with *neo* are shown at right. Note the small difference between the fragment size obtained with the 5′ *rid* flank probe (5.7kb in wild type compared to 5.3kb in replacement strains). Both the control strain X9–1 and the Δ*rid* replacement strains have two copies of *hph*. We assessed RIP frequency of *hph* in selfings of these strains (see Table 6.2).

of single spore isolates obtained from primary transformants showed that *rid* was replaced with a single copy of *hph* (e.g. KRP#8) while in other transformants, a second copy of *hph* was inserted ectopically in addition to replacing *rid* with *hph*, (e.g. KRP#9). A third class of transformants showed multiple ectopic copies of *hph*, either by tandem integration at one locus or at multiple loci without disrupting *rid* (e.g. KRP#7).

All strains with *rid* deletions and multiple copies of *hph* (e.g. FMF96, FMF97, FMF98 and KRP#9) remained as fertile as the parental or PH1 strains. RIP, however, is abolished and all progeny obtained from selfings remained resistant to hygromycin (Table 6.2). Neither RFLPs nor DNA methylation were detected in any of the progeny from *rid* selfings (data not shown). This suggests that the function of the RID homologue in *G. zeae* is akin to that of *N. crassa* rather than to that of *A. immersus* Masc1 or *A. nidulans* DmtA. Importantly, deletion of a gene that would generate a recessive mutation in *Neurospora* resulted in easily observable phenotypes in a selfing of mutant *G. zeae* strains. This experiment thus constitutes a proof-of-principle study for the likely success of future screens or selections for recessive RIP mutants in *G. zeae*.

Recognition of RIP in sequenced genomes

The reference genome of *N. crassa* OR74A contains long AT-rich tracks of inactivated TEs that appear to have been mutated by RIP (Galagan *et al.*, 2003; Selker *et al.*, 2003). In these predominantly pericentric and centromeric regions, and in dispersed AT-rich regions degenerate TEs carry not just RIP-type transition mutations when compared for example to extant Tad elements in the Adiopodoumé strain, but also many non-RIP transitions, transversions and short insertions or deletions (indels), presumably because the DNA sequence of TEs, once inactivated, will accumulate DNA changes at a rate similar to other non-coding genome sequences, i.e. 'drift'. This makes recognition of ancient or ancestral RIP by mere sequence analysis exceedingly difficult.

The availability of an ever-increasing number of genome sequences, particularly from ascomycetes, will broaden our perspective of how RIP may have affected genome evolution. Genomic repeat analysis shows evidence for repetitive AT-rich regions in many fungal genomes, many of which are likely pericentric or centromeric DNA but have not been assembled completely or remain unmapped (K.M. Smith, K.R. Pomraning and M. Freitag, unpublished data). C to T transitions are the most common class of mutations in most organisms and would leave much repetitive DNA with the TpA/ApT bias observed in *Neurospora* (Selker *et al.*, 2003). In organisms where RIP has been directly observed by experiment (Table 6.1) a bias has been found for C:T mutations in certain dinucleotide contexts such as the CpA context for *N. crassa* (Selker 1990), *P. anserina* (Graia *et al.*, 2001) and *G. zeae* (Cuomo *et al.*, 2007). CpT is preferred by *M. grisea* (Ikeda *et al.*, 2002), while *L. maculans* targets CpAs and CpGs at similar frequencies (Idnurm and Howlett 2003), as we have observed in our studies with *G. zeae* (Table 6.3). There does not seem to be a simple rule for dinucleotide context, but in most organisms an insufficient number of mutated alleles has been examined. Stepwise mutation from CpG first to CpA then to TpA, where RIP is supposed to act in two successive rounds on CpG dinucleotides is one way to explain different dinucleotide bias in some species (Braumann *et al.*, 2007; Braumann *et al.*, 2008). We do not yet understand if the CpA bias in *N. crassa* and perhaps *G. zeae* is based on the enzymatic activity of RID, although this protein is the best candidate for this mutagenic activity. Alternatively, preference for CpAs may be caused by increased deamination at methylated CpA sites relative to other sites, or by lower affinity of the DNA repair machinery for CpAs.

Bias in certain contexts has been used to analyse repeats for evidence of directed mutation (i.e. RIP) in many sequenced fungal genomes. Results from these analyses – while they should be evaluated with caution – suggest that RIP has occurred during the evolution of many fungi, both species that are currently thought to be sexual as well as those with no known sexual stage (asexual) (Clutterbuck, 2010). In the absence of direct observation of RIP it will remain difficult, if not impossible, to determine whether the AT-rich nature of repeats was caused by past RIP, spontaneous transition mutation or (most likely) a combination of both.

Repeated DNA in the published *Fusarium* genomes

Near complete sequences for the genomes of several *Fusarium* species are now assembled and publicly available, including *G. zeae* (*F. graminearum*), *F. oxysporum* f. sp. *lycopersici*, *F. verticillioides*, and *N. haematococca* (*F. solani* MPVI) (Cuomo *et al.*, 2007; Coleman *et al.*, 2009; Ma *et al.*, 2010). These genomes, together with an increasing number of sequences for additional accessions of these species, provide an incredible resource and will allow unprecedented insights into how sexuality and presumably RIP affects genome content and organization. Some hallmarks of the presence or absence of continuous RIP are easily discerned. For example, the RIP-positive *G. zeae* has the smallest, DNA repeat-poor genome, is homothallic, undergoes sexual reproduction regularly, and exhibits efficient premeiotic recombination. In stark contrast, *F. oxysporum*, which is asexual and thus presumably lacks active RIP, has a large core genome with the highest percentage of repeated DNA. The RIP-positive *F. verticillioides* and *N. haematococca* are intermediate in genome size and number of DNA repeats, heterothallic, and able to undergo sexual reproduction regularly (Selker 1990; Coleman *et al.*, 2009). Dispensable or lineage-specific chromosomes observed in *F. oxysporum* f. sp. *lycopersici* and *N. haematococca*, but lacking in *G. zeae*, are also extremely enriched for repeated DNA when compared to the core genome. In most cases the spectrum of TEs is different from the core chromosomes and the TEs are only distantly related to TEs found on the core chromosomes (Cuomo *et al.*, 2007; Coleman *et al.*, 2009; Ma *et al.*, 2010).

The absence of large gene families in *N. crassa* and other ascomycetes has been attributed to RIP (Galagan *et al.*, 2003; Galagan and Selker 2004), which presumably destroys both copies of a duplicated gene. However, RIP is hardly ever complete in each sexual cycle, and in many cases all copies of a gene duplication or repeated DNA segment will only be lightly mutated, even in selfings with homothallic species. Spores produced during early phases of the sexual cycle are less prone to RIP (Singer *et al.*, 1995a), which is expected to allow favourable gene duplications to be selected for even while purifying selection may apply to one copy of the duplication (Fig. 6.7). Examples for selection of an unmutated copy of a gene while screening for the second, duplicated copy come from experiments in *Neurospora* (Adhvaryu and Selker, 2008). In this particular case a GFP-tagged copy of the protein phosphatase 1 gene (*pp1*) was only recoverable in the presence of an unmutated copy of the wild-type gene.

We and others (Coleman *et al.*, 2009; Ma *et al.*, 2010) searched for transcripts in all published *Fusarium* genomes to identify coding regions that appear duplicated but show >60% nucleotide identity to find cases where RIP may have accelerated evolution. We have not found any obvious evidence for this scenario. In our studies with *G. zeae* we found cases in which premeiotic recombination appeared extremely active, often resulting in loss of several or (amazingly) all copies of the duplicated *hph* gene (L.R. Connolly, K.R. Pomraning, J.P. Whalen and M. Freitag, unpublished data). Thus, it seems likely that a gene may be duplicated and undergo one or successive cycles of RIP, after which one (or both) copies would be lost. Such changes are currently undetectable but may become more obvious as high-throughput sequencing of whole genomes becomes more accessible so that lab-evolved isolates from multiple generations can be analysed in detail.

RIP in nature

Evolutionary costs or benefits of RIP are still under debate. Analysis of genome sequences has uncovered AT-rich regions that are depleted of certain CpN dinucleotides, characteristic of RIP (Clutterbuck 2010), even in organisms which show no experimental evidence for extant RIP, for example *A. nidulans* (Lee *et al.*, 2008) and *Trichoderma reesei* (K.R. Pomraning and M. Freitag, unpublished data). This suggests that these fungi may have had RIP at some point in the past but have now lost this ability, possibly caused by changes in the efficiency of the RIP machinery or the sexuality of the fungus (Clutterbuck 2010).

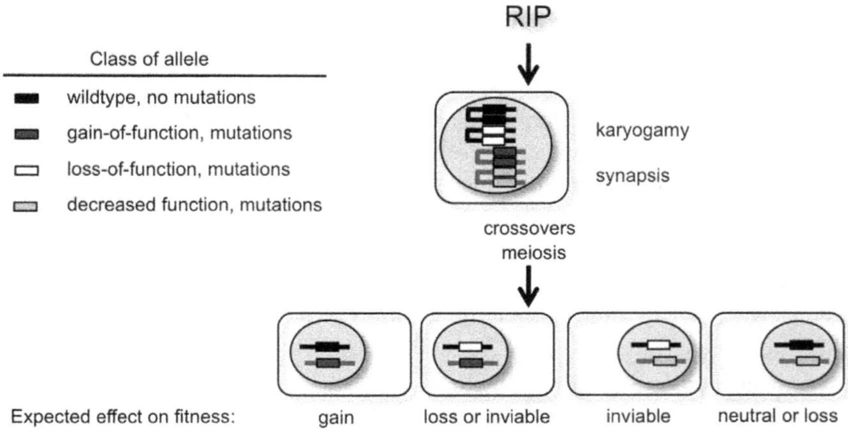

Figure 6.7 Expected effects of RIP on genetic diversity. The cartoon is similar to Figure 6.1. After RIP has occurred, meiotic divisions will segregate differentially mutated alleles of gene-sized duplications into nuclei of ascospores. In principle, four different classes of alleles are expected. Wild-type alleles may have escaped RIP, even in many rounds of premeiotic DNA replication, or they may carry only synonymous mutations. Alleles with slightly reduced function may carry some missense mutations in conserved or non-conserved regions of proteins but they may not result in easily discernible phenotypes. Loss-of-function alleles result in obvious phenotypes, even lethality. This is the class that has been most often studied by use of reporter genes, such as *hph*. Direct gain-of-function mutations have not been observed but are difficult to find. Gain-of-function (i.e. increase in pathogenicity) has been observed in field studies as a consequence of loss-of-function of an avirulence gene in *L. maculans* (Fudal *et al.*, 2009). Two important points should be stressed: (1) RIP is hardly ever complete, i.e. rarely are all copies of a duplication or repeated DNA segment mutated in all progeny, even in selfings with homothallic species. (2) All meiotic products with more than one functional copy of each duplication can drift or may be under selection, thus evolution of diversity should be accelerated by RIP-type mutator phenomena.

In *N. crassa* RIP clearly functions as a genome defence mechanism against TEs (Kinsey *et al.*, 1994; Anderson *et al.*, 2001), perhaps at the cost of slowing or halting evolution of new genes by gene duplication followed by divergence (Galagan *et al.*, 2003; Galagan and Selker 2004). Duplication of conserved essential genes that are intolerant to changes in amino acid sequence is counteracted by RIP. This evolutionary scenario appears to apply to *N. crassa* and *G. zeae*, which both have few paralogues and very little GC-rich repetitive DNA (Galagan *et al.*, 2003; Cuomo *et al.*, 2007). Accordingly, both *P. anserina* and *M. grisea*, in which RIP has been observed to be less efficient, have more gene duplications that are intact (Dean *et al.*, 2005; Espagne *et al.*, 2008). Of course there are genomes that do not fit either category, as the genome of *S. macrospora*, which does not undergo RIP, is rather repeat-poor (Nowrousian *et al.*, 2010). This leaves the possibility that RIP-dependent divergence of genes occurs quickly after duplication by creating a multitude of alleles that vary widely between unmutated and heavily mutated alleles (Fig. 6.7). Most mutations are likely to be detrimental but the few that are not should be selected for in direct competition with sibling spores containing alternative alleles. Maintenance of a new beneficial allele would be difficult to select for in subsequent sexual cycles since most additional RIP mutations would likely be detrimental, thus making a majority of progeny less fit then the original parent with the new advantageous allele. To get around this conundrum, one copy of the originally duplicated gene, likely the copy that is not providing a selective advantage, would need to be lost prior to the next fertilization event. We have some evidence for this scenario in *G. zeae*, where high premeiotic recombination rates seem to be the rule, rather than the exception. On the whole, this scenario would allow for accelerated evolution by RIP without mutating a gene to the point where it contained multiple nonsense or missense codons. Clearly, detailed lab evolution studies are warranted, and the homothallic *G. zeae* seems an ideal candidate model system for this approach.

Of course, RIP does not discriminate between functions for DNA sequences, i.e. TEs or normally expressed endogenous genes. There are, however, few examples of normally functional genes that have been subjected to RIP after duplication. In *Mycosphaerella graminicola*, a homologue of the *dim-2* gene, which is typically present as a single copy in ascomycetes, appears to have been duplicated repeatedly, to the point where 28 identifiable copies exist in the sequenced genome (Dhillon *et al.*, 2010). As mentioned above, in *Neurospora dim-2* is the sole known DNA methyltransferase and essential for cytosine methylation (Kouzminova and Selker, 2001). In *M. graminicola* all extant copies of *dim-2* bear transition mutations reminiscent of RIP. Not surprisingly, no DNA methylation has been detected in this fungus (Dhillon *et al.*, 2010).

Lastly, RIP can act to inactivate genes that are detrimental to the organism under changing environmental conditions. For example, in *L. maculans* RIP was found to be responsible for mutating an avirulence gene in a single year during field trials on disease-resistant rapeseed (*Brassica napus*), introducing nonsense codons within the gene, thus leading to novel virulence (Fudal *et al.*, 2009). Besides the well-documented destruction of TEs in *N. crassa*, this is the first direct evidence of a clearly adaptive advantage of RIP, in this case for a plant pathogen in the field. While the efficiency of RIP was overall low, especially when compared to complete deletion of the avirulence gene aided by recombination between flanking retrotransposons (Fudal *et al.*, 2009), the frequency would likely be sufficient for survival of these strains in nature. The recently published *L. maculans* genome sequence reveals that many effector genes are embedded in regions that may be heavily mutated by RIP in successive sexual cycles (Rouxel *et al.*, 2011). How exactly such alleles fare in the long run needs to be addressed by laboratory and field experiments that assess what types of alleles may allow for selective sweeps in the population.

Acknowledgements

We would like to thank Shinji Honda, Corby Kistler, Seogchan Kang, Yin-Won Lee, Eric Selker, Frances Trail, and Jin-Rong Xu for strains or plasmids. Research on *G. zeae* chromatin is funded by grants from the American Cancer Society (RSG-08-030-01-CCG) and the National Institutes of Health (R01GM097637), while research on RIP was made possible by start-up funds from the OSU Computational and Genome Biology Initiative.

References

Adhvaryu, K.K., and Selker, E.U. (2008). Protein phosphatase PP1 is required for normal DNA methylation in *Neurospora*. Genes Dev. *22*, 3391–3396.

Aleksenko, A., and Clutterbuck, A.J. (1996). The plasmid replicator AMA1 in *Aspergillus nidulans* is an inverted duplication of a low-copy-number dispersed genomic repeat. Mol. Microbiol. *19*, 565–574.

Anderson, C., Tang, and Kinsey, Q., J.A. (2001). Elimination of active tad elements during the sexual phase of the *Neurospora crassa* life cycle. Fungal Genet. Biol. *33*, 49–57.

Aramayo, R., and Metzenberg, R.L. (1996). Meiotic transvection in fungi. Cell *86*, 103–113.

Bergemann, M., Lespinet, O., M'Barek, S.B., Daboussi, M.J., and Dufresne, M. (2008). Genome-wide analysis of the *Fusarium oxysporum* mimp family of MITEs and mobilization of both native and *de novo* created *mimps*. J. Mol. Evol. *67*, 631–642.

Braumann, I., van den Berg, M., and Kempken, F. (2007). Transposons in biotechnologically relevant strains of *Aspergillus niger* and *Penicillium chrysogenum*. Fungal Genet. Biol. *44*, 1399–1414.

Braumann, I., van den Berg, M., and Kempken, F. (2008). Repeat induced point mutation in two asexual fungi, *Aspergillus niger* and *Penicillium chrysogenum*. Curr. Genet. *53*, 287–297.

Cambareri, E.B., Helber, J., and Kinsey, J.A. (1994). *Tad1-1*, an active LINE-like element of *Neurospora crassa*. Mol. Gen. Genet. *242*, 658–665.

Cambareri, E.B., Jensen, Schabtach, B.C., E., and Selker, E.U. (1989). Repeat-induced G-C to A-T mutations in *Neurospora*. Science 244(4912): 1571–1575.

Carr, P.D., Tuckwell, D., Hey, P.M., Simon, L., d'Enfert, C., Birch, M., Oliver, J.D., and Bromley, M.J. (2010). The transposon impala is activated by low temperatures: use of a controlled transposition system to identify genes critical for viability of *Aspergillus fumigatus*. Eukaryot. Cell *9*, 438–448.

Clutterbuck, A.J. (2004). MATE transposable elements in *Aspergillus nidulans*: evidence of repeat-induced point mutation. Fungal Genet. Biol. *41*, 308–316.

Clutterbuck, J.A. (2011). Genomic evidence of repeat-induced point mutation (RIP) in filamentous ascomycetes. Fungal Genet. Biol. *48*, 306–326.

Coleman, J.J., Rounsley, S.D., Rodriguez-Carres, M., Kuo, A., Wasmann, C.C., Grimwood, J., Schmutz, J., Taga, M., White, G.J., Zhou, S., *et al.* (2009). The genome of *Nectria haematococca*: contribution of supernumerary chromosomes to gene expansion. PLoS Genet. *5*, e1000618.

Colot, V., Maloisel, L., and Rossignol, J.L. (1996). Interchromosomal transfer of epigenetic states in Ascobolus: transfer of DNA methylation is mechanistically related to homologous recombination. Cell *86*, 855–864.

Colot, V., and Rossignol, J.L. (1999). Eukaryotic DNA methylation as an evolutionary device. Bioessays *21*, 402–411.

Cuomo, C.A., Guldener, U., Xu, J.R., Trail, F., Turgeon, B.G., Di Pietro, A., Walton, J.D., Ma, L.J., Baker, S.E., Rep, M., *et al.* (2007). The *Fusarium graminearum* genome reveals a link between localized polymorphism and pathogen specialization. Science *317*, 1400–1402.

Daboussi, M.J. (1997). Fungal transposable elements and genome evolution. Genetica *100*, 253–260.

Daboussi, M.J., and Capy, P. (2003). Transposable elements in filamentous fungi. Ann. Rev. Microbiol. *57*, 275–299.

Daviere, J.M., Langin, T., and Daboussi, M.J. (2001). Potential role of transposable elements in the rapid reorganization of the *Fusarium oxysporum* genome. Fungal Genet. Biol. *34*, 177–192.

Dean, R.A., Talbot, N.J., Ebbole, D.J., Farman, M.L., Mitchell, T.K., Orbach, M.J., Thon, M., Kulkarni, R., Xu, J.R., Pan, H. *et al.* (2005). The genome sequence of the rice blast fungus *Magnaporthe grisea*. Nature *434*, 980–986.

Dhillon, B., Cavaletto, J.R., Wood, K.V., and Goodwin, S.B. (2010). Accidental amplification and inactivation of a methyltransferase gene eliminates cytosine methylation in *Mycosphaerella graminicola*. Genetics *186*, 67–77.

Espagne, E., Lespinet, O., Malagnac, F., Da Silva, C., Jaillon, O., Porcel, B.M., Couloux, A., Aury, J.M., Segurens, B., Poulain, J., *et al.* (2008). The genome sequence of the model ascomycete fungus *Podospora anserina*. Genome Biol. *9*, R77.

Faugeron, G., Rhounim, L., and Rossignol, J.L. (1990). How does the cell count the number of ectopic copies of a gene in the premeiotic inactivation process acting in *Ascobolus immersus*?' Genetics *124*, 585–591.

Freedman, T., and Pukkila, P.J. (1993). De novo methylation of repeated sequences in *Coprinus cinereus*. Genetics *135*, 357–366.

Freitag, M., Hickey, P.C., Khlafallah, T.K., Read, N.D., and Selker, E.U. (2004). HP1 is essential for DNA methylation in *Neurospora*. Mol. Cell *13*, 427–434.

Freitag, M., Lee, D.W., Kothe, G.O., Pratt, R.J., Aramayo, R., and Selker, E.U. (2004). DNA methylation is independent of RNA interference in *Neurospora*. Science *304*, 1939.

Freitag, M., Williams, R.L., Kothe, G.O., and Selker, E.U. (2002). A cytosine methyltransferase homologue is essential for repeat-induced point mutation in *Neurospora crassa*. Proc. Natl. Acad. Sci. U.S.A. *99*, 8802–8807.

Fudal, I., Ross, S., Brun, H., Besnard, A.L., Ermel, M., Kuhn, M.L., Balesdent, M.H., and Rouxel, T. (2009). Repeat-induced point mutation (RIP) as an alternative mechanism of evolution toward virulence in *Leptosphaeria maculans*. Mol. Plant Microbe Interact. *22*, 932–941.

Galagan, J.E., Calvo, S.E., Borkovich, K.A., Selker, E.U., Read, N.D., Jaffe, D., FitzHugh, W., Ma, L.J., Smirnov, S., Purcell, S., *et al.* (2003). The genome sequence of the filamentous fungus *Neurospora crassa*. Nature *422*, 859–868.

Galagan, J.E., Calvo, S.E., Cuomo, C., Ma, L.J., Wortman, J.R., Batzoglou, S., Lee, S.I., Basturkmen, M., Spevak, C.C., Clutterbuck, J., Kapitonov, V., *et al.* (2005). Sequencing of *Aspergillus nidulans* and comparative analysis with *A. fumigatus* and *A. oryzae*. Nature *438*, 1105–1115.

Galagan, J.E., and Selker, E.U. (2004). RIP: the evolutionary cost of genome defense. Trends Genet. *20*, 417–423.

Goll, M.G., and Bestor, T.H. (2005). Eukaryotic cytosine methyltransferases. Ann. Rev. Biochem. *74*, 481–514.

Goyon, C., and Faugeron, G. (1989). Targeted transformation of *Ascobolus immersus* and *de novo* methylation of the resulting duplicated DNA sequences. Mol. Cell Biol. *9*, 2818–2827.

Goyon, C., Rossignol, J.L., and Faugeron, G. (1996). Native DNA repeats and methylation in Ascobolus. Nucleic Acids Res. *24*, 3348–3356.

Graia, F., Lespinet, O., Rimbault, B., Dequard-Chablat, M., Coppin, E., and Picard, M. (2001). Genome quality control: RIP (repeat-induced point mutation) comes to *Podospora*. Mol. Microbiol. *40*, 586–595.

Hammond, T.M., Andrewski, M.D., Roossinck, M.J., and Keller, N.P. (2008). *Aspergillus* mycoviruses are targets and suppressors of RNA silencing. Eukaryot. Cell *7*, 350–357.

Hammond, T.M., and Keller, N.P. (2005). RNA silencing in *Aspergillus nidulans* is independent of RNA-dependent RNA polymerases. Genetics *169*, 607–617.

Honda, S., and Selker, E.U. (2008). Direct interaction between DNA methyltransferase DIM-2 and HP1 is required for DNA methylation in *Neurospora crassa*. Mol. Cell Biol. *28*, 6044–6055.

Hua-Van, A., Daviere, J.M., Kaper, F., Langin, T., and Daboussi, M.J. (2000). Genome organization in *Fusarium oxysporum*: clusters of class II transposons. Curr. Genet. *37*, 339–347.

Idnurm, A., and Howlett, B.J. (2003). Analysis of loss of pathogenicity mutants reveals that repeat-induced point mutations can occur in the Dothideomycete *Leptosphaeria maculans*. Fungal Genet. Biol. *39*, 31–37.

Ikeda, K., Nakayashiki, H., Kataoka, T., Tamba, H., Hashimoto, Y., Tosa, Y., and Mayama, S. (2002). Repeat-induced point mutation (RIP) in *Magnaporthe grisea*: implications for its sexual cycle in the natural field context. Mol. Microbiol. *45*, 1355–1364.

Karaoglu, H., Lee, C.M., and Meyer, W. (2005). Survey of simple sequence repeats in completed fungal genomes. Mol. Biol. Evol. *22*, 639–649.

Kim, T.S., Booth, J.G., Gauch, Jr, H.G. ., Sun, Q., Park, J., Lee, Y.H., and Lee, K. (2008). Simple sequence repeats in *Neurospora* crassa: distribution, polymorphism and evolutionary inference. BMC Genomics *9*, 31.

Kinsey, J.A. (1989). Restricted distribution of the Tad transposon in strains of *Neurospora*. Curr. Genet. *15*, 271–275.

Kinsey, J.A. (1993). Transnuclear retrotransposition of the Tad element of *Neurospora*. Proc. Natl. Acad. Sci. U.S.A. *90*, 9384–9387.

Kinsey, J.A., Garrett-Engele, P.W., Cambareri, E.B., and Selker, E.U. (1994). The *Neurospora* transposon Tad is sensitive to repeat-induced point mutation (RIP). Genetics *138*, 657–664.

Kinsey, J.A., and Helber, J. (1989). Isolation of a transposable element from *Neurospora crassa*. Proc. Natl. Acad. Sci. U.S.A. *86*, 1929–1933.

Kouzminova, E., and Selker, E.U. (2001). *dim-2* encodes a DNA methyltransferase responsible for all known cytosine methylation in *Neurospora*. EMBO J. *20*, 4309–4323.

Labbe, J., Murat, C., Morin, E., Le Tacon F., and Martin, F. (2011). Survey and analysis of simple sequence repeats in the *Laccaria bicolor* genome, with development of microsatellite markers. Curr. Genet. *57*, 75–88.

Lee, D.W., Freitag, M., Selker, E.U., and Aramayo, R. (2008). A cytosine methyltransferase homologue is essential for sexual development in *Aspergillus nidulans*. PLoS One 3, e2531.

Lee, H.C., Li, L., Gu, W., Xue, Z., Crosthwaite, S.K., Pertsemlidis, A., Lewis, Z.A., Freitag, M., Selker, E.U., Mello, C.C., *et al.* (2010). Diverse pathways generate microRNA-like RNAs and Dicer-independent small interfering RNAs in fungi. Mol. Cell *38*, 803–814.

Lee, J., Lee, T., Lee, Y.W., Yun, S.H., and Turgeon, B.G. (2003). Shifting fungal reproductive mode by manipulation of mating type genes: obligatory heterothallism of *Gibberella zeae*. Mol. Microbiol. *50*, 145–152.

Lewis, Z.A., Honda, S., Khlafallah, T.K., Jeffress, J.K., Freitag, M., Mohn, F., Schubeler, D., and Selker, E.U. (2009). Relics of repeat-induced point mutation direct heterochromatin formation in *Neurospora crassa*. Genome Res. *19*, 427–437.

Li, L., Chang, S.S., and Liu, Y. (2010). RNA interference pathways in filamentous fungi. Cell Mol. Life Sci. *67*, 3849–3863.

Ma, L.J., van der Does, H.C., Borkovich, K.A., Coleman, J.J., Daboussi, M.J., Di Pietro, A., Dufresne, M., Freitag, M., Grabherr, M., Henrissat, B., *et al.* (2010). Comparative genomics reveals mobile pathogenicity chromosomes in *Fusarium*. Nature *464*, 367–373.

Macintyre, G., Atwood, C.V., and Cupples, C.G. (2001). Lowering S-adenosylmethionine levels in *Escherichia coli* modulates C-to-T transition mutations. J. Bacteriol. *183*, 921–927.

Malagnac, F., Wendel, B., Goyon, C., Faugeron, G., Zickler, D., Rossignol, J.L., Noyer-Weidner, M., Vollmayr, P., Trautner, T.A., and Walter, J. (1997). A gene essential for *de novo* methylation and development in *Ascobolus* reveals a novel type of eukaryotic DNA methyltransferase structure. Cell *91*, 281–290.

Maloisel, L., and Rossignol, J.L. (1998). Suppression of crossing-over by DNA methylation in *Ascobolus*. Genes Dev. *12*, 1381–1389.

Miao, V.P., Covert, S.F., and VanEtten, H.D. (1991). A fungal gene for antibiotic resistance on a dispensable ('B') chromosome. Science *254*, 1773–1776.

Miao, V.P., Freitag, M., and Selker, E.U. (2000). Short TpA-rich segments of the zeta-eta region induce DNA methylation in *Neurospora crassa*. J. Mol. Biol. *300*, 249–273.

Miao, V.P., Matthews, D.E., and VanEtten, H.D. (1991). Identification and chromosomal locations of a family of cytochrome P-450 genes for pisatin detoxification in the fungus Nectria haematococca. Mol. Gen. Genet. *226*, 214–223.

Murat, C., Riccioni, C., Belfiori, B., Cichocki, N., Labbe, J., Morin, E., Tisserant, E., Paolocci, F., Rubini, A., and Martin, F. (2011). Distribution and localization of microsatellites in the Perigord black truffle genome and identification of new molecular markers. Fungal Genet. Biol. *48*, 592–601.

Noubissi, F.K., McCluskey K., and Kasbekar, D.P. (2000). Repeat-induced point mutation (RIP) in crosses with wild-isolated strains of *Neurospora crassa*: evidence for dominant reduction of RIP. Fungal Genet. Biol. *31*, 91–97.

Nowrousian, M., Stajich, J.E., Chu, M., Engh, I., Espagne, E., Halliday, K., Kamerewerd, J., Kempken, F., Knab, B., Kuo, H.C., *et al.* (2010). De novo assembly of a 40 Mb eukaryotic genome from short sequence reads: *Sordaria macrospora*, a model organism for fungal morphogenesis. PLoS Genet. *6*, e1000891.

Ogasawara, H., Obata, H., Hata, Y., Takahashi, S., and Gomi, K. (2009). Crawler, a novel Tc1/mariner-type transposable element in *Aspergillus oryzae* transposes under stress conditions. Fungal Genet. Biol. *46*, 441–449.

Pasquali, M., and Kistler, C. (2006). Gibberella zeae ascospore production and collection for microarray experiments. J. Vis. Exp. *1*, 115.

Pickford, A.S., Catalanotto, C., Cogoni, C., and Macino, G. (2002). Quelling in *Neurospora crassa*. Adv. Genet. *46*, 277–303.

Pomraning, K.R., Smith, K.M., and Freitag, M. (2009). Genome-wide high throughput analysis of DNA methylation in eukaryotes. Methods 47, 142–150.

Rep, M., and Kistler, H.C. (2010). The genomic organization of plant pathogenicity in *Fusarium* species. Curr. Opin. Plant Biol. *13*, 420–426.

Rep, M., van der Does H.C., and Cornelissen, B.J. (2005). Drifter, a novel, low copy hAT-like transposon in *Fusarium oxysporum* is activated during starvation. Fungal Genet. Biol. *42*, 546–553.

Reyes-Dominguez, Y., Boedi, S., Sulyok, M., Wiesenberger, G., Stoppacher, N., Krska, R., and Strauss, J. (2012). Heterochromatin influences the secondary metabolite profile in the plant pathogen Fusarium graminearum. Fungal Genet. Biol. *49*, 39–47.

Rhounim, L., Rossignol, J.L., and Faugeron, G. (1992). Epimutation of repeated genes in *Ascobolus immersus*. EMBO J. *11*, 4451–4457.

Romano, N., and Macino, G. (1992). Quelling: transient inactivation of gene expression in *Neurospora crassa* by transformation with homologous sequences. Mol. Microbiol. *6*, 3343–3353.

Rouxel, T., Grandaubert, J., Hane. J.K., Hoede, C., van de Wouw, A.P., Couloux, A., Dominguez, V., Anthouard, V., Bally, P., Bourras, S., *et al.* (2011). Effector diversification within compartments of the *Leptosphaeria maculans* genome affected by Repeat-Induced Point mutations. Nature Commun. *2*, 202/ doi: 10.1038/ncomms1189.

Selker, E.U. (1990). Premeiotic instability of repeated sequences in *Neurospora crassa*. Ann. Rev. Genet. *24*, 579–613.

Selker, E.U., Cambareri, E.B., Jensen, B.C., and Haack, K.R. (1987). Rearrangement of duplicated DNA in specialized cells of *Neurospora*. Cell *51*, 741–752.

Selker, E.U., and Garrett, P.W. (1988). DNA sequence duplications trigger gene inactivation in *Neurospora crassa*. Proc. Natl. Acad. Sci. U S A. *85*, 6870–6874.

Selker, E.U., Tountas, N.A., Cross, S.H., Margolin, B.S., Murphy, J.G., Bird, A.P., and Freitag, M. (2003). The methylated component of the *Neurospora crassa* genome. Nature *422*, 893–897.

Seong, K., Hou, Z., Tracy, M., Kistler, H.C., and Xu, J.R. (2005). Random Insertional Mutagenesis identifies genes associated with virulence in the wheat scab fungus *Fusarium graminearum*. Phytopathology *95*, 744–750.

Shen, J.C., Rideout, W.M. 3rd, Jones, P.A. (1992). High frequency mutagenesis by a DNA methyltransferase. Cell *71*, 1073–1080.

Shiu, P.K., and Metzenberg, R.L. (2002). Meiotic silencing by unpaired DNA: properties, regulation and suppression. Genetics *161*, 1483–1495.

Shiu, P.K., Raju, N.B., Zickler, D., and Metzenberg, R.L. (2001). Meiotic silencing by unpaired DNA. Cell *107*, 905–916.

Singer, M.J., Kuzminova, E.A., Tharp, A., Margolin, B.S., and Selker, E.U. (1995a). Different frequencies of RIP among early vs. late ascospores of *Neurospora crassa*. Fungal Genetics Newsl. *42*, 74–75.

Singer, M.J., Marcotte, B.A., and Selker, E.U. (1995). DNA methylation associated with repeat-induced point mutation in *Neurospora crassa*. Mol. Cell Biol. *15*, 5586–5597.

Smith, K.M., Sancar, G., Dekhang, R., Sullivan, C.M., Li, S., Tag, A.G., Sancar, C., Bredeweg, E.L., Priest, H.D., McCormick, R.F., *et al.* (2010). Transcription factors in light and circadian clock signaling networks revealed by genomewide mapping of direct targets for *Neurospora* white collar complex. Eukaryot. Cell *9*, 1549–1556.

Smith, K.M., Phatale, P.A., Sullivan, C.M., Pomraning, K.R., and Freitag, M. (2011). Heterochromatin is required for normal distribution of *Neurospora crassa* CenH3. Mol. Cell Biol. *31*, 2528–2542.

Son, H., Min, K., Lee, J., Raju, N.B., and Lee, Y.-W. (2011). Meiotic silencing in the homothallic fungus *Gibberella zeae*. Fungal Biol. *115*, 1290–1302.

Tamaru, H., and Selker, E.U. (2001). A histone H3 methyltransferase controls DNA methylation in *Neurospora crassa*. Nature *414*, 277–283.

Tamaru, H., and Selker, E.U. (2003). Synthesis of signals for *de novo* DNA methylation in *Neurospora crassa*. Mol. Cell Biol. *23*, 2379–2394.

Tamaru, H., Zhang, X., McMillen, D., Singh, P.B., Nakayama, J., Grewal, S.I., Allis, C.D., Cheng, X., and Selker, E.U. (2003). Trimethylated lysine 9 of histone H3 is a mark for DNA methylation in *Neurospora crassa*. Nat. Genet. *34*, 75–79.

Tinoco, M.L., Dias, B.B., Dall'Astta, R.C., Pamphile, J.A., and Aragao, F.J. (2010). *In vivo* trans-specific gene silencing in fungal cells by *in planta* expression of a double-stranded RNA. BMC Biol *8*, 27.

Watters, M.K., Randall, T.A., Margolin, B.S., Selker, E.U., and Stadler, D.R. (1999). Action of repeat-induced point mutation on both strands of a duplex and on tandem duplications of various sizes in *Neurospora*. Genetics *153*, 705–714.

Wortman, J.R., Gilsenan, J.M., Joardar, V., Deegan, J., Clutterbuck, J., Andersen, M.R., Archer, D., Bencina, M., Braus, G., Coutinho, P., *et al.* (2009). The 2008 update of the *Aspergillus nidulans* genome annotation: a community effort. Fungal Genet. Biol. 46 Suppl *1*, S2–13.

Yebra, M.J., and Bhagwat, A.S. (1995). A cytosine methyltransferase converts 5-methylcytosine in DNA to thymine. Biochemistry *34*, 14752–14757.

Zolan, M.E., and Pukkila, P.J. (1986). Inheritance of DNA methylation in *Coprinus cinereus*. Mol. Cell Biol. *6*, 195–200.

The Nitrogen Regulation Network and its Impact on Secondary Metabolism and Pathogenicity

7

Philipp Wiemann and Bettina Tudzynski

Abstract

Nitrogen is essential for fungal growth because it is a component of both nucleic acids and proteins. Fungi have two predominant mechanisms to incorporate ammonium into their metabolism: 1) the NADP-dependent, glutamate-dehydrogenase-catalysed reductive amination of 2-oxoglutarate to form glutamate; and 2) the ATP-dependent, glutamine synthase-catalysed fusion of ammonium and glutamate to form glutamine. Beside ammonium, fungi can also utilize a broad variety of other nitrogen sources, such as nitrate, proteins, amino acids, uric acid, allantoin and urea. Efficient control mechanisms are needed to coordinate activation/repression of genes and their products that are involved in sensing, transporting and/or metabolizing nitrogen-containing substances. Furthermore, nitrogen availability plays a critical role in how fungi interact with plants as pathogens and endophytes. Thus, nitrogen limitation has been proposed to be a key signal for activating the expression of virulence-associated genes in plant pathogens. Additionally, quality and quantity of nitrogen also affects the formation of a broad range of secondary metabolites. These secondary metabolites often contribute to virulence on the host of the fungus and additionally can bare a threat to animal and human health when they occur as contaminants of food and feed. This chapter will review the genetic basis of the nitrogen regulation network with the focus on the genus *Fusarium* which contains some of the most devastating plant pathogens.

Introduction: the nitrogen circuit

Nitrogen is essential for life because it is a component of both nucleic acids and proteins. Although elemental nitrogen (N_2) is the most abundant gas in the earth's atmosphere, it can be utilized by organisms only after it has been reduced to ammonia (NH_3), which exists as ammonium ions (NH_4^+) in aqueous solutions. N_2 reduction is catalysed by the enzyme nitrogenase, which is produced by a limited number of prokaryotes: Proteobacteria (*Rhizobium* spp., *Azomonas* spp. and *Azotobacter* spp.), Cyanobacteria and Actinobacteria (*Frankia* spp.). Fungi can acquire ammonium via a specific transport system in their cell membrane and subsequently use it in amino acid and nucleic acid biosynthesis. Nitrate (NO_3^-), which can be formed from NH_4^+ by some Proteobacteria (*Nitrosomonas* spp. and *Nitrobacter* spp.), is another source of inorganic nitrogen. Fungi also have a highly specific system to transport NO_3^- from the environment into hyphae. Once in hyphae, NO_3^- is reduced to NH_4^+ by the activities of the enzymes nitrate reductase and nitrite reductase (Fig. 7.1).

Fungi have two predominant mechanisms to incorporate ammonium into their metabolism: (1) the NADP-dependent, glutamate dehydrogenase (GdhA)-catalysed reductive amination of 2-oxoglutarate to form glutamate (Glu); and (2) the ATP-dependent, glutamine synthase (GS)-catalysed fusion of NH_4^+ and glutamate to form glutamine (Gln). Except for proline (Pro), which is synthesized by reduction of Glu, *de novo* synthesis

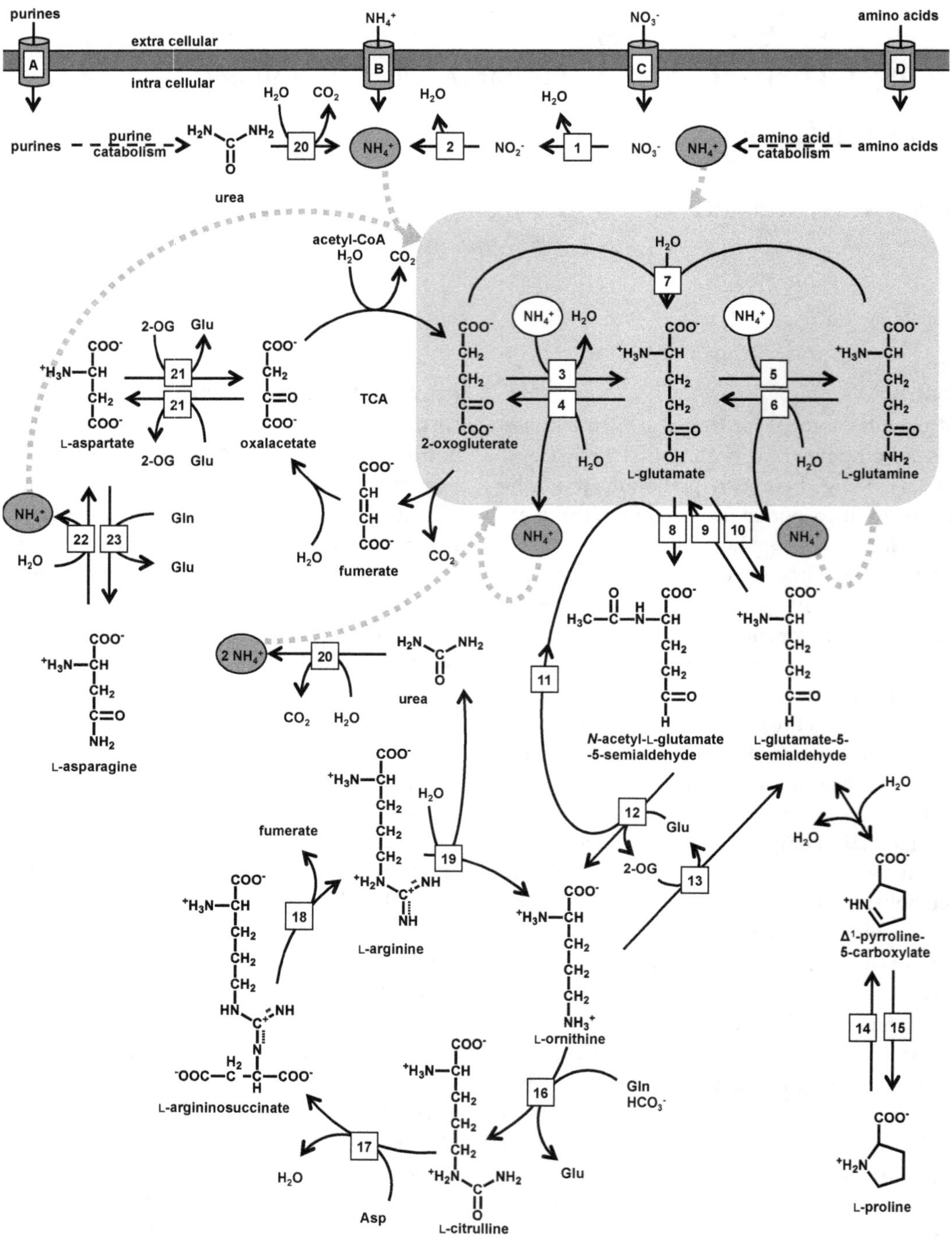

Figure 7.1 Conserved metabolic routes of nitrogen uptake, release and incorporation in amino acid metabolism: When possible, Fusarium proteins and GenBank accession numbers are stated, otherwise the *F. verticillioides* gene identifiers (Broad Institute) with the highest homology by BLASTX to characterized proteins are given in squared brackets. Metabolic routes are deduced from *N. crassa* (reviewed in Radford, 2004), *A. nidulans* and *S. cerevisiae*. TCA: Tricarboxylic acid cycle. **A:** Purine transporters. **B:** Ammonium permeases. **C:** Nitrate transporters. **D:** Amino acid permeases. **1:** Nitrate reductase, NiaD (CAA62232). **2:** Nitrite reductase, FVEG_07298.3 [NIT6 (EAA31119), NiiA (AAA33315)]. **3:** NADPdependent glutamate

of other amino acids involves incorporation of the amino group(s) by the central mechanism of transamination (Fig. 7.1). Amino acids are then used as the building blocks or precursors for all other nitrogen-containing metabolites, including proteins and nucleic acids.

Fungi can also utilize a broad variety of organic compounds as sources of nitrogen. These compounds are often the waste-products of vertebrates and include proteins, uric acid, allantoin and urea. Both uric acid and allantoin can be transported directly into hyphae where they are converted first to urea and then to NH_4^+ (Fig. 7.1). Secreted enzymes generally break down proteins to smaller peptides or individual amino acids that can be absorbed into hyphae by peptide and amino acid transporters. Amino acids can be catabolized by either transamination to produce other amino acids or deamination to form NH_4^+, which can be used to synthesize glutamate or glutamine, by GdhA or GS, respectively, and thereby close the nitrogen circuit.

It is easy to imagine that efficient control mechanisms are needed to coordinate activation and repression of the enzymatic processes introduced above. This Chapter will review the genetic basis of these mechanisms with a focus on *Fusarium*. The nitrogen metabolism network likely plays a critical role in how *Fusarium* interacts with plants as a pathogen and endophyte and how the fungus synthesizes a broad range of secondary metabolites, of which some can affect plant and human health (see Chapters 8 and 9).

Nitrogen availability controls virulence factors

When *Fusarium* and plants encounter each other in the presence of sufficient nitrogen, e.g. in soil of fertilized fields, their relationship could be described as competitive, because they compete for the same soil nutrients in order to grow. In some cases, e.g. endophytic growth of some fusaria, the relationship might even be commensalistic or mutualistic, where one or both partners benefit from the relationship, respectively. But what happens when nitrogen is limited?

Nitrogen limitation has been proposed to be a key signal for activating the expression of virulence-associated genes in plant pathogens (reviewed in Snoeijers *et al.*, 2000). Sufficiently fertilized tomato and wheat plants show a significant reduction of disease symptoms when infected with *Fusarium oxysporum* f. sp. *lycopersici* and *Fusarium graminearum*, respectively, compared to non-fertilized plants (López-Berges *et al.*, 2010; Yang *et al.*, 2010). Lack of nutrients

dehydrogenase, GdhA (CAJ21196). **4:** NAD+dependent glutamate dehydrogenase, GdhB (CAC27837). **5:** glutamine synthetase, GlnA (CAC27836). **6:** Glutaminase, FVEG_02650.3 [GtaA (BAA86935) (Koibuchi *et al.*, 2004)]. **7:** Glutamate synthase, GltA (CAJ20840). **8:** Acetylglutamate kinase / *N*acetylglutamoyl phospho reductase, FVEG_02213.3, FVEG_11104.3 and FVEG_09768.3 [ARG6 (EAA35492)]. **9:** Pyrroline5carboxylate reductase, FVEG_01267.3, FVEG_03747.3 and FVEG_10723.3 [Put2p (AAB68907) (Brandriss and Magasanik, 1979); PruA (CAA64836) (Schaap *et al.*, 1997)]. **10:** Glutamate kinase, FVEG_08231.3 [PRO4 (EAA27994)]; Glutamoyl phospho reductase, FVEG_02844.3 [PRO3 (EAA31599)]. **11:** *N*acetylglutamate synthase, FVEG_07306.3 [ARG14 (EAA33088)]; glutamate *N*acetyltransferase, FVEG_04728.3 [ARG7 (CAD70860) (Davis, 1979)]. **12:** Acetylornithine transaminase, FVEG_09656.3 [ARG5 (ARG8) (EAA34262 (Q9P3I3)) (Pauwels *et al.*, 2003)]. **13:** Ornithine transaminase, FVEG_06950.3 [OTA (EAA27181), OtaA (AAB18259)]. **14:** Proline oxidase, FVEG_03607.3 [PrnD (CAC18796)]. **15:** Pyrroline5carboxylate reductase, FVEG_05327.3 [PRO1 (EAA28126), PrnC (AAF72527)]. **16:** Small subunit carbomyol phosphate synthase, FVEG_05607.3 [ARG2 (EAA33254)]; large subunit carbomyol phospho synthase, FVEG_01344.3 [ARG3 (EAA36214)]; ornithine carbamoyltransferase, FVEG_04951.3 [ARG12 (EAA27637)]. **17:** Argininosuccinate synthase, FVEG_04732.3 and FVEG_11230.3 [ARG1 (EAA34740)]. **18:** Argininosuccinate lyase, Arg1 (BAB40769). **19:** Arginase, FVEG_11677.3 [AGA (EAA30523), AgaA (AAB05775) (Borsuk *et al.*, 1999)]. **20:** Urease, FVEG_00443.3 (Strope *et al.*, 2011); urea carboxylyase, FVEG_04571.3 (Strope *et al.*, 2011); urea amidolyase, FVEG_11593.3 [Dur1,2p / Dur80p (CAA85172) (Roon *et al.*, 1972)]. **21:** Mitochondrial aspartate aminotransferase, FVEG_11407.3 [Aat1p (CAA81946) (Morin *et al.*, 1992)]; cytosolic aspartate aminotransferase, FVEG_06348.3 [Aat2p (CAA97550) (Verleur *et al.*, 1997)]. **22:** Asparaginase, FVEG_03277.3 [Asp1p (AAB64757) (Sinclair *et al.*, 1994) and Asp3p (AAB67684) (Kim *et al.*, 1988)] and FVEG_01059.3. **23:** Asparagine synthase, FVEG_02650.3 [ASN1 (EAA31713), Asn1p (CAA88594) and Asn2p (CAA97135) (Dang *et al.*, 1996)] and FVEG_01066.3 [ASN2 (EAA32918)].

is thought to induce a switch in fungal lifestyle from non-pathogenic to pathogenic that is mediated by alterations in the transcriptome and proteome. In the wheat pathogen *F. graminearum*, distinct differences among several protein kinases and transcription factors occur upon nitrogen starvation (Rampitsch *et al.*, 2010; see Chapter 4). Studies of the plant pathogens *F. oxysporum* f. sp. *lycopersici*, *F. graminearum* and *Cladosporium fulvum* have established that several genes found to be induced during fungal growth *in planta* are also strongly up-regulated under conditions of nitrogen limitation, suggesting that nitrogen starvation mimics conditions experienced by fungi *in planta* (Coleman *et al.*, 1997; Trail *et al.*, 2003; Divon *et al.*, 2005). The up-regulated genes include those predicted to function in ammonium, amino acid and peptide transport as well as in acquisition and metabolism of alternative nitrogen sources (Divon *et al.*, 2005). There is further evidence for a connection between nitrogen metabolism and pathogenicity from the *Colletotrichum acutatum* nitrate reductase gene (*nit1*), which is preferentially expressed under nitrogen starvation and during appressoria formation. Likewise, the gene (*nir1*) encoding the transcriptional activator of the nitrate utilization genes, is essential for appressorium formation and wild-type pathogenicity on unwounded strawberry plants under nitrogen starvation conditions (Horowitz *et al.*, 2006). The *Colletotrichum gloeosporioides* gene encoding GS is expressed during infection of tropical legumes, but its role in pathogenicity has not been demonstrated (Stephenson *et al.*, 1997). Inactivation of the argininosuccinate synthase gene resulted in altered host specificity of *Botrytis cinerea* (Patel *et al.*, 2010). Similarly, deletion of the putative argininosuccinate lyase reduced pathogenicity of *F. oxysporum* f. sp. *melonis* on susceptible melon cultivars (Namiki *et al.*, 2001). A forward genetic screen for pathogenicity factors of *F. oxysporum* f. sp. *lycopersici* identified the homologue of the *Aspergillus nidulans* transcription factor MeaB (López-Berges *et al.*, 2009). In *A. nidulans*, deletion of *meaB* confers resistance to methylammonium indicating a role in nitrogen metabolism (Polley and Caddick, 1996). Interestingly, *meaB* in *A. nidulans*, *F. oxysporum* f. sp. *lycopersici* and *Fusarium fujikuroi* is expressed preferentially under nitrogen abundance, and its deletion in *F. oxysporum* f. sp. *lycopersici* caused increased vascular wilt symptoms of tomato plants when supplemented with ammonium (Wong *et al.*, 2007; López-Berges *et al.*, 2010; Wagner *et al.*, 2010). Other examples of pathogenicity-associated, nitrogen-regulated genes are the *Magnaporthe oryzae* hydrophobin (*MPG1*) and serine protease (*SPM1*) genes, and the *C. fulvum* avirulence gene *Avr9* (Talbot *et al.*, 1993; Van den Ackerveken *et al.*, 1994; Donofrio *et al.*, 2006). Taken together, the accumulated data strongly argue for an immediate connection between the ability of phytopathogenic fungi to sense ambient nitrogen availability as an input signal and a determination of non-pathogenic or pathogenic lifestyle as an output pathway.

There is also evidence that the connection between nitrogen availability and pathogenicity can be mediated through production of secondary metabolites (e.g. mycotoxins, pigments, and phytohormones) in *Fusarium*. The best example of this is production of the phytohormones gibberellic acids (GAs) by *F. fujikuroi*. GA production is strongly repressed in the presence of high levels of nitrogen and is responsible for hyperelongation (*bakanae* disease) of rice seedlings induced by *F. fujikuroi* (Borrow *et al.*, 1964; Bu'lock *et al.*, 1975; Tudzynski *et al.*, 1999; Wiemann *et al.*, 2010). Production of the mycotoxins fumonisins by *F. verticillioides* and trichothecenes by *F. graminearum* is repressed by high levels of at least some nitrogen sources (Shim and Woloshuk, 1999; Gardiner *et al.*, 2009) and contributes to the ability of the former to cause foliar symptoms on maize seedlings (Williams *et al.*, 2007; Glenn *et al.*, 2008) and the latter to cause wheat head blight (Bai *et al.*, 2002; Dyer *et al.*, 2005). In addition, production of the polyketides bikaverin, nor-bikaverin, nectriafurone, anhydrofusarubin lactol and 5-*O*-methyljavanicin is correlated with virulence in the cotton pathogen *F. oxysporum* f. sp. *vasinfectum* (Bell *et al.*, 2003). Similarly, production of bikaverin in *F. fujikuroi* and fumonisins in *Fusarium proliferatum* are also induced upon nitrogen starvation (Linnemannstöns *et al.*, 2002; Schönig *et al.*, 2008; Kohut *et al.*, 2009; Wiemann *et al.*, 2009), but whether these secondary metabolites are involved in pathogenicity remains to be elucidated.

Remarkably, several of the aforementioned genes are targets of the same GATA-type zinc finger transcription factor. Orthologues of this transcription factor have different designations in different fungi, e.g. NIT2 in *Neurospora crassa* and AreA in *A. nidulans*. In the model fungi *N. crassa* and *A. nidulans*, NIT2/AreA interact with the nitrate pathway-specific transcription factor NIT4/NirA, and both the NIT2/AreA and NIT4/NirA are required for expression of the nitrate reductase gene (*nit-3/niaD*) (Arst and Cove, 1973; Tomsett and Garrett, 1980; Burger *et al.*, 1991a,b; Yuan *et al.*, 1991; Feng and Marzluf, 1998; Muro-Pastor *et al.*, 2004). In a variety of fungi, mutation of *nit-2/areA* homologues, the gene encoding NIT2/AreA, affects both secondary metabolite synthesis and pathogenicity. For example, production of the secondary metabolites gibberellins and fumonisins is negatively affected in *areA* mutants of *F. fujikuroi* and *F. verticillioides* respectively (Tudzynski *et al.*, 1999; Kim and Woloshuk, 2008). With respect to pathogenicity, deletion of the *nit-2/areA* homologue, *Nfr1*, in the tomato pathogen *Cladosporium fulvum*, results in down-regulation of the avirulence gene *Avr9* (Pérez-García *et al.*, 2001). Deletion of the *nit-2/areA* homologues in *F. oxysporum* f. sp. *lycopersici* (*Fnr1*), *C. fulvum* (*Nfr1*) and *Colletotrichum lindemuthianum* (*CLNR1*) results in attenuated pathogenicity on their respective host plants (Divon *et al.*, 2006; Pérez-García *et al.*, 2001; Pellier *et al.*, 2003; López-Berges *et al.*, 2010). Furthermore, the gene encoding the pathogenicity factor MeaB was affected in expression in *areA* loss-of-function mutants in *F. fujikuroi*, *F. oxysporum* f. sp. *lycopersici* and *A. nidulans* (López-Berges *et al.*, 2010; Schönig *et al.*, 2008; Wagner *et al.*, 2010). Surprisingly, the *nit-2/areA* homologue, *NUT1*, in *Magnaporthe grisea* (now *M. oryzae*) is dispensable for pathogenicity on rice even though it is required for wild-type expression of the hydrophobin-encoding gene *MPG1*, which is required for pathogenicity, and the nitrate reductase-encoding gene *NIAD* during nitrogen starvation (Froeliger and Carpenter, 1996; Lau and Hamer, 1996; Soanes *et al.*, 2002). Because of NIT2/AreA-mediated effects on expression of genes encoding virulence factors and on virulence itself in multiple plant pathogenic fungi, a more thorough review of this GATA-type transcription factor is warranted.

A GATA-type transcription factor controls many nitrogen-related metabolic processes

Filamentous fungi can readily utilize the inorganic nitrogen source NO_3^- but preferentially use energetically more favoured nitrogen sources such as NH_4^+ and Gln when they are available. In order to be utilized, NO_3^- has to be transported into the cell and converted, via the nitrate–nitrite reductase system, to NH_4^+, which can then enter into the nitrogen circuit via GdhA and GS as described above. The genes involved in NO_3^- uptake and its reduction have been identified in several fungi. *N. crassa* has only one nitrate transporter, NIT10 (Gao-Rubinelli and Marzluf, 2004), whereas *A. nidulans* has two, NrtA (CrnA) and NrtB (Unkles *et al.*, 1991, 2001, Wang *et al.*, 2008). Analysis of the *F. graminearum/F. oxysporum/F. verticillioides* genome sequence database (Ma *et al.*, 2010) identified only one putative nitrate transporter per genome. Genes encoding homologues of the *N. crassa* and *A. nidulans* nitrate reductases NIT3/NiaD (Fu and Marzluf, 1987a; Maladier *et al.*, 1989) have also been characterized in *F. fujikuroi* and *F. oxysporum* (Diolez *et al.*, 1993; Tudzynski *et al.*, 1996; de Queiroz *et al.*, 1998; Fujii and Takaya, 2008). A homologue of the *N. crassa* and *A. nidulans* nitrite reductase NIT6/NiiA (Exley *et al.*, 1993; Johnston *et al.*, 1990) was characterized in *F. oxysporum* (Kobayashi and Shoun, 1995). The genes encoding CrnA, NiiA and NiaD are located adjacent to each other in *A. nidulans*, whereas the homologues of these genes are at different loci in *N. crassa* and *Fusarium* (Johnstone *et al.*, 1990; Slot and Hibbert, 2007).

Because the process of NO_3^- reduction to NH_4^+ is energy consuming and the product of the nitrate reductase, nitrite (NO_2^-), is toxic, it seems reasonable to assume that the expression of genes involved in these processes would be highly coordinated. Early biochemical studies in *N. crassa* and *A. nidulans* showed that nitrate reductase activity is induced by NO_3^- and repressed by NH_4^+ (Kinsky, 1961; Cove, 1966). Random genome mutagenesis led to the identification of numerous genes that function during reduction of nitrate to NH_4^+ in *N. crassa* and *A. nidulans* (Perkins, 1959; Cove and Pateman, 1963; Sorger and Giles, 1965). Growth and biochemical analyses revealed that one gene (*nit-3/niaD*) encodes nitrate reductase, another (*nit-6/niiA*) encodes nitrite reductase, five genes (*nit-1/cnxABC, nit-7/cnxE, nit8/cnxF, nit-9/cnxG* and *cnxH*) are involved in assembling the molybdenum-containing cofactor (MoCo) required for the nitrate reductase, and two genes (*nit-4/nirA* formerly *nit-5/niiB*, and *nit-2/areA*) encode transcriptional regulators (Pateman *et al.*, 1967; Arst and Cove, 1973; Coddington, 1976; Garrett and Cove, 1976; Tomsett and Garrett, 1980, Unkles *et al.*, 1991; 2001, Gao-Rubinelli and Marzluf, 2003). Mutagenesis of *Fusarium* led to identification of homologues of all but *nit-6/niiA* and *nit-2/areA* (Klittich and Leslie, 1988; Daboussi *et al.*, 1991; Liu and Sundheim, 1996). However, homologues of *nit-2/areA* have since been identified in multiple fusaria such as *F. fujikuroi* (Tudzynski *et al.*, 1999) and *F. verticillioides* (Kim and Woloshuk, 2008).

With the establishment of molecular genetic techniques, it was possible to clone and characterize the *nit-2* and *areA* genes in *N. crassa* and *A. nidulans* respectively (Caddick *et al.*, 1986; Fu and Marzluf, 1987b). Subsequent cross-genus complementation showed that *nit-2* and *areA* are functional orthologues (Davis and Hynes, 1987). Analyses of the predicted amino acid sequences of NIT2 and AreA revealed a high homology of the C-terminus to mammalian *trans*-acting regulators of the GATA-type zinc finger transcription factor family known to bind DNA (Fu and Marzluf, 1990a; Fu and Marzluf, 1990b; Kudla *et al.*, 1990). GATA-type transcription factors occur in fungi, metazoans and plants, where the DNA-binding domain consists of four cysteine (Cys, C) residues that coordinate one zinc atom (reviewed in Scazzocchio, 2000). In *N. crassa* and *A. nidulans*, NIT2/AreA binds preferentially to at least two 5′HGATAR DNA motifs located within 30 bp of each other (Feng *et al.*, 1993; Chiang and Marzluf, 1994; Ravagnani *et al.*, 1997). The N-terminal region of the NIT2/AreA-encoding sequence is highly variable and is dispensable for function in *A. nidulans* (Caddick et al., 1998). The significant sequence identity shared by *nit-2* and *areA* has facilitated the identification of orthologous genes in other Ascomycetes, including *Fusarium* (*areA, Fnr1* and *nnu*) (Dickmann and Leslie, 1992; Tudzynski *et al.*, 1999; Divon *et al.*, 2006; Kim and Woloshuk, 2008), *Penicillium* (*nre*) (Haas *et al.*, 1995), *Magnaporthe* (*NUT1*) (Froeliger and Carpenter, 1996), *Metarhizium* (*nrr1*) (Screen *et al.*, 1998), *Cladosporium* (*Nrf1*) (Pérez-García *et al.*, 2001), *Colletotrichum* (*CLMR1*) (Pellier *et al.*, 2003) and *Microsporium* (*dnr1*) (Yamada *et al.*, 2006). So far, the ability of *Fusarium* AreA to bind to two adjacent GATA motifs has been demonstrated experimentally only in *F. fujikuroi* (Mihlan *et al.*, 2003).

In contrast to filamentous Ascomycetes, regulation of nitrogen utilization in *Saccharomyces cerevisiae* involves four GATA-type transcription factors: Gln3, Gat1, Dal80 and Gzf3. In response to limitation of NH_4^+ or Gln, Gln3 and Gat1 (= Nil1) are responsible for activation of genes that allow utilization of energetically less favoured nitrogen sources like NO_3^-. Both proteins show homology to NIT2/AreA homologues from filamentous fungi. In the presence of NH_4^+ or Gln, two other GATA-type transcription factors, Dal80 (= Uga43) and Gzf3 (= Nil2 and Deh1) repress the same set of genes, most likely by binding to the same 5′GATA motifs and thereby preventing binding of Gln3 and Gat1 (reviewed in Magasanik and Kaiser, 2002; Fig. 7.2).

Nitrate utilization

The role of NIT2/AreA and NIT4/NirA in the nitrate assimilation pathway is understood best in *N. crassa* and *A. nidulans*. In both fungi, NO_3^- utilization requires the presence of NIT2/AreA as well as NIT4/NirA, a $Zn(II)_2Cys_6$ binuclear, or C6 zinc, cluster transcription factor (Burger *et al.*, 1991b; Yuan *et al.*, 1991). NIT2/AreA is induced by absence of favoured nitrogen sources such as NH_4^+ or Gln (Pateman *et al.*, 1967; Arst and Cove, 1973; Coddington, 1976) whereas NIT4/NirA is induced by NO_3^- (Sorger and Giles, 1965; Pateman and Cove, 1967; Hurlburt and Garrett, 1988). The gene coding for NIT4/NirA was cloned in *N. crassa* and *A. nidulans* (Burger *et al.*, 1991a; Yuan *et al.*, 1991) and complementation of a *F. oxysporum* nitrate non-utilizing mutant with the *A. nidulans nirA* suggests the presence of a functional NirA homologue in *Fusarium* (Daboussi *et al*, 1991). (Burger *et al.*, 1991b; Yuan *et al.*, 1991). Most C6 zinc cluster transcription factors bind to palindromic or repetitive DNA sequences (Schjerling and Holmberg, 1998). In *A. nidulans*, NirA binds as a homodimer to the motif 5′CTCCGHGG in the bidirectional promoter region of *niiA* and *niaD* (Punt *et al.*, 1995; Strauss *et al.*, 1998). Apart from NO_3^- as a signal that causes retention of NirA in the nucleus by disrupting its association with the nuclear export protein KapK (Bernreiter *et al.*, 2007), DNA binding by NirA also requires AreA (Narendja *et al.*, 2002; Muro-Pastor *et al.*, 2004; Berger *et al.*, 2006). Although AreA is involved in histone H3 acetylation leading to loss of positioned nucleosomes, NirA alone is able to compensate for AreA-dependent chromatin remodelling in an *areA* loss-of-function mutant when *nrtA* and *nirA* are constitutively expressed thus resulting in the ability to utilize NO_3^- (Narendja *et al.*, 2002, Berger *et al.*, 2006, Berger *et al.*, 2008). Astonishingly, the original observation of NirA acting independently of AreA to induce nitrate and nitrite reductase activities was made almost 30 years ago (Rand and Arst, 1978).

In *Fusarium*, expression of *niaD* is repressed by Gln and induced by NO_3^- in an AreA-dependent manner (Tudzynski *et al.*, 1999, Divon *et al.*, 2006; Schönig *et al.*, 2008) such that *areA* mutants exhibit severe growth defects when NO_3^- and NO_2^- are the sole nitrogen source (Kim and Woloshuk, 2008; Divon *et al.*, 2006; Teichert *et al.*, 2006; López-Berges *et al.*, 2010). The observation that a *nirA* homologue exists in *F. oxysporum* indicates that nitrate metabolism is also co-regulated by NirA and AreA in *Fusarium* (Daboussi *et al.*, 1991; Fig. 7.2C). Whether the regulatory mechanisms that control nitrate utilization in *Fusarium* are similar to those that exist in *A. nidulans* awaits further elucidation.

Involvement in purine utilization

The utilization of other non-preferred nitrogen sources related to the degradation products of purine nucleotides (e.g. hypoxanthine, xanthine, uric acid, allantoin, allantoate and urea) is often under the control of a similar system requiring substrate-inducing conditions and absence of the preferred nitrogen sources NH_4^+ or Gln. Seminal work in *A. nidulans* on the genetics, biochemistry and regulation of purine utilization revealed that the C6 zinc cluster transcription factor UaY mediates induction by uric acid (Suárez *et al.*, 1995) in conjunction with AreA in order to allow maximum gene transcription (reviewed in Scazzocchio, 1994). Briefly, *azgA, uapA, uapC* code for hypoxanthine, xanthine and uric acid transporters respectively (Gorfinkiel *et al.*, 1993; Diallinas *et al.*, 1995; Cecchetto *et al.*, 2004). After movement into the cell, these materials are degraded to allantoin by HxA, the MoCo-dependent xanthine oxidase (Glatigny and Scazzocchio, 1995), HxB, the MoCo-activator (Sealy-Lewis *et al.*, 1978; Amrani *et al.*, 1999), XanA, the MoCo-independent xanthine oxidase (Cultrone *et al.*, 2005, 2007) and UaZ, the urate oxidase (or uricase) (Oestreicher and Scazzocchio, 1993). UapC is induced and localized in the plasma membrane in response to its substrate uric acid, and it is repressed and internalized in response to NH_4^+ and Gln, all of which are dependent on UaY and AreA (Valdez-Taubas *et al.*, 2000, 2004). However, UaY and AreA-independent expression of the UapC-encoding gene *uapC* was observed during conidiospore germination in *A. nidulans*, suggesting that additional regulatory mechanisms are involved (Amillis *et al.*, 2004). In *N. crassa*, abundance and activity of uricase is also repressed by NH_4NO_3 but exhibits a basal level of activity even under nitrogen starvation (Nahm and Marzluf, 1987). NIT2 was shown to bind to double GATA motifs in the promoter region of *alc-1*, the gene encoding allantoin oxidase and that is also repressed by Gln (Lee *et al.*, 1990a,b). Interestingly, the gene encoding urease, which catalyses the last step of the purine salvage pathway resulting in free NH_4^+ ions, seems not to be regulated by NIT2 and nitrogen availability in *N. crassa* (Reinert and Marzluf, 1975) (Fig. 7.1). In *A. nidulans*, urease (UreB) activity is reduced by NH_4^+ and Gln (Mackay and Pateman, 1982), but whether AreA or UaY are involved in the regulation is not clear. The growth reduction observed in *areA* loss-of-function mutants on urea as the sole nitrogen source (Pellier *et al.*, 2003) could be explained by the dependency of *ureA*, coding for the high-affinity urea transporter, on AreA (Abreu

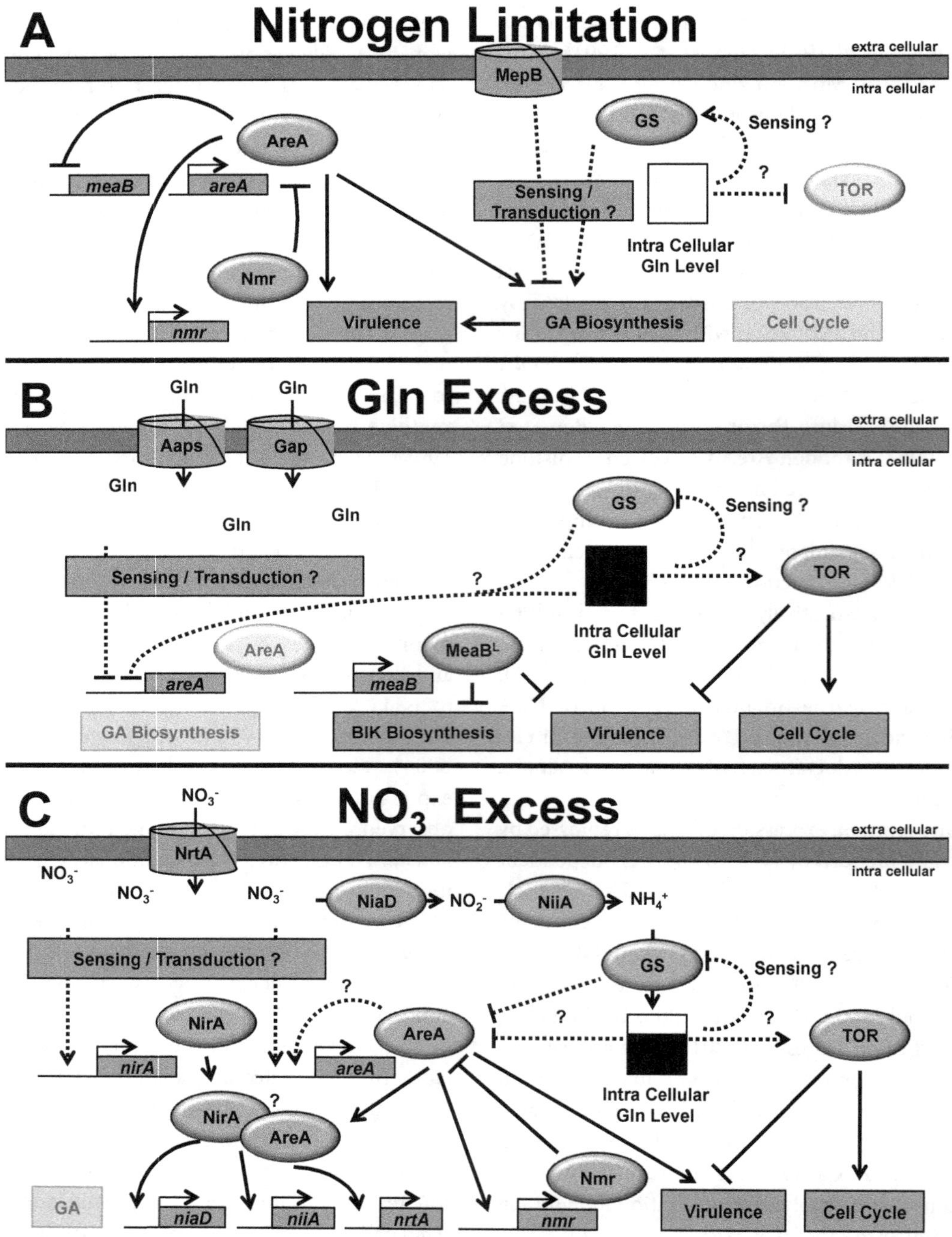

Figure 7.2 Model of key components of the nitrogen regulation network in *Fusarium fujikuroi* and *F. oxysporum* f. sp. *lycopersici* under different nitrogen regimes: **A:** Under nitrogen starvation the intracellular Gln level is low (white square). TOR is inactive which leads to a reduced cell cycle. The limited Gln concentration is putatively sensed by the active GS which putatively transduces this signal by interaction with other factors. AreA is active represses *meaB*^{L} and induces expression of a variety of genes including *nmr*, *glnA* and *mepB*. Nmr interacts with AreA thereby putatively modifying its activity. AreA leads to transcriptional activation of virulence factors (e.g. gibberellin (GA) genes). MepB is highly abundant and functions as a putative sensor that transduces signals from the environment by putative protein/protein interactions. **B:** Under Gln excess, specific amino acid permeases (Aaps) and the general amino acid permease (Gap) facilitate transport into the cell. The intracellular Gln level is high (black square). TOR is active resulting in activation of the cell cycle and a reduction of virulence. The high Gln concentration is putatively sensed by the GS and other

Table 7.1 Growth of *nit-2/areA* deletion mutants of various ascomycetes compared with the corresponding wild type on different nitrogen sources

	N. crassa[a]	*A. nidulans*[b]	*F. fujikuroi*[c]	*F. graminearum*[d]	*F. oxysporum*[e]	*F. verticillioides*[f]
Adenine		–		–		
Alanine	+/–			–		
Ammonium	+	+	–	–	–	+/–
Arginine	+	–	+	+/–		
Asparagine			–	+/–		
Aspartate	+/–	–	–			
Citrulline	+		+/–			
Glutamate	–	–	–	+	–	
Glutamine	+	+	+	+	+	+
Glycine				–		
Histidine	–		–	–		
Hypoxanthine	–	–		–		
Nitrate	–	–	–	–	–	–
Nitrite	–	–		–	–	
Ornithine	–	–	–			
Phenylalanine	–	–	–	–		
Proline	–	–	+/–	+		
Serine	–	–		–		
Threonine		–	–			
Tryptophan	–			–		
Urea	+/–	+/–	+			
Uric acid	–	–		–		
Valine	–	–		–		

+, Wild-type-like growth; +/–, restricted growth; –, no growth.
[a]Reinert and Marzluf (1975), Coddington (1976), Facklam and Marzluf (1997).
[b]Arst and Cove (1973), Pellier *et al.* (2003).
[c]Tudzynski *et al.* (1999), Wagner *et al.* (2010); our unpublished data
[d]Dickman and Leslie (1992).
[e]Divon *et al.* (2006), López-Berges *et al.* (2010).
[f]Kim and Woloshuk (2008).

sensors and results in low *areA* mRNA levels. The absence of an active AreA leads to reduced GA (and fumonisin) gene expression and induction of *meaB*[L] expression. MeaB[L] itself is involved in repression of bikaverin (BIK) gene expression and virulence. **C:** Under NO_3^- excess, the nitrate transporter NrtA provides NO_3^- influx into the cell. The high NO_3^- concentration is sensed by so far unidentified sensors that transduce this signal resulting in *nirA* and *areA* expression. NirA and AreA activate expression of genes essential for NO_3^- reduction to NH_4+ (e.g. *niaD*, *niiA*) as well as *nmr*. As the level of Nmr protein increases AreA is putatively modulated by direct interaction. Additionally, AreA activates virulence genes, e.g. GA genes. NiaD and NiiA convert NO_3^- into NH_4+ which is finally incorporated into Gln by the GS, building up an increasing intracellular Gln level (black and white square). This intracellular level perceived and leads to activation of TOR resulting in cell cycle activation and repression of virulence. Furthermore, increasing Gln concentration might be sensed by the GS which directly or indirectly inactivates AreA activity.

et al., 2010). The slight reduction in intracellular Glu levels in *areA* loss-of-function mutants grown with urea compared to the wild type (Berger *et al.*, 2008) does not support the hypothesis that AreA is responsible for UreA and UreB activity.

In *F. oxysporum* f. sp. *lycopersici* the uricase-encoding homologue to *uaZ* was induced under nitrogen starvation conditions in an AreA-dependent manner (Divon *et al.*, 2005; Divon *et al.*, 2006). Additionally, the expression of two distinct homologues of *hxA* in *F. fujikuroi* (*xdh1* and *xdh2*) and a gene (*ncs1*) with homology to putative allantoin transporters were induced under nitrogen starvation and were AreA dependent (Schönig *et al.*, 2008). Recent phylogenetic analyses of four sequenced *Fusarium* species revealed the existence of two putative urease encoding genes. Genes putatively encoding enzymes for another salvage pathway of urea (urea carboxylase and urea amidolyase) are present in the genus *Fusarium* (Strope *et al.*, 2011) (Fig. 7.1). Since *N. crassa* and *A. nidulans* have only one urease-encoding gene, to utilize urea as the sole nitrogen source (Strope *et al.*, 2011), this could account for the observed growth differences on urea of the respective *nit-2*/*areA* mutants when compared to *Fusarium areA* mutants (Table 7.1). So far, no specific C6 zinc cluster transcription factor homologous to the *A. nidulans* UaY has been characterized in *Fusarium*, which in addition to AreA, might be involved in regulation of purine salvage pathways. Initial investigations on urease, urea carboxylase and urea amidolyase regulation should be expanded, as these enzymes catalyse crucial step(s) in NH_4^+ acquisition from urea derived from purines or degradation of arginine (Arg), aspartate (Asp), Pro, Glu and asparagine (Asn) (Fig. 7.1).

Involvement in amino acid metabolism

Although *areA* loss-of-function mutants of *A. nidulans*, *N. crassa* as well as various *Fusarium* species have severe growth defects on multiple amino acids (see Table 7.1 and references therein), the involvement of AreA at a molecular level has been investigated best for proline utilization in *A. nidulans*. Here, five clustered genes are involved in proline catabolism (Hull *et al.*, 1989). The gene *prnA* encodes a C6 zinc cluster transcription factor that binds to 5′CCGG repeats in the bidirectional promoter of *prnB* and *prnC* (Gómez *et al.*, 2002). The genes *prnB*, *prnD* and *prnC* encode a proline transporter, a proline oxidase and a pyrroline-5-carboxylate reductase, respectively (Jones *et al.*, 1981; Sophianopoulou and Scazziocchio, 1989). Expression of all genes requires proline and a functional PrnA protein (Sharma and Arst, 1985). Repression of *prnB* occurs under nitrogen and carbon sufficiency mediated by the absence of AreA and the presence of CreA, a Cys2/His2 zinc finger transcription factor, respectively (Arst and Cove, 1973; Arst *et al.*, 1980; Gómez *et al.*, 2003), whereas subsequent repression of *prnC* and *prnD* is due to exclusion of the inducer Pro resulting from down-regulation of the permease encoding gene *prnB* (Cubero *et al.*, 2000). Furthermore, chromatin remodelling is involved in induction and repression of *prnB* (García *et al.*, 2004; Reyes-Dominguez *et al.*, 2008). A link between the metabolic pathways of Pro, Arg and Glu is represented by the ornithine transaminase that catalyses the conversion between ornithine and glutamate-5-semialdehyde. In this metabolic route, Arg can be converted to Pro and Glu. In *A. nidulans* a similar involvement of AreA in regulation of *otaA*, coding for the ornithine transaminase, as seen in *prnB* gene expression under carbon repressing and nitrogen inducing conditions, is suggested (Dzikowska *et al.*, 1999).

In *F. fujikuroi*, a putative *prnD* homologue was repressed in the *areA* deletion mutant compared to the wild type under nitrogen starvation in the presence of glucose, indicating that AreA also plays a role in proline metabolism in *Fusarium* (Schönig *et al.*, 2008). Apart from the highly complex effects of AreA on Pro and Arg catabolism in *A. nidulans*, an intriguing observation can be made, when comparing growth tests on different amino acids as sole nitrogen sources (see Table 7.1 and references therein). The major growth defect on several amino acids in *Fusarium* spp. could be explained by data obtained from microarray results comparing wild type and *areA* deletion mutants in *F. oxysporum* f. sp. *lycopersici* and *F. fujikuroi* that identified several amino acid and peptide transporters as well as all three NH_4^+ permeases as AreA target genes (Divon *et al.*,

2006; Teichert *et al.*, 2008; Schönig *et al.*, 2008). The observation that all *areA* mutants show wild-type-like growth on Gln indicate that anabolism of all other amino acids from Gln and Gln uptake into the cells is functional. In contrast, growth on other nitrogen sources is strongly impaired. What is so unique about Gln compared to the other amino acids? One answer to this question is that there is only one enzyme, GS, to synthesize Gln from Glu and free NH_4^+ in an ATP-dependent reaction. All other amino acids can be derived from their respective α-keto acid by transamination from Glu or Gln. As there is strong growth impairment of *F. fujikuroi* and *F. oxysporum areA* mutants on Glu (see Table 7.1 and references therein), there are three possible explanations: (1) Glu cannot be transported into cells, (2) GS is inactive and/or (3) there are no free NH_4^+ ions available as substrate for GS. Results obtained in *F. fujikuroi* showing that the GS-encoding gene *glnA* is a target gene of AreA support our second hypothesis. However, transcriptional levels of *glnA* are not fully repressed in the *areA* deletion mutant, indicating an effect of AreA on *glnA* translation and\or post-translational activation of the GS (Teichert *et al.*, 2004; Teichert *et al.*, 2006; Schönig *et al.*, 2008). In the basidiomycete *Agaricus bisporus*, a post-translational modification of the GS is suggested, as enzyme activity does not correlate with mRNA levels upon Gln or NH_4^+ addition (Kersten *et al.*, 1997). The involvement of NIT2/AreA homologue in this process is unknown. In *A. nidulans*, elegant work measuring the intracellular Gln level in an *areA* loss-of-function mutant also indicates impairment of the GS, as the Gln level is drastically reduced (Berger *et al.*, 2008).

In *N. crassa*, the transcription factor CPC1 is a central component mediating the transcriptional up-regulation of genes involved in amino acid metabolism under conditions of amino acid imbalance (Carsiotis and Jones, 1974; Barthelmess and Kolanus, 1990; Tian *et al.*, 2007). In *F. fujikuroi*, the *cpc1* homologue is up-regulated when amino acids are imbalanced (i.e. inhibition or deletion of GS) but is down-regulated under nitrogen starvation (Teichert *et al.*, 2004; Schönig *et al.*, 2009). Furthermore, expression of several genes involved in amino acid metabolism are dependent on *cpc1* transcription indicating that it has similar functions as CPC1 in *N. crassa* (Schönig *et al.*, 2009). Remarkably, under nitrogen starvation conditions, *cpc1*, but none of the Cpc1 target genes, is highly over-expressed in the *areA* deletion mutant in *F. fujikuroi* (Schönig *et al.*, 2008, 2009). This suggests that amino acid imbalance occurs in the *areA* deletion mutant (as discussed above), and that a functional AreA is required for induction of expression of Cpc1-target genes. Interestingly, in *A. nidulans*, a binding sequence motif for CPC1 is present in the *prnB* promoter that is also regulated by AreA (Tazebay *et al.*, 1997). Elucidating the mechanisms by which AreA is interconnected with Cpc1 and GS will lead to a major advance in understanding the role of AreA in amino acid metabolism.

Involvement in secondary metabolism

A characteristic feature of many fungi is their ability to produce secondary metabolites, small molecules that are not absolutely required for growth (reviewed in Keller *et al.*, 2005). The products of secondary metabolism in *Fusarium* are diverse and can contribute to virulence on host plants. *F. fujikuroi* is known for its ability to produce gibberellins, a group of diterpenoid plant hormones (reviewed in Bömke and Tudzynski, 2009) that significantly contribute to virulence during plant infection (Carter *et al.*, 2008; Karov *et al.*, 2009; Wiemann *et al.*, 2010). The growth-promoting function of gibberellins has been exploited extensively by agri- and horticulture biotechnology industries (Rademacher, 1997). Fermentation of *F. fujikuroi* for gibberellin production delivers highest yields under nitrogen limiting conditions (Borrow *et al.*, 1964; Bu'lock *et al.*, 1975). Remarkably, the GATA-type transcription factor AreA is essential for gibberellin production under nitrogen starvation (Tudzynski *et al.*, 1999; Mihlan *et al.*, 2003). These findings are unusual, as most AreA-dependent genes are involved in nitrogen utilization and gibberellins cannot serve as nitrogen source for the fungus. In *F. verticillioides*, fumonisin production is also dependent on AreA under nitrogen limiting conditions (Shim and Woloshuk, 1999; Kim and Woloshuk, 2008) although the direct promotion of fumonisin cluster genes transcription by AreA has not yet been demonstrated. In *P. chrysogenum*, penicillin production is repressed under nitrogen excessive conditions (Kolar *et al.*, 1991; Feng *et al.*, 1994). Although the AreA homologue NRE is able to bind to promoter regions of the penicillin biosynthetic genes (Haas and Marzluf, 1995), the direct involvement of the GATA-type transcription factor in penicillin production remains to be proven experimentally.

Surprisingly, when NO_3^- is present in high concentrations as the only nitrogen source, gibberellin biosynthetic gene repression is not abrogated, although AreA must be functional in order to utilize NO_3^- (Wiemann *et al.*, 2010). These data suggest that (1) an essential co-activator is missing, (2) different conformations of AreA can have distinct functions and/or (3) a repressing element circumvents AreA binding to gibberellin gene promoters. Astonishingly, gene expression and concomitant production of the red polyketide pigment bikaverin is also induced under nitrogen starvation conditions, but independently of AreA (Wiemann *et al.*, 2009). Taken together, these findings substantiate the hypothesis that nitrogen response regulatory elements in addition to AreA are involved in regulation of secondary metabolism in *Fusarium*.

Regulation of NIT2/AreA in response to nitrogen conditions

Now that the gene targets that respond to nitrogen quality and quantity in dependence of NIT2/AreA have been defined, an important question is: how is NIT2/AreA activity itself regulated? The diversity of NIT2/AreA dependent out-put pathways mentioned before already suggests that this regulation is complex, as active forms of the transcriptional activator must be present under different conditions including the complete absence of nitrogen as well as the presence of specific nitrogen sources. General modulation of activity can be achieved by several mechanisms including, transcriptional control, transcript stability, post-translational modification, protein allocation and protein–protein interaction.

Transcriptional control and mRNA/protein stability

In *A nidulans*, strong transcriptional induction of several distinct transcripts of *areA* under nitrogen starvation conditions is partially dependent on autogenous activation (Langdon *et al.*, 1995). Recent studies show that *areA* transcription is induced by NO_3^- and is inversely correlated with intracellular Gln levels (Schinko *et al.*, 2010). In *N. crassa*, a similar but far less pronounced induction of transcription of the *areA* homologue *nit-2* is not autogenously controlled by its product under nitrogen starvation conditions (Fu and Marzluf, 1987b). In *F. fujikuroi* three distinct *areA* transcripts are present under nitrogen starvation and NO_3^- abundant conditions, but like in *A. nidulans*, the distinct transcripts cannot be detected under nitrogen repressive conditions (Wiemann *et al.*, 2010). When NO_3^- is present as the sole nitrogen source, induction of *areA* transcription also occurs in *F. oxysporum* f. sp. *lycopersici* (López-Berges *et al.*, 2010). Whether *areA* in *Fusarium* is subject to autogenous activation is so far unknown. Other essential questions to be answered are (1) How can the fungi differentiate between the different nitrogen sources (e.g. expression of *areA* is induced with NO_3^- and repressed by Gln)? (2) What mediates repression under Gln conditions; (3) How can the fungi sense the availability of nitrogen or the complete absence of a nitrogen source, and (4) How are these different signals correctly coordinated?

In *A. nidulans*, *areA* transcript stability is dependent on Gln and NH_4^+ availability and is mediated by the 3′ untranslated region (3′UTR) (Platt *et al.*, 1996; Morozov *et al.*, 2000). Interestingly, Gln-mediated *areA* degradation is dependent on a functional AreA product, while, in contrast, NH_4^+-mediated degradation occurs independently of AreA (Morozov *et al.*, 2001). In both *N. crassa* and *F. fujikuroi*, transcripts of *areA* seems to be stable under NO_3^- and Gln conditions over time (Tao and Marzluf, 1999; Tudzynski *et al.*, 1999). In *F. fujikuroi*, transcription of *niaD* is undetectable upon Gln addition, despite a stable *areA* transcript (Tudzynski *et al.*, 1999), indicating either a direct destabilizing effect of Gln on the *niaD* transcript or an indirect effect mediated by a post-translationally modified AreA protein in response to Gln (Fig. 7.2C). In *A. nidulans, niaD* transcript is destabilized by deadenylation following addition of Gln, but whether this is dependent on *areA* destabilization itself or a modified AreA protein remains to be determined (Caddick *et al.*, 2006). In *N. crassa* and *A. nidulans*, stability of the NIT2/AreA protein appears to be unaffected by addition of Gln and NH_4^+, suggesting another level of regulation of NIT2/AreA activity on output pathways (Tao and Marzluf, 1999; Todd *et al.*, 2005). Whether protein stability of AreA is effected under different nitrogen conditions is a question still to be answered for *Fusarium*.

Modulation of NIT2/AreA by NMR1/NmrA

In *N. crassa*, mutants, referred to as MS5, were identified that exhibit an increased nitrate reductase activity under repressing conditions mediated by NH_4^+, Gln and Glu (Premakumar *et al.*, 1980). NO_3^- and NO_2^- induction is still needed for nitrate and nitrite reductase activity, and although repression by NH_4^+, Gln and Glu still occurs, the repression is alleviated and NIT3 was present in higher concentrations, suggesting that the mutation acts post-translationally rather than on the level of transcription (Dunn-Coleman, *et al.*, 1981; Tomsett *et al.*, 1981). Complementation and subsequent sequence analyses revealed that two independent MS5 mutants had a single point mutation leading to one amino acid substitution in the same gene, *nmr-1* (Jarai and Marzluf, 1990; Young *et al.*, 1990). Transcriptional regulation of *nmr-1* seems to be constitutive in different nitrogen conditions and in *nit-2* and *nmr-1* mutants (Fu *et al.*, 1988). Furthermore, the *nmr-1* mutation in strain MS5 does not lead to altered *nit-2* expression and only slightly derepresses *nit-3* transcription in the presence of Gln (Fu and Marzluf, 1988).

Yeast two hybrid analyses showed that NIT2 and NMR1 interact with each other at two specific C-terminal regions of NIT2 and that binding of NIT2 to DNA *in vitro* is most likely inhibited by interaction with NMR1 (Xiao *et al.*, 1995). Deletion of a C-terminal region of NIT2 resulted in enhanced NIT3 activity in the presence of NO_3^- and Glu, and does not show interaction with NMR1 (Pan *et al.*, 1997). Deletion and overexpression of the *A. nidulans nmr-1* homologues gene, *nmrA*, had a similar affects on some output pathways of AreA under NH_4^+ and Gln repressing conditions, respectively, e.g. nitrate reductase activity and *amdS-lacZ* reporter gene expression (Adrianopoulos *et al.*, 1998; Wong *et al.*, 2007; Wagner *et al.*, 2010). In *A. nidulans*, expression of *nmrA* is derepressed in the *areA* mutant under nitrogen starvation and generally more pronounced when NH_4^+ or Gln is present, but does not influence expression of *areA* (Wong *et al.*, 2007; Wagner *et al.*, 2010). Nuclear localization of NmrA was more pronounced under NH_4^+ conditions than in nitrogen starved cells (Zhao *et al.*, 2010). From *in vitro* experiments there is evidence that NmrA binds to AreA, whereas the C-terminal region of NmrA is more crucial for binding than the nine extreme C-terminal residues of AreA, and that the affinity of AreA to 5′GATA motif-containing DNA is stronger than to NmrA (Lamb *et al.*, 2004). Furthermore, NmrA, which belongs to the short-chain dehydrogenase/reductase (SDR) superfamily, was shown to bind NAD^+ and $NADP^+$, but not NH_4^+ or Gln, challenging its role as the direct sensor of intracellular nitrogen (Stammers *et al.*, 2001; Lamb *et al.*, 2003). Also, neither binding of NAD^+ nor $NADP^+$ changes interaction of NmrA with AreA *in vitro*, but binding of NAD^+ suppresses proteolytic cleavage of NmrA, thereby stabilizing AreA-binding capability (Lamb *et al.*, 2004; Zhao *et al.*, 2010). Surprisingly, in *F. fujikuroi* deletion of the *nmr-1/nmrA* homologues gene *nmr* does not result in a derepression of gibberellin gene expression under nitrogen sufficient conditions (60 mM Gln), although *nmr* complements the respective mutants of *N. crassa* and *A. nidulans* (Mihlan *et al.*, 2003, Wagner *et al.*, 2010). This might also be explained by significantly reduced *areA* expression under these conditions independently of the presence or absence of Nmr (Wiemann *et al.*, 2010; unpublished data). When Gln concentrations are lower (6 mM Gln) expression of gibberellin genes starts earlier in the *nmr* deletion mutant compared to the wild type (Schönig *et al.*, 2008). Overexpression of *nmr* results in enhanced resistance to ClO_3^- in the presence of 60 mM Gln, suggesting a reduced nitrate reductase activity (Mihlan *et al.*, 2003; Schönig *et al.*, 2008; Wagner *et al.*, 2010; Fig. 7.2). Similar to the situation in *N. crassa*, NO_3^--dependent induction of *niaD* was not overcome in the *nmr* deletion mutant in *F. fujikuroi*, but *niaD* expression was even slightly increased (Schönig *et al.*, 2008). Although Nmr and AreA in *F. fujikuroi* are able to interact in yeast two hybrid experiments (Schönig *et al.*, 2008), the accumulated data suggest that there is a discrepancy between different AreA out-put pathways in *F. fujikuroi* (Fig. 7.2). Whereas the regulation of the nitrate utilization pathway seems to be similar to regulation in *N. crassa* and *A. nidulans*, with respect to AreA and Nmr, the gibberellin production pathway seems to require additional regulator(s) and is only partially dependent on NmrA-mediated nitrogen metabolite repression of AreA. As transcript stability of *areA* seems not to be affected by NO_3^- and Gln addition (Tudzynski *et al.*, 1999), the protein stability of AreA under these conditions and the potential role of NmrA therein should be investigated in the future. A striking difference between *F. fujikuroi* and *A. nidulans* is that expression of *nmr* in *F. fujikuroi* is not inverse to *areA* expression. In addition, *nmr* expression in *F. fujikuroi* is not constitutive under different nitrogen conditions as in *N. crassa* but is abundant under nitrogen starvation conditions in an AreA-dependent manner (Wagner *et al.*, 2010).

The accumulated data from filamentous fungi on the interconnection between NIT2/AreA and NMR1/NmrA are so far insufficient to form a general model for a role in a fungal response to environmental nitrogen signals. Although the genetic components seem to be conserved in each fungus, the response to the same in-put is different among fungi. However, in response to NO_3^- and its subsequent utilization, *Neurospora*, *Aspergillus* and *Fusarium* all have a common regulatory mechanism: the conserved transcription factor NIT2/AreA acts as positive element when NO_3^- is available and NMR1/NmrA acts as a negative regulator under Gln or NH_4^+ sufficiency. Unusual results from *N. crassa*, demonstrating that nitrate reductase activity oscillates independently of the clock regulators White Collar 1 (WC1) and Frequency (FRQ), suggest that during NO_3^- utilization a negative feedback loop takes place in which the nitrate reductase activates its own repressor (Christensen *et al.*, 2004). Obviously, NO_3^- triggers *nit-2/areA* expression/activity, which is required for NO_3^- utilization to subsequent amino acids, specifically Gln. Expression/activation of *nmr-1/nmrA* seems to be either dependent on AreA (in *Fusarium*) or repressing nitrogen sources, such as Gln (in *Aspergillus*). By binding to NIT2/AreA, NMR1/NmrA is suggested to inactivate its activity, thereby shutting down its own positive signal (Fig. 7.2). This of course, is a highly speculative and generalized model that needs experimental proof in the future. And furthermore, the identification of sensors that mediate NO_3^- and Gln signals is of crucial interest in order to be able to explain the observed responses. The fact that the AreA-dependent gibberellin out-put pathway in *F. fujikuroi* is not similarly affected by Nmr as the NO_3^- utilization out-put pathway, argues for additional nitrogen response regulators that are integrated into the transduction network.

Additional factors involved in the nitrogen transduction network

Resistance to toxic methylammonium ions (MA) in *A. nidulans* in the presence of NO_3^-, referred to as *mea*^{R}, can arise through mutations at one of two genetic loci (Arst and Cove, 1969). Analysis of mutations at *meaA* suggested that it plays a role in transport (Arst and Page, 1973) while mutation of *meaB* leads to derepression of nitrogen repressed genes indicating a role in regulation (Arst and Cove, 1973; Polley and Caddick, 1996). Almost 30 years later the protein encoded by *meaA* was characterized as a high capacity NH_4^+ transporter which is expressed AreA-dependently under NH_4^+-sufficient and starvation conditions (Monahan *et al.*, 2002; 2005). In *F. fujikuroi* the high capacity NH_4^+ transporter encoding gene, *mepB*, is preferentially expressed under starvation conditions in an AreA-dependent manner. In contrast to *A. nidulans*, MepB also serves as a NH_4^+ sensor which is able to transduce repressing signals towards nitrogen-dependent secondary metabolite production pathways (see below) (Teichert *et al.*, 2008).

Sequence analyses of *meaB* revealed that it

likely encodes a protein belonging to the bZIP transcription factor family. In *A. nidulans*, *meaB* is preferentially expressed under nitrogen sufficient and repressed under nitrogen starvation conditions in an AreA-dependent manner (Wagner *et al.*, 2010). The observation that *meaB/areA* double mutants display additional phenotypes in *F. fujikuroi* as compared to an *areA* single mutant indicate that *meaB* is involved in processes independently of AreA (Wagner *et al.*, 2010; Fig. 7.2B). Efforts to delineate these processes have been met with mixed success. Recent work in *A. nidulans* revealed contradictory results on whether MeaB binds to the *nmrA* promoter, thereby inducing transcript synthesis under nitrogen surplus, which would lead to decreased AreA activity. It has been reported by Wong *et al.* (2007) that *meaB* binds to the *nmrA* promoter, thereby activating *nmrA* transcription, while only slight or no effect of MeaB on *nmrA* expression in *A. nidulans* and *F. fujikuroi*, respectively was observed by Wagner *et al.* (2010). Furthermore, *meaB* was shown to be almost constitutively expressed independently of AreA in *A. nidulans* (Wong *et al.*, 2007) while two transcripts are produced in an AreA- and nitrogen-dependent manner in *F. fujikuroi*. The reason for these contrasting results remain unclear.

The complicated role played by *meaB* in *F. fujikuroi* is partially mediated by the synthesis of two distinct transcripts, depending on the nitrogen status. When nitrogen is depleted, a second, smaller *meaB* transcript (*meaB*^{S}) is initiated from a second transcriptional start that resides within the first intron of the full length *meaB* open reading frame (ORF). Evidence suggests that *meaB*^{S} is not translated into a protein. The production of *meaB*^{S} has been proposed to be induced by AreA and/or one of its downstream effectors which concomitantly blocks transcription of the large, full-length transcript (*meaB*^{L}). The latter occurs under Gln sufficiency in the wild type or in the *areA* mutant independent of nitrogen availability. In contrast to *nmr* mutants, *meaB* mutants exhibit increased sensitivity towards ClO_3^- and NO_2^- in the presence of NH_4^+, and show reduced expression of *glnA*, suggesting independent functions for these two proteins. Strikingly, the *F. fujikuroi meaB/areA* double deletion mutant shows severe growth defects when Gln is the only nitrogen source, whereas single deletion mutants can grow wild-type-like, respectively (Wagner *et al.*, 2010). These data suggest that despite their preferred transcription under different nitrogen conditions, AreA and MeaB coordinately affect amino acid metabolism. Furthermore, expression of the AreA-independently/nitrogen-dependently controlled bikaverin genes is absolutely derepressed in the *meaB/areA* double mutant even under Gln surplus conditions (Wagner *et al.*, 2010). Whether this is a direct affect or simply a general stress response due to the restricted growth caused by a Gln utilization deficit is unresolved.

The distinct role of MeaB in NO_3^- utilization and secondary metabolism is also so far unresolved in *Fusarium*. Growth defects of *F. oxysporum* f. sp. *lycopersici meaB* mutants on NO_3^- suggest an activating effect on AreA or NirA (López-Berges *et al.*, 2010). On the other hand, the *meaB* deletion mutant revealed increased expression of AreA-dependent gibberellin and AreA-independent bikaverin genes in *F. fujikuroi* under nitrogen limiting conditions suggesting a repressing effect on a regulator different from AreA (Wagner *et al.*, 2010). Further experiments will be essential to allow an unambiguous integration of MeaB into the established regulation network.

Another locus involved in nitrogen regulation in *A. nidulans* was identified based on the discovery of mutants that suppress the *areA* mutant phenotype (Arst *et al.*, 1989). This locus, called *areB*, is predicted to encode three different proteins, all containing an N-terminal GATA zinc finger domain and a C-terminal leucine zipper domain with moderate homology to the negative acting nitrogen response regulators Dal80p/Uga43p and Gzf3p/Nil2p/Deh1p from *S. cerevisiae* (Conlon *et al.*, 2001). *AreB* loss-of-function mutants exhibit derepression of the *otaA* gene involved in Arg metabolism under certain carbon limiting conditions, suggesting a role as a repressing element (Dzikowska *et al.*, 2003). Deletion and overexpression of *areB* in an *A. nidulans fmdS-lacZ* reporter strain support this role due to enhanced and repressed β-galactosidase activity under nitrogen starvation in the respective strains (Wong *et al.*, 2009). However, deletion of *areB* did not increase growth on unfavoured nitrogen sources such as NO_3^- and alanine (Ala) (Wong *et al.*, 2009). Analyses of the *P. chrysogenum areB* homologue, *nreB*, found that its transcription is repressed by the favoured nitrogen source Gln and that *nreB* overexpression leads to reduced *niaD* and *niiA* transcription, but does not alter *nre* (*areA*) expression (Haas *et al.*, 1997). In *F. fujikuroi*, expression of *areB* is repressed by high amounts of Gln and NO_3^-. This is in contrast to *areA*, which is repressed by Gln but not by NO_3^- (Wiemann *et al.*, 2010). This difference in repression of AreA and AreB indicates that AreB does not participate in NO_3^- utilization. Elucidation of how AreB functions within the nitrogen regulation network will be challenging but necessary. Defining output pathways and investigating influence on and by other known and so far cryptic factors will aid in defining the role of AreB more precisely.

Despite all the information regarding the factors known to regulate the different out-put pathways upon varying nitrogen situations that fungi can face in their lifestyles, one essential question remains: How do fungal cells sense different nitrogen qualities and quantities in order to achieve a coordinate response? To examine this question, three potential sensory systems under investigation in *Fusarium* will be presented: nitrogen permeases, the TOR kinase and GS.

Are ammonium permeases and other nitrogen permeases potential sensors?

The nature of the chemical signals that define the nitrogen status of the fungal cell as well as the character of sensors are not well understood. In *S. cerevisiae*, several plasma membrane-localized sensors obtain information from the extracellular environment, including the availability of amino acids, ammonium and glucose (Forsberg and Ljungdahl, 2001; Bahn *et al.*, 2007). Thus, within certain fungal species, one of the highly conserved ammonium permeases has evolved a regulatory function and probably induces a set of genes via a yet unidentified signal transduction cascade. Examples of homologues of this permease include Mep2p (*S. cerevisiae* and *Candida albicans*), Amt1 (*Hebeloma cylindrosporum*), and Ump2 (*Ustilago maydis*), all of which are required for the induction of filamentous growth under low-nitrogen conditions (Biswas and Morschhauser, 2005; Holsbeeks *et al.*, 2004; Marini *et al.*, 2003; Lorenz and Heitman, 1998; Van Nuland *et al.*, 2006; Rutherford *et al.*, 2008). One general hypothesis is that these members of the Amt/Mep/Rh protein family act as transceptors that regulate downstream effector molecules (Lorenz and Heitman, 1998). Transceptors bind and/or transport nutrients as part of their sensing mechanism, and mutations that affect transport would also affect growth phenotypes, such as pseudohyphal differentiation, that are influenced by nitrogen status. In contrast, if the absence of ammonium were the signal for Mep2 to act as a sensor, then mutations that prevent Mep2 from binding or transporting ammonium might activate pseudohyphal growth. These opposing models were tested by mutational analysis of residues D186, H194, and H348 in Mep2 of *S. cerevisiae* which are involved in transport function and are conserved throughout the Mep/Amt/Rh family (Javelle *et al.*, 2004; Boeckstaens *et al.*, 2007; Rutherford *et al.*, 2008). While the $Mep2^{H194A}$ protein has a significantly reduced ability to transport ammonium, the $Mep2^{H348A}$ protein shows full transport capacity. Although both $Mep2^{H194A}$ and $Mep2^{H348A}$ proteins were stably expressed and correctly localized, neither transport-defective cells expressing the $Mep2^{H194A}$ protein nor transport-proficient cells expressing H348A proteins were able to undergo pseudohyphal growth in response to low ammonium, indicating that binding but not transport is sufficient for signalling (Rutherford *et al.*, 2008).

Another interesting finding is that the ability of Mep2 to induce pseudohyphal growth is dependent on its expression levels (Biswas and Morschhäuser, 2005; Rutherford *et al.*, 2008). Thus, *MEP2* expression from a high copy plasmid resulted in more robust pseudohyphal growth compared with expression from a low copy plasmid. *MEP2* overexpression enhances differentiation even under normally repressive conditions and induces a transcriptional profile that is consistent with activation of the mitogen-activated protein (MAP) kinase pathway and the downstream acting Ste12 transcription factor (Rutherford *et al.*, 2008). Together, these data strengthen the model that the lack of ammonium in the environment is not the initiating signal for pseudohyphal growth as has been proposed, and that the importance of Mep2 does not relate to its influence on the internal levels of nitrogen.

F. fujikuroi has three ammonium permeases (MepA, MepB and MepC) that were functionally characterized via phenotypic and methylamine uptake studies with *S. cerevisiae* mutants (Teichert *et al.*, 2008). The genome sequences of *F. graminearum*, *F. oxysporum* and *F. verticillioides* include genes encoding closely related homologues of MepA, MepB and MepC. MepB is probably the high capacity ammonium permease in *F. fujikuroi*, and deletion of the corresponding gene, *mepB*, results in significant growth reduction on media with low concentrations of ammonium as well as in derepression of multiple genes that are normally repressed at higher concentrations of ammonium. For example, the GA and bikaverin biosynthetic genes were strongly up-regulated in Δ*mepB* mutants in early stages of culture when ammonium levels were still high (Fig. 7.2A). The peptide transporter gene *mtd1* and the amino acid permease gene *aap8*, which are normally repressed by high levels of ammonium, were also derepressed at early stages of growth of Δ*mepB* mutants (Teichert *et al.*, 2008). In addition, some genes (e.g. *ddr48* and *cipC*) that are down regulated in the glutamine synthetase mutant (Δ*glnA*) were significantly up-regulated in the Δ*mepB* mutants (Teichert *et al.*, 2004; 2008). To determine whether this derepression was due to a nitrogen-sensing defect rather than reduced ammonium transport capacity, the second high capacity transporter, MepC, was overexpressed in the Δ*mepB* background. Although the strong growth defect was partially overcome, the derepression of the nitrogen-repressed genes mentioned above was not affected. These findings suggest a sensing or regulatory role of MepB in addition to a permease function. However, downstream signalling components mediating the signal from MepB have not yet been identified.

Another question to be answered is the role of amino acid permeases (AAPs), e.g. that of the general amino acid permease (GAP), in nitrogen sensing. As members of the Mep/Amt/Rh family, AAPs are located at the surface of cells to relay nutritional information. In *S. cerevisiae*, one of the 20 AAPs, GAP, has a dual function as amino acid transporter and receptor (transceptor) that is able to transduce a signal reflecting amino acid availability. The detection of extracellular amino acid deficiency leads to up-regulation of proteins involved in biosynthesis and transport of the deficient amino acid(s) (Donaton *et al.*, 2003). Recent studies of amino acid transporters in fruit flies and mammalian cell lines revealed similar transceptor functions: AAPs sense and signal amino acid availability to the target of rapamycin (TOR) pathway in a manner similar to yeast (Hundahl and Taylor, 2009). Not much is yet known about a dual role of AAPs in filamentous fungi. Among the 68 amino acid permease identified in the recently sequenced genome of *F. fujikuroi*, only one, the GAP homologue has been identified by complementation of the respective *S. cerevisiae* mutant (B. Schönig and B. Tudzynski, unpublished; Fig. 7.2B).

In *S. cerevisiae*, the activity of at least six distinct nitrogen permeases is regulated by a nutrient-regulated protein kinase, the nitrogen permease reactivator protein Npr1. Under conditions of nitrogen sufficiency, Npr1 promotes post-Golgi sorting of GAP1 and the three ammonia permeases, MEP1/2/3, to the plasma membrane and their retention in the membrane. Inversely, Npr1 promotes ubiquitination and subsequent degradation of the GAP, but not of MEP1/2/3, under nitrogen sufficiency (De Craene *et al.*, 2001; Crespo *et al.*, 2004). There seems to be another way for Npr1 to affect the activity of these proteins. Thus, expression and membrane localization of the ammonium permease Mep2 is not altered in Δ*npr1* mutants of *S. cerevisiae* and *Candida albicans*, while the transport activity is strongly inhibited, possibly through loss of Npr1-mediated conformational changes (Neuhäuser *et al.*, 2011; Rutherford *et al.*, 2008). In *F. fujikuroi*, three Npr1 homologues have recently been identified, and the functional characterization of the single and double deletion mutants is currently under investigation (B. Schönig, D. Wagner and B. Tudzynski, unpublished).

It is not clear what nitrogen sources *Fusarium* can sense and how the fungus can differentiate between different nitrogen sources. To address the question of whether ammonium acts as an effector of nitrogen metabolite repression in *Fusarium* or if repression of nitrogen catabolite repression (NCR) target genes by ammonium is due to its conversion to glutamate and/or glutamine, the *F. fujikuroi gdhA* deletion mutant, which is unable to assimilate ammonium through reductive synthesis of glutamate from 2-oxoglutarate, was treated with the specific GS inhibitor L-methionine-DL-sulfoximine (MSX) to abolish GS-mediated ammonium assimilation. Northern blot analyses demonstrate that ammonium itself can be sensed, probably by MepB, and causes strong nitrogen metabolite repression of NCR target genes without being metabolized to glutamate or glutamine (Teichert *et al.*, 2008).

The special role of the glutamine synthetase (GS) in the nitrogen regulation network

Assimilation of different nitrogen sources, such as nitrate, urea, or am1ino acids, results in formation of NH_4^+ which is subsequently converted first to Glu via glutamate dehydrogenase and then to Gln via glutamine synthetase (GS). There are three distinct GS enzyme families which can easily be distinguished by length: GSI with 360 amino acids on average, GSII with 450 and GSIII with 730 (Van Rooyen *et al.*, 2011). All three form multimeric proteins containing double-ringed quaternary structures composed of identical units: GSI- and GSIII-type enzymes contain 12 identical subunits of which the two hexameric rings of GSIII-type GS face each other the opposite way compared to GSI-type enzymes, whereas GSII- enzymes consist of an octamer composed of two tetramers placed back-to-back (Llorca *et al.*, 2006). However, recent crystal structure analyses of mammalian and *S. cerevisiae* GSs indicate that the eukaryotic GSII has a decamer structure of two pentameric rings showing the same orientation as the GSI rings (Krajevski *et al.*, 2008; He *et al.*, 2009). Recent molecular studies and genome projects have shown that the genes of GSI, GSII and GSIII families are broadly distributed among prokaryotes and eukaryotes suggesting that the GS family members arose prior to the divergence of prokaryotes and eukaryotes.

Based on experimental data in various fungi, it has been suggested that Gln, and not NH_4^+, NO_3^- or Glu, is the key effector of nitrogen metabolite repression that ensures preferential utilization of reduced nitrogen sources, such as NH_4^+, over more complex ones, such as NO_3^-, urea, purines, and proteins, when multiple nitrogen resources are available (Arst and MacDonald, 1975; Dunn- Coleman and Garrett, 1980, 1981; Wiame *et al.*, 1985; Caddick, 1994; Platt *et al.*, 1996). For plant pathogenic fungi, nitrogen may be a limiting nutrient during early stages of infection, such as during spore germination and penetration. In support of this hypothesis, GS-encoding genes are highly expressed in *C. gloeosporioides* and *Blumeria graminis* during plant infection (Stephenson *et al.*, 1997; Zhang *et al.*, 2005). Further, an increase in cytosolic levels of Gln and GS-encoding mRNAs in bean leaves upon infection by *C. lindemuthianum* suggest that nitrogen mobilization also plays a key role in the host during plant–pathogen interactions (Pageau *et al.*, 2006). Two isoenzymes of GS in arbuscular mycorrhiza (AM) fungi of the phylum *Glomeromycota* were highly up-regulated during symbiotic interactions and appear to be important for nitrogen uptake and translocation by the fungi and for fungus-to-plant transfer of nitrogen (Tian *et al.*, 2010).

Recent studies with *A. nidulans* showed a 50% drop in Gln concentration within 5 min of transfer of mycelia into a nitrogen-free medium as compared to media with ammonium (Berger *et al.*, 2008). The levels of most other amino acids were not significantly different in the two media. Accordingly, both the Gln level and GS activity are strongly regulated to maintain a constant nitrogen status. Taken together, these data suggest that Gln is indeed the marker for the nitrogen status (Berger *et al.*, 2008). The question to be asked is whether Gln itself is the key corepressor of many metabolic processes as proposed by Premakumar *et al.* (1979) and Margelis *et al.* (2003), or whether GS has an important role not only in providing Gln and regulating intracellular Gln levels, but also as key control point in the nitrogen regulatory network as shown for *N. crassa* (Mora, 1990; Dunn-Coleman and Garrett, 1980, 1981). In a *S. cerevisiae* mutant with reduced GS activity due to a mutation in the GS structural gene, GLN1, expression of both the nitrogen-regulated gene

GDH2 (NAD-linked glutamate dehydrogenase) and the defective GLN1 gene was similar to that in cells lacking the repressor Ure2 of the positively acting GATA transcription factor Gln3. Similarly, treatment with the GS inhibitor MSX leads to increased expression of nitrogen-regulated genes in cells grown on a preferred nitrogen source (Crespo *et al.*, 2002; Magasanik, 2005). It is, however, not clear if the drop in the intracellular concentration of Gln or reduced GS activity is responsible for the release of Gln3 from its association with the repressor Ure2.

A regulatory role for GS has been also postulated for *F. fujikuroi* as a result of efforts to increase GA production by lowering intracellular Gln levels by deletion of *glnA*, the GS-encoding gene (Teichert *et al.*, 2004). Deletion of *glnA* was expected to up-regulate both GA and bikaverin biosynthetic genes but instead significantly down-regulated the genes, even under nitrogen-limiting conditions, and blocked production of both secondary metabolites (Fig. 7.2). Inhibition of GS activity with MSX also down-regulated GA and bikaverin biosynthetic genes. Together, these results support the hypothesis that GS plays a regulatory role (Teichert *et al.*, 2004). HPLC analysis of the wild type and the Δ*glnA* mutant confirmed that *glnA* deletion significantly reduced intracellular glutamine levels even when exogenous Gln was fed to the mutant (G. Mey and B. Tudzynski, unpublished data). Under nitrogen starvation, the expression of GA and bikaverin genes was induced in the wild type but not in the mutant even though the intracellular Gln pool was exhausted to the same extent in both strains. These data further support our hypothesis that GS plays an important role in nitrogen regulation.

To determine whether loss of starvation-induced expression of nitrogen-repressed genes (e.g. GA and BIK – bikaverin biosynthetic genes) in the *glnA* mutant is mediated by the cross pathway control regulator, CPC1, gene expression was examined in a Δ*cpc1* mutant of *F. fujikuroi* that was treated with MSX to inhibit GS activity (Schönig *et al.*, 2009). If up-regulation of CPC1 under glutamine starving conditions was involved in down-regulation of secondary metabolism, this effect would be abolished in the Δ*cpc1* mutant. However, the absence of CPC1 did not restore wild-type like expression of GA and BIK genes in MSX-treated cultures, indicating that CPC1 is not involved in this GS-specific down-regulation of secondary metabolite production (Schönig *et al.*, 2009). In contrast, amino acid starvation and subsequent up-regulation of CPC1 in *A. nidulans* results in down-regulation of penicillin (secondary metabolite) production and favours metabolic flux of precursors towards amino acid biosynthesis (Busch *et al.*, 2003).

To determine whether *glnA* deletion specifically affects genes involved in biosynthesis of secondary metabolites, a macroarray approach was used to compare transcriptional profiles of wild-type *F. fujikuroi* with the *glnA* mutant. This analysis revealed a set of additional genes that are strongly up or down-regulated in the mutant, including genes involved in nitrogen metabolism (uricase, CPC1), in translation control (eEF1α; eIH5a), stress response (*ddr48*; *cipC*), ribosome biogenesis, glyoxylate cycle (isocitrate lyase), and histone modification (histone acetyltransferase) (Teichert *et al.*, 2004) indicating that the GS plays a major role in regulating different processes beside secondary metabolism

Interestingly, sequence and structural analyses indicated that prokaryotic GSI and eukaryotic GSII enzymes are highly conserved and share a similar catalytic mechanism suggesting a common evolutionary origin (He *et al.*, 2009). This hypothesis is supported by the presence of GSI and GSII enzymes in some prokaryotes, such as *Streptomyces*, *Rhizobium* and *Agrobacterium* (Reuther and Wohlleben, 2007). Complementation of the *F. fujikuroi glnA* deletion mutant with either the GSI and GSII genes (*glnA* and *glnII* respectively) from *Streptomyces coelicolor* (kindly provided by W. Wohlleben, University of Tübingen) further supports a common catalytic mechanism for GSI and GSII. Restoration of both glutamine synthesis and production of secondary metabolites in the *F. fujikuroi glnA* mutant with bacterial GS genes was surprising since regulation of GS activity in *S. coelicolor* is regulated by bacteria-specific posttranslational modifications such as adenylation in response to ammonium concentration, and by interacting transcription factors. Despite the fact that the corresponding adenylyltransferase GlnE as well as homologues of those transcription factors are missing in fungal systems, the bacterial GS proteins were active in the fungal background.

In *Bacillus subtilis,* GSI has a regulatory role in nitrogen metabolism and transmits the nitrogen signal to two transcription factors, TnrA and GlnR (Wray and Fisher, 2008, 2010). In this case, GS activity is regulated by a Gln feed-back inhibition, whereby Gln-bound GS is inhibited and regulates DNA binding of GlnR and TnrA. Mutations in GS that disrupt a seven-residue loop or replace Glu304 resulted in high levels of resistance to glutamine and to the GS-specific inhibitor MSX and lead to constitutive expression of genes regulated by GlnR and TnrA (Wray and Fisher, 2010). Alignment of amino acid sequences of GSs from bacteria, plants, mammals and fungi revealed a high level of sequence conservation of the seven-residue loop. Site-directed mutagenesis of these conserved residues in the *F. fujikuroi* GS are currently under investigation.

The role of the TOR pathway

How fungi sense their intracellular nitrogen status and what kinds of nitrogen sources they sense has been best studied in *S. cerevisiae,* where a drop in intracellular levels of Gln provides the stimulus for expression of genes regulated by the AreA homologue Gln3 (Crespo *et al.*, 2002; Magasanik, 2005). Likewise, in plants, internal pools of amino acids, mainly Gln, may provide a signal for nitrogen status and allow for regulation of the uptake of nitrate, ammonium, and amino acids (Miller *et al.*, 2007). In bacteria, Gln provides a signal for nitrogen status under nitrogen sufficient conditions, whereas 2-oxoglutarate can provide a signal for nitrogen starvation conditions because it accumulates when ammonium is limited (Wisedchaisri *et al.*, 2010).

The ability to sense the intracellular nitrogen status ensures that the energetically demanding process of reducing nitrate to ammonium occurs only when a limited amount of ammonium is available. Besides transmembrane localized permeases which have transport and sensing functions, intracellular sensors that can sense the amount and quality of nitrogen in intracellular pools must also exist. All living organisms sense and respond to nutrient-derived signals to adapt their physiology to promote survival. In eukaryotic organisms ranging from yeasts to humans, the TOR (target of rapamycin) kinase and the downstream TOR signalling pathway are global regulators of cell growth. The central components of this signalling cascade are the TOR protein kinases, which are activated by amino acid-derived signals and stimulate a set of anabolic processes, including translation, transcription, and ribosome biogenesis (for review see Rohde *et al.*, 2008).

Rapamycin is an antitumour and immunosuppressant macrocyclic lactone that has the ability to bind the peptidylisoprolyl isomerase FK506-binding protein (FKBP12). When bound to FKBP12, rapamycin causes inactivation of TOR kinases, which in turn triggers a series of events such as G1 cell cycle arrest, protein synthesis inhibition, glycogen accumulation, and autophagy that also occur in response to stress and nutrient starvation (Rubio-Texeira, 2007).

In *S. cerevisiae,* TOR proteins form two distinct, evolutionary conserved multimeric protein complexes known as TOR complex 1 and 2 (TORC1 and TORC2). Each complex mediates distinct physiological processes in response to nutrient availability: the rapamycin-sensitive TORC1 regulates temporal processes of growth, whereas the rapamycin-insensitive TORC2 is thought to control spatial aspects of growth such as actin polarization (Rohde *et al.*, 2008).

Genome-wide transcriptional analysis revealed a remarkable set of TORC1 controlled genes, including transcriptional regulators of rapamycin-sensitive genes (Rohde and Cardenas, 2004). However, the mechanisms by which signals of nutrient availability are sensed by TORC1 and transduced to transcription factors to elicit a response are not fully understood. In *S. cerevisiae,* TORC1 regulates two highly conserved phosphatases, Sit4 and PP2A, which in turn regulate cellular localization of the two positively acting GATA type transcription factors, Gln3 and Gat1. Both, Gln3 and Gat1, are responsible for expression of nitrogen catabolite repressed (NCR) genes needed for utilization and metabolism of alternative nitrogen sources when Gln or NH_4^+ are not available. Under nitrogen sufficient conditions, Gln3 and Gat1 are cytoplasmic, and NCR genes are not expressed. Under nitrogen starvation or rapamycin treatment, both transcription factors are translocated to the nucleus and transcription of NCR genes is derepressed. However, Sit4 and PP2A affect subcellular localization of Gln3 and Gat1 differently (Tate *et al.*, 2010). While nuclear localization of Gat1 is absolutely dependent on PP2A, Gln3 dephosphorylation and subsequent expression of NCR-sensitive genes is dependent mainly on Sit4.

Much less is known on the role of TOR kinase in filamentous fungi. In contrast to yeast, the sequenced genomes of filamentous fungi contain only one *TOR* homologue. Genes (e.g. *torA*, *fprA*, *jipA*, *sitA*, and *tapA*) encoding putative components of the TOR pathway have been identified in the genomes of *A. nidulans* and other fungi (Fitzgibbon *et al.*, 2005). Despite the presence of these putative homologues, there is currently no clear evidence that TOR signalling acts through regulating subcellular localization of the GATA factors AreA and AreB (homologues of Gln3 and Gat1) by phosphorylation/dephosphorylation in response to nitrogen availability or rapamycin treatment. Although phenotypes of *A. nidulans* mutants suggest that the TOR pathway plays only a minor role in regulating nitrogen metabolism, transcriptome analysis in *F. fujikuroi* revealed a role of TOR in regulating expression of NCR genes. A first indication for the presence of a functional TOR protein in *F. fujikuroi* was the high level of sensitivity to rapamycin (100 ng/ml) of a wild-type strain of the fungus, suggesting an interaction of the rapamycin–FKBP12 complex with an active TOR homologue in this fungus (Teichert *et al.*, 2006). Addition of rapamycin to wild-type cultures resulted in partial deregulation of GA and bikaverin biosynthetic genes, but also of some other NCR genes, e.g. the ammonium permease gene *mepC*. This partial deregulation was obtained only with nitrogen concentrations less than 10 mM (Fig. 7.2). In contrast, genes encoding the ribosome biogenesis protein 40S26E and the translation initiation factor eIF4A, which are highly expressed under nitrogen sufficiency, were down-regulated by rapamycin. Deletion of the FKBP12-encoding gene, *fpr1*, in *F. fujikuroi* resulted in rapamycin resistance and no change in gene expression profiles in the presence and absence of rapamycin, indicating that the interaction of rapamycin and FKBP12 observed in yeast also occurs in *Fusarium* (Teichert *et al.*, 2006). Macroarray experiments comparing growth in 10 mM ammonium nitrate with or without rapamycin treatment identified additional genes whose expression is affected by rapamycin treatment. Among rapamycin-affected genes were those involved in stress response, ribosome biogenesis, translation initiation/elongation, autophagy, cross-pathway control, tricarboxylic acid and glyoxylate cycles, and secondary metabolism. Like in yeast, ribosome biogenesis genes (e.g. encoding both 40S and 60S subunits) and translation initiation (e.g. *eIF1A* and *eIF4A*) are down-regulated by rapamycin whereas genes involved in autophagy (*idi4* and *idi7*) and protection against oxidative stress (thioredoxin- and peroxiredoxin-encoding genes) are up-regulated by rapamycin. Interestingly, the FKBP12-encoding gene *fpr1* was also shown to be up-regulated by rapamycin.

In contrast to yeast, the nitrogen repression of NCR genes in *F. fujikuroi* cannot be overcome by rapamycin-mediated TOR inhibition at high Gln or NH_4^+ concentrations (100 mM), probably due to an alternative nitrogen-sensing mechanism (Teichert *et al.*, 2006; Fig. 7.2). It is also not clear whether TORC1 regulates the subcellular localization and thereby the activity of the GATA factors AreA and/or AreB in response to nitrogen availability or rapamycin through the action of TOR-regulated phosphatases as it has been shown for Gln3 and Gat1 in *S. cerevisiae* (Tate *et al.*, 2009).

In the tomato pathogen *F. oxysporum* cellophane penetration, a virulence-related process, also requires an active TOR cascade. Although NH_4^+ represses cellophane penetration in wild-type *F. oxysporum*, this repression is reversed by rapamycin, presumably through inhibition of TOR (López-Berges *et al.*, 2010). Interestingly, NH_4^+ did not repress cellophane penetration by a Δ*areA* mutant of *F. oxysporum*, indicating that AreA is not required for cellophane penetration but instead may repress penetration. Nitrogen source and TOR also control other virulence-related functions, such as vegetative hyphal fusion and root adhesion in *F. oxysporum* (López-Berges *et al.*, 2010). Similar to *F. fujikuroi*, rapamycin treatment increased transcript levels of three AreA target genes (*niaD*, *niiA* and *mepB*) in *F. oxysporum*. This up-regulation was also observed in a Δ*meaB* mutant, indicating that TOR may act independently of MeaB during nitrogen repression. By contrast, rapamycin failed to up-regulate these three NCR-sensitive genes in the Δ*areA* mutant, suggesting that TOR mediates nitrogen catabolite repression through inhibition of AreA (López-Berges *et al.*, 2010).

Conclusions

Fungi have acquired an extensive network of anabolic and catabolic pathways to utilize structurally diverse nitrogen sources. They are able to respond to quantitative and qualitative changes in nitrogen through complex regulatory mechanisms that include (1) sensing nitrogen availability, (2) transducing the signal to transcription factors and/or other regulatory proteins, (3) interactions between transcription factors, (4) formation of regulatory protein complexes, and, finally, (5) binding transcription factors to DNA to affect transcription. The most prominent nitrogen regulator is the conserved GATA factor AreA. This protein is responsible for activating the transcription of a set of nitrogen-regulated genes when preferred nitrogen sources, such as glutamine or ammonia, are limited or when only non-preferred nitrogen sources, such as nitrate, are available. Although significant progress has been made in understanding how AreA functions, much remains unknown. For example, although AreA likely interacts with numerous proteins, only one such protein, Nmr, has been identified. In addition, AreA phosphorylation sites and the protein kinases that regulate AreA activity have not yet been identified. Complicating the story even further is the observation that regulation of subcellular localization of AreA in filamentous fungi is different than for the yeast AreA homologue Gln3. For future studies, it will be important to define output pathways for each of the aforementioned factors in order to elucidate their role in the nitrogen regulation network. The role of other factors, such as MeaB and AreB, in these out-put pathways is just emerging.

Some years ago, Boris Magasanik, who studied nitrogen regulation for most of his scientific career, wrote 'Although we understand many of the molecular details of how transcription of nitrogen-regulated genes is controlled, we know relatively little about the nature and origin of the intracellular signals that govern the activity of the transcription factors' (Magasanik, 2005). Work on *S. cerevisiae* has provided insights into fundamental aspects of the nitrogen regulation network beginning with nitrogen sensing, signal transduction, and subsequent transcriptional response, particularly with respect to the role of the TOR complex in the network (reviewed in Georis *et al.*, 2009). Comparable knowledge in filamentous fungi is decidedly lacking, and a direct link of the TOR complex and AreA activity in *Fusarium* and other filamentous fungi has not yet been demonstrated. However, there is evidence that the *Fusarium* GS acts as an intracellular nitrogen sensor and plays an important role in regulation of secondary metabolism (Teichert *et al.*, 2004; and unpublished data) (Fig. 7.2).

Except for the sensing function of MepB in *F. fujikuroi*, little is known in *Fusarium* about the role of nitrogen permeases in sensing their substrates and about regulation of permease stability by Npr1-like protein kinases as shown in *S. cerevisiae* (De Craene *et al.*, 2001). Identifying intra- and extracellular nitrogen sensors and connecting them to downstream processes that result in coordinated responses of out-put pathways will be a major topic of future research. Such efforts will be aided by the ever-increasing number of *Fusarium* genome sequences and the application of functional genomics, proteomics, transcriptomics, metabolomics and bioinformatics. Already, microarray analysis has been used to identify AreA and nitrogen-regulated genes in *F. fujikuroi* and *F. oxysporum* (Schönig *et al.*, 2008; Divon *et al.*, 2005), a phosphoproteome has been used to identify *F. graminearum* proteins expressed under nitrogen-limiting conditions (Rampitsch *et al.*, 2010), and a genome-wide bioinformatic approach has been used to identify known and novel regulatory motifs in gene promoter regions of four *Fusarium* genome sequences (Kumar *et al.*, 2010). Future application of these and other technologies should provide a clear understanding not only of the functions of individual components of the nitrogen regulation network in *Fusarium* but also of the entire network, from nitrogen signal sensing to changes in gene expression and protein activity, as well as cross-talk between the nitrogen network and other metabolic networks such as carbon metabolism and stress responses (Fig. 7.2). Because of the importance of nitrogen regulation in pathogenicity and secondary metabolism, a thorough understanding of the nitrogen metabolic network should provide insights into how to control plant diseases and mycotoxin contamination problems caused by *Fusarium*.

References

Abreu, C., Sanguinetti, M., Amillis, S., and Ramon, A. (2010). AreA, the major urea/H+ symporter in *Aspergillus nidulans*. Fungal Genet. Biol. *47*, 1023–1033.

Amillis, S., Cecchetto, G., Sophianopoulou, V., Koukaki, M., Scazzocchio, C., and Diallinas, G. (2004). Transcription of purine transporter genes is activated during the isotropic growth phase of *Aspergillus nidulans* conidia. Mol. Microbiol. *52*, 205–216.

Amrani, L., Cecchetto, G., Scazzocchio, C., and Glatigny, A. (1999). The *hxB* gene, necessary for the post-translational activation of purine hydroxylases in *Aspergillus nidulans*, is independently controlled by the purine utilization and the nicotinate utilization transcriptional activating systems. Mol. Microbiol. *31*, 1065–1073.

Andrianopoulos, A., Kourambas, S., Sharp, J.A., Davis, M.A., and Hynes, M.J. (1998). Characterization of the *Aspergillus nidulans nmrA* gene involved in nitrogen metabolite repression. J. Bacteriol. *180*, 1973–1977.

Arst, Jr, H.N. , and Cove, D.J. (1969). Methylammonium resistance in *Aspergillus nidulans*. J. Bacteriol. *98*, 1284–1293.

Arst, Jr, H.N. ., and Cove, D.J. (1973). Nitrogen metabolite repression in *Aspergillus nidulans*. Mol. Gen. Genet. *126*, 111–141.

Arst, Jr, H.N. ., and Page, M.M. (1973). Mutants of *Aspergillus nidulans* altered in the transport of methylammonium and ammonium. Mol. Gen. Genet. *121*, 239–245.

Arst, H.N., Jr., and MacDonald, D.W. (1975). A gene cluster in *Aspergillus nidulans* with an internally located cis-acting regulatory region. Nature. *254*, 26–31.

Arst, Jr, H.N. ., MacDonald, D.W., and Jones, S.A. (1980). Regulation of proline transport in *Aspergillus nidulans*. J. Gen. Microbiol. *116*, 285–294.

Arst, H.N., Jr, Tollervey, D., and Caddick, M.X. (1989). A translocation associated, loss-of-function mutation in the nitrogen metabolite repression regulatory gene of *Aspergillus nidulans* can revert intracistronically. Mol. Gen. Genet. *215*, 364–367.

Bahn, Y.S., Xue, C., Idnurm, A., Rutherford, J.C., Heitman, J., and Cardenas, M.E. (2007). Sensing the environment: lessons from fungi. Nat. Rev. Microbiol. *5*, 57–69.

Bai, G.H., Desjardins, A.E., and Plattner, R.D. (2002). Deoxynivalenol-nonproducing *Fusarium graminearum* causes initial infection, but does not cause disease spread in wheat spikes. Mycopathologia *153*, 91–98.

Barthelmess, I.B., and Kolanus, J. (1990). The range of amino acids whose limitation activates general amino-acid control in *Neurospora crassa*. Genet. Res. *55*, 7–12.

Bell, A.A., Wheeler, M.H., Liu, J., Stipanovic, R.D., Puckhaber, L.S., and Orta, H. (2003). United States Department of Agriculture-Agricultural Research Service studies on polyketide toxins of *Fusarium oxysporum* f sp *vasinfectum*, potential targets for disease control. Pest Manag. Sci. *59*, 736–747.

Berger, H., Pachlinger, R., Morozov, I., Goller, S., Narendja, F., Caddick, M., and Strauss, J. (2006). The GATA factor AreA regulates localization and *in vivo* binding site occupancy of the nitrate activator NirA. Mol. Microbiol. *59*, 433–446.

Berger, H., Basheer, A., Bock, S., Reyes-Dominguez, Y., Dalik, T., Altmann, F., and Strauss, J. (2008). Dissecting individual steps of nitrogen transcription factor cooperation in the *Aspergillus nidulans* nitrate cluster. Mol. Microbiol. *69*, 1385–1398.

Bernreiter, A., Ramon, A., Fernández-Martínez, J., Berger, H., Araùjo-Bazan, L., Espeso, E.A., Pachlinger, R., Gallmetzer, A., Anderl, I., Scazzocchio, C., and Strauss, J. (2007). Nuclear export of the transcription factor NirA is a regulatory checkpoint for nitrate induction in *Aspergillus nidulans*. Mol. Cell Biol. *27*, 791–802.

Biswas, K., and Morschhauser, J. (2005). The Mep2p ammonium permease controls nitrogen starvation-induced filamentous growth in *Candida albicans*. Mol. Microbiol. *56*, 649–669.

Boeckstaens, M., André, B., and Marini, A.M. (2007). The yeast ammonium transport protein Mep2 and its positive regulator, the Npr1 kinase, play an important role in normal and pseudohyphal growth on various nitrogen media through retrieval of excreted ammonium. Mol. Microbiol. *64*, 534–546.

Bömke, C., and Tudzynski, B. (2009). Diversity, regulation, and evolution of the gibberellin biosynthetic pathway in fungi compared to plants and bacteria. Phytochemistry *70*, 1876–1893.

Borrow, A., Brown, S., Jefferys, E.G., Kessell, R.H., Lloyd, E.C., Lloyd, P.B., Rothwell, A., Rothwell, B., and Swait, J.C. (1964). The kinetics of metabolism of *Gibberella fujikuroi* in stirred culture. Can. J. Microbiol. *10*, 407–444.

Borsuk, P., Dzikowska, A., Empel, J., Grzelak, A., Grześkowiak, R., and Weglenski, P. (1999). Structure of the arginase coding gene and its transcript in *Aspergillus nidulans*. Acta Biochim. Pol. *46*, 391–403.

Boudekou, S., and Andre, B. (1999). A co-activator of nitrogen-regulated transcription in *Saccharomyces cerevisiae*. Mol. Microbiol. *31*, 753–762.

Brandriss, M.C., and Magasanik, B. (1979). Genetics and physiology of proline utilization in *Saccharomyces cerevisiae*: Enzyme induction by proline. J. Bacteriol. *140*, 498–503.

Brown, S.H., Yarden, O., Gollop, N., Chen, S., Zveibil, A., Belausov, E., and Freeman, S. (2008). Differential protein expression in *Colletotrichum acutatum*: changes associated with reactive oxygen species and nitrogen starvation implicated in pathogenicity on strawberry. Mol. Plant Pathol. *9*, 171–190.

Bu'lock, J., Detroy, R., Hostalek, Z., and Munin-al-Shakarchi, A. (1974). Regulation of secondary biosynthesis in *Gibberella fujikuroi*. Trans. Br. Mycol. Soc. *62*, 377–389.

Burger, G., Tilburn, J., and Scazzocchio, C. (1991a). Molecular cloning and functional characterization of the pathwayspecific regulatory gene *nirA*, which controls nitrate assimilation in *Aspergillus nidulans*. Mol. Cell Biol. *11*, 795–802.

Burger, G., Strauss, J., Scazzocchio, C., and Lang, B.F. (1991b). *nirA*, the pathway-specific regulatory gene of nitrate assimilation in *Aspergillus nidulans*, encodes a putative GAL4-type zinc finger protein and contains four introns in highly conserved regions. Mol. Cell Biol. *11*, 5746–5755.

Busch, S., Bode, H.B., Brakhage, A.A., and Braus, G.H. (2003). Impact of the cross-pathway control on the regulation of lysine and penicillin biosynthesis in *Aspergillus nidulans*. Curr. Genet. *42*, 209–219. Epub 2002 Dec 3.

Caddick, M.X., Arst, H.N., Jr, Taylor, L.H., Johnson, R.I., and Brownlee, A.G. (1986). Cloning of the regulatory gene *areA* mediating nitrogen metabolite repression in *Aspergillus nidulans*. EMBO J. *5*, 1087–1090.

Caddick, M.X., Peters, D., and Platt, A. (1994). Nitrogen regulation in fungi. Antonie Van Leeuwenhoek. *65*, 169–177.

Caddick, M.X., and Arst, H.N., Jr. (1998). Seletion of the 389 N-terminal residues of the transcriptional activator AREA does not result in nitrogen metabolite derepression in *Aspergillus nidulans*. J. Bacteriol. *180*, 5762–1764.

Caddick, M.X., Jones, M.G., van Tonder, J.M., Le Cordier, H., Narendja, F., Strauss, J., and Morozov, I.Y. (2006). Opposing signals differentially regulate transcript stability in *Aspergillus nidulans*. Mol. Microbiol. *62*, 509–519.

Calderon, J., Martinez, L.M., and Mora, J. (1990). Isolation and characterization of a *Neurospora crassa* mutant altered in the alpha polypeptide of glutamine synthetase. J. Bacteriol. *172*, 4996–5000.

Carsiotis, M., and Jones, R.F. (1974). Cross-pathway regulation, tryptophan-mediated control of histidine and arginine biosynthetic enzymes in *Neurospora crassa*. J. Bacteriol. *119*, 889–892.

Carter, L.L., Leslie, J.F., and Webster, R.K. (2008). Population structure of *Fusarium fujikuroi* from California rice and water grass. Phytopathology *98*, 992–998.

Cecchetto, G., Amillis, S., Diallinas, G., Scazzocchio, C., and Drevet, C. (2004). The AzgA purine transporter of *Aspergillus nidulans*. Characterization of a protein belonging to a new phylogenetic cluster. J. Biol. Chem. *279*, 3132–3141.

Chiang, T.Y., and Marzluf, G.A. (1994). DNA recognition by the NIT2 nitrogen regulatory protein, importance of the number, spacing, and orientation of GATA core elements and their flanking sequences upon NIT2 binding. Biochemistry *33*, 576–582.

Christensen, M.K., Falkeid, G., Loros, J.J., Dunlap, J.C., Lillo, C., and Ruoff, P. (2004). A nitrate-induced *frq*-less oscillator in *Neurospora crassa*. J Biol. Rhythms *19*, 280–286.

Coddington, A. (1976). Biochemical studies on the *nit* mutants of *Neurospora crassa*. Mol. Gen. Genet. *145*, 195–206.

Coleman, M., Henricot, B., Arnau, J., and Oliver, R.P. (1997). Starvation-induced genes of the tomato pathogen *Cladosporium fulvum* are also induced during growth *in planta*. Mol. Plant–Microbe Interact. *10*, 1106–1109.

Conlon, H., Zadra, I., Haas, H., Arst, H.N., Jr, Jones, M.G., and Caddick, M.X. (2001). The *Aspergillus nidulans* GATA transcription factor gene areB encodes at least three proteins and features three classes of mutation. Mol. Microbiol. *40*, 361–375.

Cove, D.J. (1966). The induction and repression of nitrate reductase in the fungus *Aspergillus nidulans*. Biochim. Biophys. Acta. *113*, 51–56.

Cove, D.J., and Pateman, J.A. (1963). Independently segregating genetic loci concerned with nitrate reductase activity in *Aspergillus nidulans*. Nature *198*, 262–263.

Cove, D.J., and Pateman, J.A. (1969). Autoregulation of the synthesis of nitrate reductase in *Aspergillus nidulans*. J. Bacteriol. *97*, 1374–1378.

Cove, D.J. (1979). Genetic studies of nitrate assimilation in *Aspergillus nidulans*. Biol Rev Camb. Philos. Soc. *54*, 291–327.

Crespo, J.L., Powers, T., Fowler, B., and Hall, M.N. (2002). The TOR-controlled transcription activators GLN3, RTG*1*, and RTG3 are regulated in response to intracellular levels of glutamine. Proc. Natl. Acad. Sci. U.S.A.*99*, 6784–6789.

Crespo, J.L., Helliwell, S.B., Wiederkehr, C., Demougin, P., Fowler, B., Primig, M., and Hall, M.N. (2004). NPR1 kinase and RSP5-BUL1/2 ubiquitin ligase control GLN3-dependent transcription in *Saccharomyces cerevisiae*. J. Biol. Chem. *279*, 37512–37517.

Cubero, B., Gomez, D., and Scazzocchio, C. (2000). Metabolite repression and inducer exclusion in the proline utilization gene cluster of *Aspergillus nidulans*. J. Bacteriol. *182*, 233–235.

Cultrone, A., Scazzocchio, C., Rochet, M., Montero-Morán, G., Drevet, C., and Fernández-Martin, R. (2005). Convergent evolution of hydroxylation mechanisms in the fungal kingdom, molybdenum cofactor-independent hydroxylation of xanthine via a-ketoglutarate-dependent dioxygenases. Mol. Microbiol. *57*, 276–290.

Cultrone, A., Domínguez, Y.R., Drevet, C., Scazzocchio, C., and Fernández-Martín, R. (2007). The tightly regulated promoter of the *xanA* gene of *Aspergillus nidulans* is included in a helitron. Mol. Microbiol. *63*, 1577–1587.

Daboussi, M.J., Langin, T., Deschamps, F., Brygoo, Y., Scazzocchio, C., and Burger, G. (1991). Heterologous expression of the *Aspergillus nidulans* regulatory gene *nirA* in *Fusarium oxysporum*. Gene *109*, 155–160.

Dang, V.D., Valens, M., Bolotin-Fukuhara, M., and Daignan-Fornier, B. (1996). Cloning of the *ASN1* and *ASN2* genes encoding asparagine synthetases in *Saccharomyces cerevisiae*: differential regulation by the CCAAT-box-binding factor. Mol. Microbiol. *22*, 681–692.

Davis, R.H. (1979). Genetics of arginine biosynthesis in *Neurospora crassa*. Genetics. *93*, 557–575.

Davis, M.A., and Hynes, M.J. (1987). Complementation of *areA*-regulatory gene mutations of *Aspergillus nidulans* by the heterologous regulatory gene *nit-2* of *Neurospora crassa*. Proc. Natl. Acad. Sci. U.S.A.*84*, 3753–3757.

De Craene, J.-O., Soetens, O., and Andre, B. (2001). The Npr1 kinase controls biosynthetic and endocytic sorting of the yeast Gap1 permease. J. Biol. Chem. *276*, 43939–43948.

de Queiroz, M.V., Barros, A.O., de Barros, E.G., Guimarães, V., and de Araújo, E.F. (1998). Transformation of *Penicillium griseoroseum* nitrate reductase mutant with the *nia* gene from *Fusarium oxysporum*. Can. J. Microbiol. *44*, 487–489.

De Virgilio, C., and Loewith, R. (2006). Cell growth control: little eukaryotes make big contributions. Oncogene. *25*, 6392–6415.

Diallinas, G., Gorfinkiel, L., Arst, H.N., Jr, Cecchetto, G., and Scazzocchio, C. (1995). Genetic and molecular characterization of a gene encoding a wide specificity purine permease of *Aspergillus nidulans* reveals a novel family of transporters conserved in prokaryotes and eukaryotes. J. Biol. Chem. *270*, 8610–8622.

Dickman, M.B., and Leslie, J.F. (1992). The regulatory gene *nit-2* of *Neurospora crassa* complements a nnu mutant of *Gibberella zeae* (*Fusarium graminearum*). Mol. Gen. Genet. *235*, 458–462.

Diolez, A., Langin, T., Gerlinger, C., Brygoo, Y., and Daboussi, M.J. (1993). The *nia* gene of *Fusarium oxysporum*, isolation, sequence and development of a homologous transformation system. Gene *131*, 61–67.

Divon, H.H., Rothan-Denoyes, B., Davydov, O., Di Pietro, A., and Fluhr, R. (2005). Nitrogen-responsive genes are differentially regulated in planta during *Fusarium oxysporum* f. sp. *lycopersici* infection. Mol. Plant Pathol. *6*, 459–470.

Divon, H., Ziv, C., Davydov, O., Yarden, O., and Fluhr, R. (2006). The global nitrogen regulator, FNR*1*, regulates fungal nutrition-genes and fitness during *Fusarium oxysporum* pathogenesis. Mol. Plant Pathol. *7*, 485–497.

Divon, H.H., and Fluhr, R. (2007). Nutrition acquisition strategies during fungal infection of plants. FEMS Microbiol. Lett. *266*, 65–74.

Donaton, M.C., Holsbeeks, I., Lagatie, O., Van Zeebroeck, G., Crauwels, M., Winderickx, J., and Thevelein, J.M. (2003). The Gap1 general amino acid permease acts as an amino acid sensor for activation of protein kinase A targets in the yeast Saccharomyces cerevisiae. Mol. Microbiol. *50*, 911–929.

Donofrio, N.M., Oh, Y., Lundy, R., Pan, H., Brown, D.E., Jeong, J.S., Coughlan, S., Mitchell, T.K., and Dean, R.A. (2006). Global gene expression during nitrogen starvation in the rice blast fungus, *Magnaporthe grisea*. Fungal Genet. Biol. *43*, 605–617.

Dunn-Coleman, N.S., and Garrett, R.H. (1980). The role of glutamine synthetase and glutamine metabolism in nitrogen metabolite repression, a regulatory phenomenon in the lower eukaryote *Neurospora crassa*. Mol. Gen. Genet. *179*, 25–32.

Dunn-Coleman, N.S., Tomsett, A.B., and Garrett, R.H. (1981). The regulation of nitrate assimilation in *Neurospora crassa*, biochemical analysis of the *nmr-1* mutants. Mol. Gen. Genet. *182*, 234–239.

Dyer, R.B., Plattner, R.D., Kendra, D.F., and Brown, D.W. (2005). *Fusarium graminearum* TRI14 is required for high virulence and DON production on wheat but not for DON synthesis *in vitro*. J. Agric. Food Chem. *53*, 9281–9287.

Dzikowska, A., Swianiewicz, M., Talarczyk, A., Wisniewska, M., Goras, M., Scazzocchio, C., and Weglenski, P. (1999). Cloning, characterization and regulation of the ornithine transaminase (*otaA*) gene of *Aspergillus nidulans*. Curr. Genet. *35*, 118–126.

Dzikowska, A., Kacprzak, M., Tomecki, R., Koper, M., Scazzocchio, C., and Weglenski, P. (2003). Specific induction and carbon/nitrogen repression of arginine catabolism gene of *Aspergillus nidulans* – functional *in vivo* analysis of the *otaA* promoter. Fungal Genet. Biol. *38*, 175–186.

Exley, G.E., Colandene, J.D., and Garrett, R.H. (1993). Molecular cloning, characterization, and nucleotide sequence of *nit-6*, the structural gene for nitrite reductase in *Neurospora crassa*. J. Bacteriol. *175*, 2379–2392.

Facklam, T.J., and Marzluf, G.A. (1978). Nitrogen regulation of amino acid catabolism in *Neurospora crassa*. Biochem. Genet. *16*,:343–354.

Feng, B., Xiao, X., and Marzluf, G.A. (1993). Recognition of specific nucleotide bases and cooperative DNA binding by the *trans*-acting nitrogen regulatory protein NIT2 of *Neurospora crassa*. Nucleic Acids Res. *21*, 3989–3996.

Ferguson, A.R., and Sims, A.P. (1971). Inactivation *in vivo* of glutamine synthetase and NAD-specific glutamate dehydrogenase: its role in the regulation of glutamine synthesis in yeasts. J. Gen. Microbiol. *69*, 423–427.

Ferguson, A.R., and Sims, A.P. (1974a). The regulation of glutamine metabolism in *Candida utilis*: the role of glutamine in the control of glutamine synthetase. J. Gen. Microbiol. *80*, 159–171.

Ferguson, A.R., and Sims, A.P. (1974b). The regulation of glutamine metabolism in *Candida utilis*: the inactivation of glutamine synthetase. J. Gen. Microbiol. *80*, 173–185.

Fitzgibbon, G.J., Morozov, Jones, I.Y., M.G., and Caddick, M.X. (2005). Genetic analysis of the TOR pathway in *Aspergillus nidulans*. Eukaryot. Cell *4*, 1595–1828.

Forsberg, H., and Ljungdahl, P.O. (2001). Sensors of extracellular nutrients in *Saccharomyces cerevisiae*. Curr. Genet. *40*, 91–109.

Froeliger, E.H., and Carpenter, B.E. (1996). *NUT1*, a major nitrogen regulatory gene in *Magnaporthe grisea*, is dispensable for pathogenicity. Mol. Gen. Genet. *251*, 647–656.

Fu, Y.-H., and Marzluf, G.A. (1987a). Molecular cloning and analysis of the regulation of nit-3, the structural gene for nitrate reductase in *Neurospora crassa*. Proc. Natl. Acad. Sci. U.S.A.*84*, 8243–8247.

Fu, Y.-H., and Marzluf, G.A. (1987b). Characterization of Nit-2, the major nitrogen regulatory gene of *Neurospora crassa*. Mol. Cell. Biol. *7*, 1691–1696.

Fu, Y.-H., and Marzluf, G.A.(1988). Metabolic control and autogenous regulation of *nit-3*, the nitrate reductase structural gene of *Neurospora crassa*. J. Bacteriol. *170*, 657–661.

Fu, Y.-H., and Marzluf, G.A. (1990). *nit-2*, the major nitrogen regulatory gene of *Neurospora crassa*, encodes a protein with a putative zinc finger DNA-binding domain. Mol. Cell. Biol. *10*, 1056–1065.

Fu, Y.-H., Young, J.L., and Marzluf, G.A. (1988). Molecular cloning and characterization of a negative-acting nitrogen regulatory gene of *Neurospora crassa*. Mol. Gen. Genet. *214*, 74–79.

Fujii, T., and Takaya, N. (2008). Denitrification by the fungus *Fusarium oxysporum* involves NADH-nitrate reductase. Biosci. Biotechnol. Biochem. *72*, 412–420.

Gao-Rubinelli, F., and Marzluf, G.A. (2004). Identification and characterization of a nitrate transporter gene in *Neurospora crassa*. Biochem. Genet. *42*, 21–34.

Garcia, I., Gonzalez, R., Gomez, D., and Scazzocchio, C. (2004). Chromatin rearrangements in the *prnD-prnB* bidirectional promoter, dependence on transcription factors. Eukaryot. Cell 3, 144–156.

Gardiner, D.M., Kazan, K., and Manners, J.M. (2009). Nutrient profiling reveals potent inducers of trichothecene biosynthesis in *Fusarium graminearum*. Fungal Genet. Biol. *46*, 604–613.

Garrett, R.H., and Cove, D.J. (1976). Formation of NADPH-nitrate reductase activity *in vitro* from *Aspergillus nidulans niaD* and *cnx* mutants. Mol. Gen. Genet. *149*, 179–186.

Georis, I., Feller, A., Tate, J.J., Cooper, T.G., and Dubois, E. (2009). Nitrogen catabolite repression-sensitive transcription as a readout of Tor pathway regulation: the genetic background, reporter gene and GATA factor assayed determine the outcomes. Genetics *181*, 861–874.

Glatigny, A., and Scazzocchio, C. (1995). Cloning and molecular characterization of *hxA*, the gene coding for the xanthine dehydrogenase (purine hydroxylase I) of *Aspergillus nidulans*. J. Biol. Chem. *270*, 3534–3550.

Glenn, A.E., Zitomer, N.C., Zimeri, A.M., Williams, L.D., Riley, R.T., and Proctor, R.H. (2008). Transformation-mediated complementation of a FUM gene cluster deletion in *Fusarium verticillioides* restores both fumonisin production and pathogenicity on maize seedlings. Mol. Plant Microbe Interact. *21*, 87–97.

Gomez, D., Cubero, B., Cecchetto, G., and Scazzocchio, C. (2002). PrnA, a Zn2Cys6 activator with a unique DNA recognition mode, requires inducer for *in vivo* binding. Mol. Microbiol. *44*, 585–597.

Gomez, D., Garcia, I., Scazzocchio, C., and Cubero, B. (2003). Multiple GATA sites, protein binding and physiological relevance for the regulation of the proline transporter gene of *Aspergillus nidulans*. Mol. Microbiol. *50*, 277–289.

Gorfinkiel, L., Diallinas, G., and Scazzocchio, C. (1993). Sequence and regulation of the *uapA* gene encoding a uric acid-xanthine permease in the fungus *Aspergillus nidulans*. J. Biol. Chem. *268*, 23376–23381.

Haas, H., and Marzluf, G.A. (1995). NRE, the major nitrogen regulatory protein of *Penicillium chrysogenum*, binds specifically to elements in the intergenic promoter regions of nitrate assimilation and penicillin biosynthetic gene clusters. Curr. Genet. *28*, 177–183.

Haas, H., Angermayr, K., Zadra, I., and Stoffler, G. (1997). Overexpression of *nreB*, a new GATA factor-encoding gene of *Penicillium chrysogenum*, leads to repression of the nitrate assimilatory gene cluster. J. Biol. Chem. *272*, 22576–22582.

Haas, H., Bauer, B., Redl, B., Stoffler, G., and Marzluf, G.A. (1995). Molecular cloning and analysis of *nre*, the major nitrogen regulatory gene of *Penicillium chrysogenum*. Curr. Genet. *27*, 150–158.

He, Y.X., Gui, L., Liu, Y.Z., Du, Y., Zhou, Y., Li, P., and Zhou, C.Z. (2009). Crystal structure of *Saccharomyces cerevisiae* glutamine synthetase Gln1 suggests a nanotube-like supramolecular assembly. Proteins *76*, 249–254.

Holsbeeks, I., Lagatie, O., Van Nuland A., Van de Velde, S., and Thevelein, J.M. (2004). The eukaryotic plasma membrane as a nutrient-sensing device. Trends Biochem. Sci. *29*, 556–564.

Horowitz, S., Freeman, S., Zveibil, A., and Yarden, O. (2006). A defect in *nir1*, a *nirA*-like transcription factor, confers morphological abnormalities and loss of pathogenicity in *Colletotrichum acutatum*. Mol. Plant Pathol. *7*, 341–354.

Hull, E.P., Green, P.M., Arst, H.N., Jr, and Scazzocchio, C. (1989). Cloning and physical characterization of the L-proline catabolism gene cluster of *Aspergillus nidulans*. Mol. Microbiol. *3*, 553–559.

Hundal, H.S., and Taylor, P.M. (2009). Amino acid transceptors: gate keepers of nutrient exchange and regulators of nutrient signaling. Am. J. Physiol. Endocrinol. Metab. *296*, E603–613.

Hurlburt, B.K., and Garrett, R.H. (1988). Nitrate assimilation in *Neurospora crassa*, enzymatic and immunoblot analysis of wild-type and nit mutant protein products in nitrate-induced and glutamine-repressed cultures. Mol. Gen. Genet. *211*, 35–40.

Jarai, G., and Marzluf, G.A. (1990). Analysis of conventional and *in vitro* generated mutants of *nmr*, the negatively acting nitrogen regulatory gene of *Neurospora crassa*. Mol. Gen. Genet. *222*, 233–240.

Javelle, A., Severi, E., Thornton, J., and Merrick, M. (2004). Ammonium sensing in *Escherichia coli*. Role of the ammonium transporter AmtB and AmtB–GlnK complex formation. J. Biol. Chem. *279*, 8530–8538.

Johnstone, I.L., McCabe, P.C., Greaves, P., Gurr, S.J., Cole, G.E., Brow, M.A., Unkles, S.E., Clutterbuck, A.J., Kinghorn, J.R., and Innis, M.A. (1990). Isolation and characterisation of the *crnA-niiA-niaD* gene cluster for nitrate assimilation in *Aspergillus nidulans*. Gene. *90*, 181–192.

Jones, S.A., Arst, H.N., Jr, and MacDonald, D.W. (1981). Gene roles in the *prn* cluster of *Aspergillus nidulans*. Curr. Genet. *27*, 150–158.

Karov, I.K., Mitrev, S.K., and Kostadinovska, E.D. (2009). *Gibberella fujikuroi* (Sawada) Wollenweber, the new parasitical fungus on rice in the Republic of Macedonia. Zbornik Matice srpske za prirodne nauke *116*, 175–182.

Keller, N.P., Turner, G., and Bennett, J.W. (2005). Fungal secondary metabolism – from biochemistry to genomics. Nat. Rev. Microbiol. 3, 937–947.

Kersten, M.A., Müller, Y., op den Camp, H.J., Vogels, G.D., van Griensven, L.J., Visser, J., and Schaap, P.J. (1997). Molecular characterization of the *glnA* gene encoding glutamine synthetase from the edible mushroom *Agaricus bisporus*. Mol. Gen. Genet. *256*, 179–186.

Kim, K.W., Kamerud, J.Q., Livingston, D.M., and Roon, R.J. (1988). Asparaginase II of *Saccharomyces cerevisiae*. Characterization of the *ASP3* gene. J. Biol. Chem. *263*, 11948–11953.

Kim, H., and Woloshuk, C.P. (2008). Role of AREA, a regulator of nitrogen metabolism, during colonization of maize kernels and fumonisin biosynthesis in *Fusarium verticillioides*. Fungal Genet. Biol. *45*, 947–953.

Kinsky, S.C. (1961). Induction and repression of nitrate reductase in *Neurospora crassa*. J. Bacteriol. *82*, 898–904.

Klittich, C., and Leslie, J.F. (1988). Nitrate reduction mutants of *Fusarium moniliforme* (*Gibberella fujikuroi*). Genetics *118*, 417–423.

Kobayashi, M., and Shoun, H. (1995). The copper-containing dissimilatory nitrite reductase involved in the denitrifying system of the fungus *Fusarium oxysporum*. J. Biol. Chem. *270*, 4146–4151.

Kohut, G., Adám, A.L., Fazekas, B., and Hornok, L. (2009). N-starvation stress induced FUM gene expression and fumonisin production is mediated via the HOG-type MAPK pathway in *Fusarium proliferatum*. Int. J. Food Microbiol. *130*, 65–69.

Koibuchi, K., Nagasaki, H., Yuasa, A., Kataoka, J., and Kitamoto, K. (2000). Molecular cloning and characterization of a gene encoding glutaminase from *Aspergillus oryzae*. Appl. Microbiol. Biotechnol. *54*, 59–68.

Kolar, M., Holzmann, K., Weber, G., Leitner, E., and Schwab, H. (1991). Molecular characterization and functional analysis in *Aspergillus nidulans* of the 5′-region of the *Penicillium chrysogenum* isopenicillin N synthetase gene. J. Biotechnol. *17*, 67–80.

Krajewski, W.W., Collins, R., Holmberg-Schiavone, L., Jones, T.A., Karlberg, T., and Mowbray, S.L. (2008). Crystal structures of mammalian glutamine synthetases illustrate substrate-induced conformational changes and provide opportunities for drug and herbicide design. J. Mol. Biol. 375, 317–328.

Kudla, B., Caddick, M.X., Langdon, T., Martinez-Rossi, N.M., Bennett, C.F., Sibley, S., Davies, R.W., and Arst, H.N., Jr (1990). The regulatory gene *areA* mediating nitrogen metabolite repression in *Aspergillus nidulans*. Mutations affecting specificity of gene activation alter a loop residue of a putative zinc finger. EMBO J. *9*, 1355–1364.

Kumar, L., Breakspear, A., Kistler, C., Ma, L.J., and Xie, X. (2010). Systematic discovery of regulatory motifs in *Fusarium graminearum* by comparing four *Fusarium* genomes. BMC Genomics. *11*, 208.

Lamb, H.K., Leslie, K., Dodds, A.L., Nutley, M., Cooper, A., Johnson, C., Thompson, P., Stammers, D.K., and Hawkins, A.R. (2003). The negative transcriptional regulator NmrA discriminates between oxidized and reduced dinucleotides. J. Biol. Chem. *278*, 32107–32114.

Lamb, H.K., Ren, J., Park, A., Johnson, C., Leslie, K., Cocklin, S., Thompson, P., Mee, C., Cooper, A., Stammers, D.K., and Hawkins, A.R. (2004). Modulation of the ligand binding properties of the transcription repressor NmrA by GATA-containing DNA and site-directed mutagenesis. Protein Sci. *13*, 3127–3138.

Langdon, T., Sheerins, A., Ravagnani, A., Gielkens, M., Caddick, M.X., and Arst, H.N., Jr (1995). Mutational analysis reveals dispensability of the N-terminal region of the *Aspergillus* transcription factor mediating nitrogen metabolite repression. Mol. Microbiol. *17*, 877–888.

Lara, M., Blanco, L., Campomanes, M., Calva, E., Palacios, R., and Mora, J. (1982). Physiology of ammonium assimilation in *Neurospora crassa*. J. Bacteriol. *150*, 105–112.

Lau, G., and Hamer, J.E. (1996). Regulatory genes controlling *MPG1* expression and pathogenicity in the rice blast fungus *Magnaporthe grisea*. Plant Cell *8*, 771–781.

Lee, H., Fu, Y.-H., and Marzluf, G.A. (1990a). Molecular cloning and characterization of *alc* the gene encoding allantoicase of *Neurospora crassa*. Mol. Gen. Genet. *222*, 140–144.

Lee, H., Fu, Y.-H., and Marzluf, G.A. (1990b). Nucleotide sequence and DNA recognition elements of alc, the structural gene which encodes allantoicase, a purine catabolic enzyme of *Neurospora crassa*. Biochemistry *29*, 8779–8787.

Legrain, C., Vissers, S., Dubois, E., Legrain, M., and Wiame, J.M. (1982). Regulation of glutamine synthetase from *Saccharomyces cerevisiae* by repression, inactivation and proteolysis. Eur. J. Biochem. *123*, 611–616.

Linnemannstöns, P., Schulte, J., del Mar Prado, M., Proctor, R.H., Avalos, J., and Tudzynski, B. (2002). The polyketide synthase gene *pks4* from *Gibberella fujikuroi* encodes a key enzyme in the biosynthesis of the red pigment bikaverin. Fungal Genet. Biol. *37*, 143–148.

Liu, W., and Sundheim, L. (1996). Nitrate nonutilizing mutants and vegetative compatibility groups in *Fusarium poae*. Fungal Genet. Biol. *20*, 12–17.

Llorca, O., Betti, M., González, J.M., Valencia, A., Márquez, A.J., and Valpuesta, J.M. (2006). The three-dimensional structure of an eukaryotic glutamine synthetase: functional implications of its oligomeric structure. J. Struct. Biol. *156*, 469–479.

López-Berges, M.S., Pietro, M.D.I., Daboussi, M.J., Wahab, H.A., Vasnier, C., Roncero, M.I., Dufresne, M., and

Hera, C. (2009). Identification of virulence genes in *Fusarium oxysporum* f. sp. *lycopersici* by large-scale transposon tagging. Mol. Plant Pathol. *10*, 95–107.

López-Berges, M.S., Rispail, N., Prados-Rosales, R.C., and Di Pietro, A. (2010). A nitrogen response pathway regulates virulence functions in *Fusarium oxysporum* via the protein kinase TOR and the bZIP protein MeaB. Plant Cell *22*, 2459–2475.

Lorenz, M.C., and Heitman, J. (1998). The MEP2 ammonium permease regulates pseudohyphal differentiation in *Saccharomyces cerevisiae*. EMBO J. *17*, 1236–1247.

Ma, L.J., van der Does, H.C., Borkovich, K.A., Coleman, J.J., Daboussi, M.J., Di Pietro, A., Dufresne, M., Freitag, M., Grabherr, M., Henrissat, B., *et al.* (2010). Comparative genomics reveals mobile pathogenicity chromosomes in *Fusarium*. Nature *464*, 367–373.

Mackay, E.M., and Pateman, J.A. (1982). The regulation of urease activity in *Aspergillus nidulans*. Biochem. Genet. *20*, 763–776.

Magasanik, B., and Kaiser, C.A. (2002). Nitrogen regulation in *Saccharomyces cerevisiae*. Gene *290*, 1–18.

Magasanik, B. (2005). The transduction of the nitrogen regulation signal in *Saccharomyces cerevisiae*. Proc. Natl. Acad. Sci. U.S.A.*102*, 16537–16538.

Malardier, L., Daboussi, M.J., Julien, J., Roussel, F., Scazzocchio, C., and Brygoo, Y. (1989). Cloning of the nitrate reductase gene (*niaD*) of *Aspergillus nidulans* and its use for transformation of *Fusarium oxysporum*. Gene *78*, 147–156.

Margelis, S., D'Souza, C., Small, A.J., Hynes, M.J., Adams, T.H., and Davis, M.A. (2001). Role of glutamine synthetase in nitrogen metabolite repression in *Aspergillus nidulans*. J. Bacteriol. *183*, 5826–5833.

Marini, A.M., Brun, A., and Chalot, M. (2003). Molecular characterization, function and regulation of ammonium transporters (Amt) and ammonium-metabolizing enzymes (GS, NADP-GDH) in the ectomycorrhizal fungus *Hebeloma cylindrosporum*. Mol. Microbiol. *47*, 411–430.

Marzluf, G.A. (1997). Genetic regulation of nitrogen metabolism in the fungi. Microbiol. Mol. Biol. Rev. *61*, 17–32.

Mihlan, M., Homann, V., Liu, T.-W.D., and Tudzynski, B. (2003). AREA directly mediates nitrogen regulation of gibberellin biosynthesis in *Gibberella fujikuroi*, but its activity is not affected by NMR. Mol. Microbiol. *47*, 975–991.

Miller, A.J., Fan, X., Shen, Q., and Smith, S.J. (2007). Amino acids and nitrate as signals for the regulation of nitrogen acquisition. J. Exp. Bot. *59*, 111–119.

Mitchell, A.P. (1985). The GLN1 locus of *Saccharomyces cerevisiae* encodes glutamine synthetase. Genetics *111*, 243–258.

Mitchell, A.P., and Magasanik. B. (1984a). Regulation of glutamine-repressible gene products by the GLN3 function in *Saccharomyces cerevisiae*. Mol. Cell Biol. *4*, 2758–2766.

Mitchell, A.P., and Magasanik, B. (1984b). Biochemical and physiological aspects of glutamine synthetase inactivation in *Saccharomyces cerevisiae*. J. Biol. Chem. *259*, 12054–12062.

Mitchell, A.P., and Magasanik, B. (1984c). Three regulatory systems control production of glutamine synthetase in *Saccharomyces cerevisiae*. Mol. Cell Biol. *4*, 2767–2773.

Monahan, B.J., Fraser, J.A., Hynes, M.J., and Davis, M.A. (2002). Isolation and characterization of two ammonium permease genes, *meaA* and *mepA*, from *Aspergillus nidulans*. Eukaryot. Cell *1*, 85–94.

Monahan, B.J., Askin, M.C., Hynes, M.J., and Davis, M.A. (2006). Differential expression of *Aspergillus nidulans* ammonium permease genes is regulated by GATA transcription factor AreA. Eukaryot. Cell *5*, 226–237.

Mora, J. (1990). Glutamine metabolism and cycling in *Neurospora crassa*. Microbiol. Rev. *54*, 293–304.

Morin, P.J., Subramanian, G.S., and Gilmore, T.D. (1992). *AAT1*, a gene encoding a mitochondrial aspartate aminotransferase in *Saccharomyces cerevisiae*. Biochim. Biophys. Acta. *1171*, 211–214.

Morozov, I.Y., Martinez, M.G., Jones, M.G., and Caddick, M.X. (2000). A defined sequence within the 3′ UTR of the *areA* transcript is sufficient to mediate nitrogen metabolite signalling via accelerated deadenylation. Mol. Microbiol. *37*, 1248–1257.

Morozov, I.Y., Galbis-Martinez, M., Jones, M.G., and Caddick, M.X. (2001). Characterization of nitrogen metabolite signalling in *Aspergillus* via the regulated degradation of *areA* mRNA. Mol. Microbiol. *42*, 269–277.

Muro-Pastor, M.I., Gonzalez, R., Strauss, J., Narendja, F., and Scazzocchio, C. (1999). The GATA factor AreA is essential for chromatin remodelling in a eukaryotic bidirectional promoter. EMBO J. *18*, 1584–1597.

Muro-Pastor, M.I., Strauss, J., Ramón, A., and Scazzocchio, C. (2004). A paradoxical mutant GATA factor. Eukaryot. Cell 3, 393–405.

Nahm, B.H., and Marzluf, G.A. (1987). Induction and *de novo* synthesis of uricase, a nitrogen-regulated enzyme in *Neurospora crassa*. J. Bacteriol. *169*, 1943–1948.

Namiki, F., Matsunaga, M., Okuda, M., Inoue, I., Nishi, K., Fujita, Y., and Tsuge, T. (2001). Mutation of an arginine biosynthesis gene causes reduced pathogenicity in *Fusarium oxysporum* f. sp. *melonis*. Mol. Plant Microbe Interact. *14*, 580–584.

Narendja, F., Goller, S.P., Wolschek, M., and Strauss, J. (2002). Nitrate and the GATA factor AreA are necessary for *in vivo* binding of NirA, the pathway-specific transcriptional activator of *Aspergillus nidulans*. Mol. Microbiol. *44*, 573–583.

Neuhäuser, B., Dunkel, N., Satheesh, S.V., and Morschhauser, J. (2011). Role of the Npr1 kinase in ammonium transport and signaling by the ammonium permease Mep2 in *Candida albicans*. Eukaryot. Cell *10*, 332.

Oestreicher, N., and Scazzocchio, C. (1993). Sequence, regulation, and mutational analysis of the gene encoding urate oxidase in *Aspergillus nidulans*. J. Biol. Chem. *268*, 23382–23389.

Pageau, K., Reisdorf-Cren, M., Morot-Gaudry, J.F., and Masclaux-Daubresse, C. (2006). The two senescence-related markers, GS1 (cytosolic glutamine synthetase) and GDH (glutamate dehydrogenase), involved in nitrogen mobilization, are differentially regulated during pathogen attack and by stress hormones and reactive oxygen species in *Nicotiana tabacum* L. leaves. J. Exp. Bot. *57*, 547–557.

Pan, H., Feng, B., and Marzluf, G.A. (1997). Two distinct protein–protein interactions between NIT2 and NMR regulatory proteins are required to establish nitrogen metabolite repression in *Neurospora crassa*. Mol. Microbiol. *26*, 721–729.

Patel, R.M., van Kan, J.A., Bailey, A.M., and Foster, G.D. (2010). Inadvertent gene silencing of argininosuccinate synthase (*bcass1*) in *Botrytis cinerea* by the pLOB1 vector system. Mol. Plant Pathol. *11*, 613–624.

Pateman, J.A., and Cove, D.J. (1967). Regulation of nitrate reduction in *Aspergillus nidulans*. Nature *215*, 1234–1237.

Pateman, J.A., Rever, B.M., and Cove, D.J. (1967). Genetic and biochemical studies of nitrate reduction in *Aspergillus nidulans*. Biochem. J. *104*, 103–111.

Pellier, A.L., Lauge, R., Veneault-Fourrey, C., and Langin, T. (2003). CLNR*1*, the AREA/NIT2-like global nitrogen regulator of the plant fungal pathogen *Colletotrichum lindemuthianum* is required for the infection cycle. Mol. Microbiol. *48*, 639–655.

Pérez-García, A., Snoeijers, S.S., Joosten, M.H., Goosen, T., and De Wit, P.J. (2001). Expression of the Avirulence gene *Avr9* of the fungal tomato pathogen *Cladosporium fulvum* is regulated by the global nitrogen response factor NRF1. Mol. Plant–Microbe. Interact. *14*, 316–325.

Perkins, D.D. (1959). New markers and multiple point linkage data in *Neurospora*. Genetics *44*, 1185–1208.

Platt, A., Langdon, T., Arst, H.J., Jr, Kirk, D., Tollervey, D., Sanchez, J.M., and Caddick, M.X. (1996). Nitrogen metabolite signalling involves the C-terminus and the GATA domain of the *Aspergillus* transcription factor AREA and the 3′ untranslated region of its mRNA. EMBO J. *15*, 2791–2801.

Polley, S.D., and Caddick, M.X. (1996). Molecular characterisation of *meaB*, a novel gene affecting nitrogen metabolite repression in *Aspergillus nidulans*. FEBS Lett. *388*, 200–205.

Premakumar, R., Sorger, G.J., and Gooden, D.(1979). Nitrogen metabolite repression of nitrate reductase in *Neurospora crassa*. J. Bacteriol. *137*, 1119–1126.

Premakumar, R., Sorger, G.J., and Gooden, D. (1980). Physiological characterization of a *Neurospora crassa* mutant with impaired regulation of nitrate reductase. J. Bacteriol. *144*, 542–551.

Proctor, R.H., Hohn, T.M., McCormick, S.P., and Desjardins, A.E. (1995). Tri6 encodes an unusual zinc finger protein involved in regulation of trichothecene biosynthesis in *Fusarium sporotrichioides*. Appl. Environ. Microbiol. *61*, 1923–1930.

Punt, P.J., Strauss, J., Smit, R., Kinghorn, J.R., van den Hondel, C.A., and Scazzocchio, C. (1995). The intergenic region between the divergently transcribed *niiA* and *niaD* genes of *Aspergillus nidulans* contains multiple NirA binding sites which act bidirectionally. Mol. Cell Biol. *15*, 5688–5699.

Rademacher, W. (1997). Gibberellins. In Fungal Biotechnology, Anke, T., ed. (Chapman & Hall, London, England). pp. 193–205.

Radford, A. (2004). Metabolic highways of *Neurospora crassa* revisited. Adv. Genet. *52*, 165–207.

Rampitsch, C., Subramaniam, R., Djuric-Ciganovic, S., and Bykova, N.V. (2010). The phosphoproteome of *Fusarium graminearum* at the onset of nitrogen starvation. Proteomics *10*, 124–140.

Rand K.N., and Arst, H.N., Jr (1978). Mutations in *nirA* gene of *Aspergillus nidulans* and nitrogen metabolism. Nature *272*, 732–734.

Ravagnani, A., Gorfinkiel, L., Langdon, T., Diallinas, G., Adjadj, E., Demais, S., Gorton, D., Arst, H.N., Jr, and Scazzocchio, C. (1997). Subtle hydrophobic interactions between the seventh residue of the zinc finger loop and the first base of an HGATAR sequence determine promoterspecific recognition by the *Aspergillus nidulans* GATA factor AreA. EMBO J. *16*, 3974–3986.

Reinert, W.R., and Marzluf, G.A. (1975). Genetic and metabolic control of the purine catabolic enzymes of *Neurospora crasse*. Mol. Gen. Genet. *139*, 39–55.

Reuther, J., and Wohlleben, W. (2007). Nitrogen metabolism in *Streptomyces coelicolor*: Transcriptional and post-translational regulation. J. Mol. Microbiol. Biotechnol. *12*, 139–146.

Reyes-Dominguez, Y., Narendja, F., Berger, H., Gallmetzer, A., Fernandez-Martin, R., Garcia, I., Scazzocchio, C., and Strauss, J. (2008). Nucleosome positioning and histone H3 acetylation are independent processes in the *Aspergillus nidulans prnD-prnB* bidirectional promoter. Eukaryot. Cell 7, 656–663.

Rohde, J.R., and Cardenas, M.E. (2004). Nutrient signaling through TOR kinases controls gene expression and cellular differentiation in fungi. Curr. Top. Microbiol. Immunol. *279*, 53–72.

Rohde, J.R., Bastidas, R., Puria, R., and Cardenas, M.E. (2008). Nutritional Control via Tor signaling in *Saccharomyces cerevisiae*. Curr. Opin. Microbiol. *11*, 153–160.

Rubio-Texeira, M. (2007). Urmylation controls Nil1p and Gln3p-dependent expression of nitrogen-catabolite repressed genes in *Saccharomyces cerevisiae*. FEBS Lett. *581*, 541–550.

Roon, R.J., Hampshire, J., and Levenberg, B. (1972). Urea amidolyase. The involvement of biotin in urea cleavage. J. Biol. Chem. *247*, 7539–7545.

Rutherford, J.C., Chua, G., Hughes, T., Cardenas, M.E., and Heitman, J. (2008). A Mep2-dependent transcriptional profile links permease function to gene expression during pseudohyphal growth in *Saccharomyces cerevisiae*. Mol. Biol. Cell. *19*, 3028–3039.

Sanchez, F., Calva, E., Campomanes, M., Blanco, L., Guzman, J., Saborio, J.L., and Palacios, R. (1980). Heterogeneity of glutamine synthetase polypeptides in *Neurospora crassa*. J. Biol. Chem. *255*, 2231–2234.

Scazzocchio, C. (1994). The purine degradation pathway, genetics, biochemistry and regulation. Prog. Ind. Microbiol. *29*, 221–257.

Scazzocchio, C. (2000). The fungal GATA factors. Curr. Opin. Microbiol. 3, 126–131.

Schaap, P.J., Müller, Y., Sonnenberg, A.S., van Griensven, L.J., and Visser, J. (1997). The *Agaricus bisporus pruA* gene encodes a cytosolic delta

1-pyrroline-5-carboxylate dehydrogenase which is expressed in fruit bodies but not in gill tissue. Appl. Environ. Microbiol. *63*, 57–62.

Schinko, T., Berger, H., Lee, W., Gallmetzer, A., Pirker, K., Pachlinger, R., Buchner, I., Reichenauer, T., Güldener, U., and Strauss, J. (2010). Transcriptome analysis of nitrate assimilation in *Aspergillus nidulans* reveals connections to nitric oxide metabolism. Mol. Microbiol. *78*, 720–738.

Schjerling, P., and Holmberg, S. (1996). Comparative amino acid sequence analysis of the C6 zinc cluster family of transcriptional regulators. Nucleic Acids Res. *24*, 4599–4607.

Schönig, B., Brown, D.W., Oeser, B., and Tudzynski, B. (2008). Cross–species hybridization with *Fusarium verticillioides* microarrays reveals new insights into *Fusarium fujikuroi* nitrogen regulation and the role of AreA and NMR. Eukaryot. Cell 7, 1831–1846.

Schönig, B., Vogel, S., and Tudzynski, B. (2009). Cpc1 mediates cross-pathway control independently of Mbf1 in *Fusarium fujikuroi*. Fungal Genet. Biol. *46*, 898–908.

Screen, S., Bailey, A., Charnley, K., Cooper, R., and Clarkson, J. (1998). Isolation of a nitrogen response regulator gene (*nrr1*) from *Metarhizium anisopliae*. Gene *221*, 17–24.

Sealy-Lewis, H.M., Scazzocchio, C., and Lee, S. (1978). A mutation defective in the xanthine alternative pathway of *Aspergillus nidulans*, its use to investigate the specificity of *uaY* mediated induction. Mol. Gen. Genet. *164*, 303–308.

Sharma, K.K., and Arst, H.N., Jr (1985). The product of the regulatory gene of the proline catabolism gene cluster of *Aspergillus nidulans* is a positive-acting protein. Curr. Genet. *9*, 299–304.

Shim, W.B., and Woloshuk, C.P. (1999). Nitrogen repression of fumonisin B1 biosynthesis in *Gibberella fujikuroi*. FEMS Microbiol. Lett. *177*, 109–116.

Sims, A.P., and Ferguson, A.R. (1974). The regulation of glutamine metabolism in *Candida utilis*: studies with $^{15}NH_3$ to measure *in vivo* rates of glutamine synthesis. J. Gen. Microbiol. *80*, 143–158.

Sims, A.P., Toone, J., and Box, V. (1974a). The regulation of glutamine synthesis in the food yeast *Candida utilis*: the purification and subunit structure of glutamine synthetase and aspects of enzyme deactivation. J. Gen. Microbiol. *80*, 485–499.

Sims, A.P., Toone, J., and Box, V. (1974b). The regulation of glutamine metabolism in *Candida utilis*: mechanisms of control of glutamine synthetase. J. Gen. Microbiol. *84*, 149–162.

Sinclair, K., Warner, J.P., and Bonthron, D.T. (1994). The *ASP1* gene of *Saccharomyces cerevisiae*, encoding the intracellular isozyme of L-asparaginase. Gene. *144*, 37–43.

Slot, J.C., and Hibbett, D.S. (2007). Horizontal transfer of a nitrate assimilation gene cluster and ecological transitions in fungi: a phylogenetic study. PLoS One 2, e1097.

Snoeijers, S.S., Pérez-García, A., Joosten, M.H.A.J., and De Wit, P.J.G.M. (2000). The effect of nitrogen on disease development and gene expression in bacterial and fungal plant pathogens. Eur. J. Plant Pathol. *106*, 493–506.

Soanes, D.M., Kershaw, M.J., Cooley, R.N., and Talbot, N.J. (2002). eRgulation of the *MPG1* hydrophobin gene in the rice blast fungus *Magnaporthe grisea*. Mol. Plant Microbe Interact. *15*, 1253–1267.

Sophianopoulou, V., and Scazzocchio, C. (1989). The proline transport protein of *Aspergillus nidulans* is very similar to amino acid transporters of *Saccharomyces cerevisiae*. Mol. Microbiol. *3*, 705–714.

Sorger, G.J., and Giles, N.H. (1965). Genetic control of nitrate reductase in *Neurospora crassa*. Genetics *52*, 777–788.

Soussi-Boudekou, S., and Andre, B. (1999). A co-activator of nitrogen-regulated transcription in *Saccharomyces cerevisiae*. Mol. Microbiol. *31*, 753–762.

Stammers, D.K., Ren, J., Leslie, K., Nichols, C.E., Lamb, H.K., Cocklin, S., Dodds, A., and Hawkins, A.R. (2001). The structure of the negative transcriptional regulator NmrA reveals a structural superfamily which includes the short-chain dehydrogenase/reductases. EMBO J. *20*, 6619–6626.

Stephenson, S.A., Green, J.R., Manners, J.M., and Maclean, D.J. (1997). Cloning and characterisation of glutamine synthetase from *Colletotrichum gloeosporioides* and demonstration of elevated expression during pathogenesis on *Stylosanthes guianensis*. Curr. Genet. *31*, 447–454.

Strauss, J., Muro-Pastor, M.I., and Scazzocchio, C. (1998). The regulator of nitrate assimilation in ascomycetes is a dimer which binds a nonrepeated, asymmetrical sequence. Mol. Cell Biol. *18*, 1339–1348.

Strope, P.K., Nickerson, K.W., Harris, S.D., and Moriyama, E.N. (2011). Molecular evolution of urea amidolyase and urea carboxylase in fungi. BMC Evol. Biol. *11*, 80.

Suarez, T., de Queiroz, M.V., Oestreicher, N., and Scazzocchio, C. (1995). The sequence and binding specificity of UaY, the specific regulator of the purine utilization pathway in *Aspergillus nidulans*, suggest an evolutionary relationship with the PPR1 protein of *Saccharomyces cerevisiae*. EMBO J. *14*, 1453–1467.

Talbot, N., Ebbole, D., and Hamer, J. (1993). Identification and characterization of *MPG1*, a gene involved in pathogenicity from the rice blast fungus *Magnaporthe grisea*. Plant Cell 5, 1575–1590.

Tao, Y., and Marzluf, G.A. (1999). The NIT2 nitrogen regulatory protein of *Neurospora*, expression and stability of *nit-2* mRNA and protein. Curr. Genet. *36*, 153–158.

Tate, J.J., Georis, I., Feller, A., Dubois, E., and Cooper, T.G. (2009). Rapamycin-induced Gln3 dephosphorylation is insufficient for nuclear localization: Sit4 and PP2A phosphatases are regulated and function differently. J. Biol. Chem. *284*, 2522–2534.

Tate, J.J., Georis, I., Feller, A., Dubois, E., and Cooper, T.G. (2010). Distinct phosphatase requirements and gata factor responses to nitrogen catabolite repression and rapamycin treatment in *Saccharomyces cerevisiae*. J. Biol. Chem. *285*, 17880–17895.

Tazebay, U.H., Sophianopoulou, V., Scazzocchio, C., and Diallinas, G. (1997). The gene encoding the major proline transporter of *Aspergillus nidulans* is upregulated during conidiospore germination and in response to proline induction and amino acid starvation. Mol. Microbiol. *24*, 105–117.

Teichert, S., Schönig, B., Richter, S., and Tudzynski, B. (2004). Seletion of the *Gibberella fujikuroi* glutamine synthetase gene has significant impact on transcriptional control of primary and secondary metabolism. Mol. Microbiol. *53*, 1661–1675.

Teichert, S., Wottawa, M., Schönig, B., and Tudzynski, B. (2006). Role of the *Fusarium fujikuroi* TOR kinase in nitrogen regulation and secondary metabolism. Eukaryot. Cell *5*, 1807–1819.

Teichert, S., Rutherford, J.C., Wottawa, M., Heitman, J., and Tudzynski, B. (2008). The impact of the ammonium permeases MepA, MepB, and MepC on nitrogen-regulated secondary metabolism in *Fusarium fujikuroi*. Eukaryot. Cell *7*, 187–201.

Tian, C., Kasiborski, B., Koul, R., Lammers, P.J., Bücking, H., and Shachar-Hill, Y. (2010). Regulation of the nitrogen transfer pathway in the arbuscular mycorrhizal symbiosis: gene characterization and the coordination of expression with nitrogen flux. Plant Physiol. *153*, 1175–1187.

Tian, C., Kasuga, T., Sachs, M.S., and Glass, N.L. (2007). Transcriptional profiling of cross pathway control in *Neurospora crassa* and comparative analysis of the Gcn4 and CPC1 regulons. Eukaryot. Cell *6*, 1018–1029.

Todd, R.B., Fraser, J.A.,Wong, K.H., Davis, M.A., and Hynes, M.J. (2005). Nuclear accumulation of the GATA factor AreA in response to complete nitrogen starvation by regulation of nuclear export. Eukaryot. Cell *4*, 1646–1653.

Tomsett, A.B., and Garrett, R.H. (1980). The isolation and characterization of mutants defective in nitrate assimilation in *Neurospora crassa*. Genetics *95*, 649–660.

Tomsett, A.B., Dunn-Coleman, N.S., and Garrett, R.H. (1981). The regulation of nitrate assimilation in *Neurospora crassa*, the isolation and genetic analysis of *nmr-1* mutants. Mol. Gen. Genet. *182*, 229–233.

Trail, F., Xu, J.R., San Miguel, P., Halgren, R.G., and Kistler, H.C. (2003). Analysis of expressed sequence tags from *Gibberella zeae* (anamorph *Fusarium graminearum*). Fungal Genet. Biol. *38*, 187–197.

Tudzynski, B., Homann, V., Feng, B., and Marzluf, G.A. (1999). Isolation, characterization and disruption of the *areA* nitrogen regulatory gene of *Gibberella fujikuroi*. Mol. Gen. Genet. *261*, 106–114.

Tudzynski, B., Mende, K., Weltring, K.M., Kinghorn, J.R., and Unkles, S.E. (1996). The *Gibberella fujikuroi niaD* gene encoding nitrate reductase, isolation, sequence, homologous transformation and electrophoretic karyotype location. Microbiology *142*, 533–539.

Unkles, S.E., Hawker, K.L., Grieve, C., Campbell, E.I., Montague, P., and Kinghorn, J.R. (1991). *crnA* encodes a nitrate transporter in *Aspergillus nidulans*. Proc. Natl. Acad. Sci. U.S.A.*88*, 204–208.

Unkles, S.E., Zhou, D., Siddiqi, M.Y., Kinghorn, J.R., and Glass, A.D. (2001). Apparent genetic redundancy facilitates ecological plasticity for nitrate transport. EMBO J. *20*, 6246–6255.

Valdez-Taubas, J., Diallinas, G., Scazzocchio, C., and Rosa, A.L. (2000). Protein expression and subcellular localization of the general purine transporter UapC from *Aspergillus nidulans*. Fungal Genet. Biol. *30*, 105–113.

Valdez-Taubas, J., Harispe, L., Scazzocchio, C., Gorfinkiel, L., and Rosa, A.L. (2004). Ammonium-induced internalisation of UapC, the general purine permease from *Aspergillus nidulans*. Fungal Genet. Biol. *41*, 42–51.

Van den Ackerveken, G.F.J.M., Dunn, R.M., Cozijnsen, T.J., Vossen, P., Van den Broek, H.W.J., and De Wit, P.J.G.M. (1994). Nitrogen limitation induces expression of the avirulence gene *avr9* in the tomato pathogen *Cladosporum fulvum*. Mol. Gen. Genet. *243*, 277–285.

Van Nuland A.,P., Vandormael, M., Donaton, M., Alenquer, A., Lourenco, E., Quintino, M., Versele, and Thevelein, J.M. (2006). Ammonium permease-based sensing mechanism for rapid ammonium activation of the protein kinase A pathway in yeast. Mol. Microbiol. *59*, 1485–1505.

Van Rooyen, J.M., Abratt, V.R., Belrhali, H., and Sewell, T. (2011). Crystal structure of type III glutamine synthetase: Surprising reversal of the inter-ring interface. Structure *19*, 471–483.

Verleur, N., Elgersma, Y., van Roermund, C.W., Tabak, H.F., and Wanders, R.J. (1997). Cytosolic aspartate aminotransferase encoded by the *AAT2* gene is targeted to the peroxisomes in oleate-grown *Saccharomyces cerevisiae*. Eur. J. Biochem. *247*, 972–980.

Wagner, D., Schmeinck, A., Mos, M., Morozov, I.Y., Caddick, M.X., and Tudzynski, B. (2010). The bZIP transcription factor MeaB mediates nitrogen metabolite repression at specific loci. Eukaryot. Cell *9*, 1588–1601.

Wang, Y., Li, W., Siddiqi, Y., Symington, V.F., Kinghorn, J.R., Unkles, S.E., and Glass, A.D. (2008). Nitrite transport is mediated by the nitrite-specific high-affinity NitA transporter and by nitrate transporters NrtA, NrtB in *Aspergillus nidulans*. Fungal Genet. Biol. *45*, 94–102.

Wiame, J.M., Grenson, M., and Arst, Jr, H.N. . (1985). Nitrogen catabolite repression in yeasts and filamentous fungi. Adv. Microb. Physiol. *26*, 1–88.

Wiemann, P., Brown, D.W., Kleigrewe, K., Bok, J.W., Keller, N.P., Humpf, H.-U., and Tudzynski, B. (2010). FfVel1 and FfLae*1*, components of a velvet-like complex in *Fusarium fujikuroi*, affect differentiation, secondary metabolism and virulence. Mol. Microbiol. *77*, 972–994.

Wiemann, P., Willmann, A., Straeten, M., Kleigrewe, K., Beyer, M., Humpf, H.-U., and Tudzynski, B. (2009). Biosynthesis of the red pigment bikaverin in *Fusarium fujikuroi*, genes, their function and regulation. Mol. Microbiol. 72, 931–946.

Williams, L.D., Glenn, A.E., Zimeri, A.M., Bacon, C.W., Smith, M.A., and Riley, R.T. (2007). Fumonisin disruption of ceramide biosynthesis in maize roots and the effects on plant development and *Fusarium verticillioides*-induced seedling disease. J. Agric. Food Chem. *55*, 2937–2946.

Wisedchaisri, G., Dranow, D.M., Lie, T.J., Bonanno, J.B., Patskovsky, Y., Ozyurt, S.A., Sauder, J.M., Almo, S.C., Wasserman, S.R., Burley, S.K., *et al.* (2010). Structural underpinnings of nitrogen regulation by the prototypical nitrogen-responsive transcriptional factor NrpR. Structure. *18,* 1512–1521.

Wong, K.H., Hynes, M.J., Todd, R.B., and Davis, M.A. (2007). Transcriptional control of *nmrA* by the bZIP transcription factor MeaB reveals a new level of nitrogen regulation in *Aspergillus nidulans.* Mol. Microbiol. *66,* 534–551.

Wong, K.H., Hynes, M.J., Todd, R.B., and Davis, M.A. (2009). Deletion and overexpression of the *Aspergillus nidulans* GATA factor AreB reveals unexpected pleiotropy. Microbiology *155,* 3868–3880.

Wray, Jr, L.V., and Fisher, S.H. (2008). *Bacillus subtilis* GlnR contains an autoinhibitory C-terminal domain required for the interaction with glutamine synthetase. Mol. Microbiol. *68,* 277–285.

Wray, Jr, L.V., and Fisher, S.H. (2010). Functional roles of the conserved Glu304 loop of *Bacillus subtilis* glutamine synthetase. Mol. Microbiol. *192,* 5018–5025.

Xiao, X., Fu, Y.-H., and Marzluf, G. (1995). The negative-acting NMR regulatory protein of *Neurospora crassa* binds to and inhibits the DNAbinding activity of the positive-acting nitrogen regulatory protein NIT2. Biochemistry *34,* 8861–8868.

Yamada, T., Makimura, K., and Abe, S. (2006). Isolation, characterization, and disruption of *dnr1,* the *areA/nit-2*-like nitrogen regulatory gene of the zoophilic dermatophyte, *Microsporum canis.* Med. Mycol. *44,* 243–252.

Yang, F., Jensen, J.D., Svensson, B., Jørgensen, H.J., Collinge, D.B., and Finnie, C. (2010). Analysis of early events in the interaction between *Fusarium graminearum* and the susceptible barley (*Hordeum vulgare*) cultivar Scarlett. Proteomics *10,* 3748–3755.

Young, J.L., Jarai, G., Fu, Y.-H., and Marzluf, G.A. (1990). Nucleotide sequence and analysis of *nmr,* a negative-acting regulatory gene in the nitrogen circuit of *Neurospora crassa.* Mol. Gen. Genet. *222,* 120–128.

Yuan, G.F., Fu, Y.-H., and Marzluf, G.A. (1991). *nit-4,* a pathway-specific regulatory gene of *Neurospora crassa,* encodes a protein with a putative binuclear zinc DNA-binding domain. Mol. Cell Biol. *11,* 5735–5745.

Zhang, Z., Henderson, C., Perfect, E., Carver, T.L., Thomas, B.J., Skamnioti, P., and Gurr, S.J. (2005). Of genes and genomes, needles and haystacks: *Blumeria graminis* and functionality. Mol. Plant Pathol. *6,* 561–575.

Zhao, X., Hume, S.L., Johnson, C., Thompson, P., Huang, J., Gray, J., Lamb, H.K., and Hawkins, A.R. (2010). The transcription repressor NmrA is subject to proteolysis by three *Aspergillus nidulans* proteases. Protein Sci. *19,* 1405–1419.

Diversity of Polyketide Synthases in *Fusarium*

8

Daren W. Brown and Robert H. Proctor

Abstract

Fusarium can produce a structurally diverse array of secondary metabolites (SMs) with a range of biological activities, including pigmentation, plant growth regulation, and toxicity to humans and other animals. Contamination of grain-based food and feed with toxic SMs produced by *Fusarium* is associated with a variety of diseases in plants and animals and results in loss of millions of dollars in grain commodities each year. Many SMs are formed via the activities of a family of large enzymes called polyketide synthases (PKSs) that consist of between five and eight functional domains. This Chapter reviews the structures and functions of *Fusarium* PKSs present in four species, *Fusarium verticillioides, Fusarium graminearum, Fusarium solani* and *Fusarium oxysporum*. Each genome has between 11 and 16 PKS genes. Re-examination of inferred phylogenetic relationships of deduced amino acid sequences provides insight into how this gene family evolved. The geneaologies suggest that collectively the *Fusarium* PKSs represent 36 distinct sets of PKS homologues, where each set catalyses synthesis of a structurally distinct polyketide. Variation in *Fusarium* PKS genes is due to both ancient and recent gene duplications, gene loss events, gain-of-function due to the acquisition of new domains, and of loss-of-function due to nucleotide mutations. The significant number and variety of evolutionary changes reflects a vast biosynthetic potential this gene family provides fungi and that may help them adapt to changing environmental conditions. Understanding how fungal polyketides are synthesized should lead to better methods to control their production and thereby reduce their negative impact on human endevors.

Introduction

Collectively, fungi produce thousands of structurally diverse natural products (Fig. 8.1) with a wide range of biological activities. Some have beneficial pharmaceutical activities, others are toxic to animals and/or plants, others can alter plant growth, and still others are brightly coloured pigments (Cole *et al.*, 2003). Because most of these natural products are not required for life, they are considered to be secondary metabolites (SMs) (Keller *et al.*, 2005). Some of the earliest research on fungal SMs employed mycelial and spore pigments as taxonomic characteristics to distinguish between fungi. However, the fungal SM with the greatest impact on humanity during the 20th century was the antibiotic penicillin, produced by various *Penicillium* species (Fig. 8.1, insert). The medical and commercial success of this 'miracle drug' resulted in a remarkable increase in SM research on fungi in order to identify SMs with pharamceutical properties. Two of the most important of these are the cholesterol-lowering drug lovastatin and the immunosuppresing drug cyclosporin.

In contrast to the fungal SMs noted above, the most well known *Fusarium* SMs (Fig. 8.1) have a detrimental impact on humanity because of their toxicity and, as a result, are referred to as mycotoxins. Consumption of grain-based food and feed contaminated with the *Fusarium* mycotoxins fumonisins, zearalenone and deoxynivalenol (DON) have been associated with a variety of diseases in humans and other animals

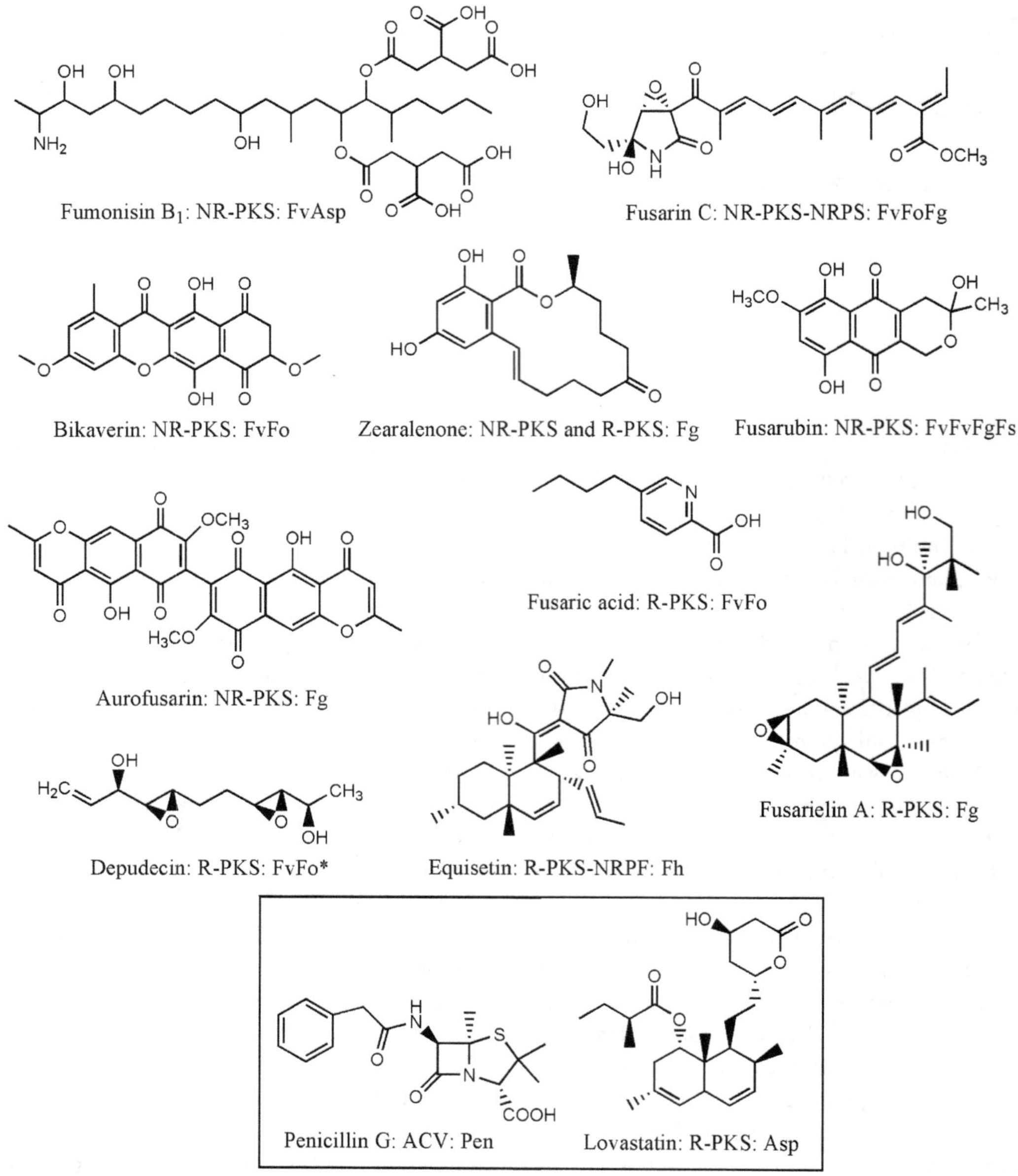

Figure 8.1 Chemical structures of polyketide-derived secondary metabolites produced by *Fusarium*. The NR-PKS and R-PKS abbreviations next to the metabolite names indicate whether a reducing or non-reducing PKS, respectively, synthesize the polyketide precursor of the metabolite. R-PKS-NRPS indicates that the R-PKS includes an NRPS module, while NRPS indicates that an NRPS, rather than a PKS, is required for synthesis of the metabolite. The two metabolites shown in the box inset are produced by fungi other than *Fusarium*. Abbreviations for the *Fusarium* species or other fungi that can produce the metabolites are also indicated next to the metabolite names: Fv=*F. verticillioides*; Fg=*F. graminearum*; Fo=*F. oxysporum*; Fh=*F. heterosporum*; Fl=*F. langsethiae*; Asp=*Aspergillus*; Pen=*Penicillium*.

(Glenn, 2007; Morgavi and Riley, 2007). The economic impact of these mycotoxins on health of humans and livestock as well as on international trade is estimated to be in the 100s of millions of US dollars each year (Glenn, 2007; Morgavi and Riley, 2007; Wu, 2007). Fumonisin-induced diseases are linked to the ability of these mycotoxins to inhibit ceramide synthase, a critical enzyme in sphingolipid metabolism. Fumonisin consumption can result in acute toxicity in some mammals (e.g. leucoencephalomalacia in horses), whereas more chronic exposure can lead to cancer in other mammals (e.g. laboratory rodents) (Howard *et al.*, 2001; Sydenham *et al.*, 1990). In humans, ingestion of fumonisin-contaminated food has been epidemiologically associated with neural tube defects and oesophageal cancer (Marasas *et al.*, 2004; Seefelder *et al.*, 2003; Voss *et al.*, 2007). The multiple harmful health effects (e.g. vomiting and gastrointestinal haemorrhaging) caused by consumption of trichothecene-contaminated grain likely result from the ability of these mycotoxins to inhibit protein synthesis (Alexander *et al.*, 2009). The detrimental health effects caused by consumption of zearalenone-contaminated grain include impaired reproduction and altered sexual development in livestock and result from the estrogenic properties of this mycotoxin (Fink-Gremmels and Malekinejad, 2007; Morgavi and Riley, 2007).

In most cases, the role of SMs in the ecology of fungi that produce them is poorly understood. The energy required for production, maintenance of genetic material required for synthesis, regulation and transport, and structural diversity suggest that SMs provide a selective advantage to the producing organisms. Nevertheless, roles of some SMs have been determined in some organisms and include contributions to virulence on plants and to competition with bacteria, insects or other fungi (see Chapter 9) (Srobarova *et al.*, 2002; Wicklow *et al.*, 2009; Wicklow and Poling, 2009; Xu *et al.*, 2008). Briefly, trichothecene production is critical for the ability of *Fusarium graminearum* to cause the wheat disease head blight, or scab, presumably by inhibiting protein synthesis and thereby interfering with plant defence responses (Proctor *et al.*, 1995). Mutants of *F. graminearum* in which trichothecene production is blocked can infect a single floret (wheat flower) and cause disease but are unable to spread to neighbouring florets in the same head. In *Fusarium fujikuroi*, gibberellin production contributes to the ability of the fungus to cause bakanae disease of rice by inducing hyper-elongation of stems and abnormal development, which result in reduced grain yield (Bomke and Tudzynski, 2009). A number of heretofore unknown *Fusarium verticillioides* SMs have been suggested to reduce disease severity in maize caused by *Ustilago maydis* (Lee *et al.*, 2009) and *A. flavus* (Wicklow *et al.*, 1988).

The majority of fungal SMs characterized to date are derived from a family of enzymes called polyketide synthases (PKSs). Fungal PKSs are large proteins with multiple domains, each of which may function in the sequential or iterative condensation of simple carboxylic acids to form a carbon chain. This Chapter reviews *Fusarium* PKSs and polyketide-derived SMs and examines the phylogenetic relationships of 62 PKS genes present in publicly available genome sequences of four fusaria: *F. graminearum*, *Fusarium oxysporum*, *Fusarium solani* and *F. verticillioides*. Accumulation of knowledge of fungal PKSs has provided insights into how to predict the structures of polyketides synthesized by novel PKSs and insights into the diversity and evolution of this important class of enzymes in a fungal genus that is of significant concern to agricultural production as well as food and feed safety. Because of the potential role of polyketide-derived SMs in plant pathogenesis and because mycotoxin contamination is often most severe in diseased crops, knowledge of *Fusarium* PKSs could lead to the identification of SMs that contribute to the ability of *Fusarium* to cause disease. Such SMs are potential targets for development of novel strategies that limit both crop disease and mycotoxin contamination caused by *Fusarium*.

PKS domain structure and function

The multidomain structure and enzymatic mechanism of fungal PKSs are similar to mammalian fatty acid synthases (FASs). FASs initiate synthesis by catalysing the condensation of an acetyl group (the starter or primer unit) and a malonyl group (the extender unit) to form a four-carbon chain with carbonyl (keto) functions at carbons 1 and 3 (C-1 and C-3) and with the release of carbon dioxide (Wakil, 1989). Three FAS domains are required for the condensation: an acyl carrier protein (ACP) domain to which all acyl intermediates bind; an acyl transferase (AT) domain that transfers the starter unit (acetyl group) from acetyl-CoA to the ACP and the extender unit (malonyl group) from malonyl-CoA to the ACP; and a β-ketoacyl synthase (KS) domain that first receives the acetyl group from the ACP and then catalyses the condensation reaction with the malonyl group bound to ACP. While the condensation product, β-ketoacyl, is bound to the ACP, three other FAS domains catalyse the successive reduction of the C-3 carbonyl (β-keto) to a covalent carbon-carbon single bond: the keto reductase (KR) domain reduces the β-keto to a hydroxyl; the dehydratase (DH) domain reduces the resulting hydroxyl to an enoyl (carbon–carbon double bond); and the enoyl reductase (ER) domain reduces the enoyl to a carbon-carbon single bond. Fatty acid synthesis then continues when the four-carbon acyl derivative, with a keto only on C-1, is then transferred from the ACP to the KS where it undergoes condensation with another ACP-bound malonyl group followed by reduction of the resulting β-keto to a carbon-carbon single bond. FASs involved in constitutive fatty acid synthesis repeat the cycle of condensation and β-keto reduction eight or nine times to form the saturated fatty acids palmitic acid (16-carbon chain) or stearic acid (18-carbon chain). In fungi, synthesis is terminated via hydrolysis by the FAS thioesterase (TE) domain.

Fungal PKSs function in the same manner as FASs in that they include an AT, KS and ACP domain and catalyse the condensation of simple carboxylic acids to form carbon chains. The enormous structural diversity of polyketides, in contrast to fatty acids, arise from a huge degree of choices made by the PKS that are tightly programmed by each enzyme. First and foremost, PKSs often contain additional domains either internal (e.g. methyltransferase domain) or at the amino or carboxyl terminus (e.g. non-ribosomal peptide synthase domains). Second, different fungal PKSs catalyse between two and nine condensations. Third, some PKSs utilize different length starter units and finally, some PKSs regulate the reduction of the β-keto group (Hopwood and Khosla, 1992; Hopwood and Sherman, 1990).

Fungal PKSs can be divided into two groups based on whether they reduce or do not reduce the β-keto of condensation products: reducing (R-PKS) and non-reducing (NR-PKS) PKSs (Fig. 8.2) (Kroken *et al.*, 2003). R-PKSs average between 2,200 to >4,000 amino acids and, like FASs, generally have KR, DH and ER domains that catalyse reduction of each β-keto to form a saturated carbon chain. The saturated polyketide backbone of fumonisins (Fig. 8.1) is synthesized by an R-PKS. Some R-PKSs have a non-functional ER domain and synthesize products with an unsaturated carbon chain. The unsaturated carbon backbone or skeleton of fusarins is synthesized by an R-PKS with a non-functional ER domain (Fig. 8.1). Fungal R-PKSs can also differ from FASs by not having a TE domain and by the presence of

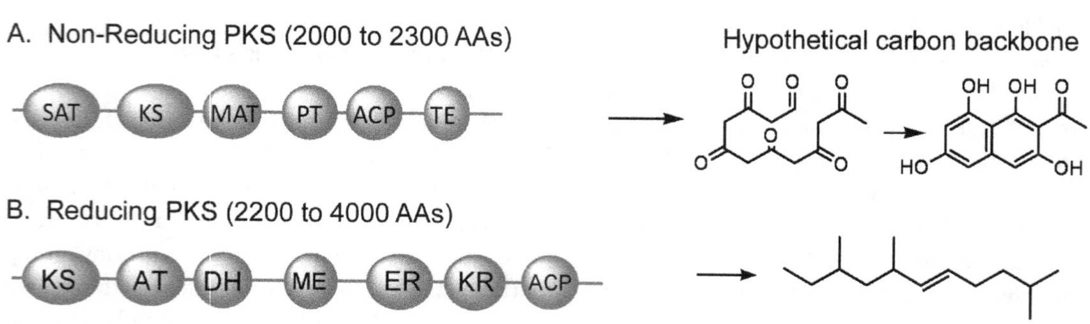

Figure 8.2 Domain organization of typical fungal nonreducing (NR-PKS) and reducing (R-PKS) PKSs. Domain abbreviations are as follows: SAT=starter unit acetyltransferase, KS=keto synthase, MAT=malonlytransferase, PT=product template domain, ACP=acyl carrier protein, TE=thioesterase, AT=acyltransferase, DH=dehydratase, ME=methyltransferase, ER=enoyl reductase, and KR=ketoreductase. Some R-PKSs include all or a portion of non-ribosomal peptide synthase (NRPS) module, while others have a carnitine acyltransferase (CAT) domain at the carboxy terminus. AA values indicate the range in number of amino acids for each group of PKS proteins.

functional domains that are not present in FASs. For example, all fungal R-PKSs characterized to date have an *S*-adenosyl methionine-dependent methyltransferase (ME) domain that catalyses methylation of the polyketide chain at one or more positions (Kroken *et al.*, 2003). The methyl groups present in fumonisins and fusarins are formed via the activity of R-PKS ME domains. Although all known fungal R-PKSs include conserved sequences indicative of a ME domain, some ME domains lack amino acid residues critical for ME activity and as a result are non-functional. Phylogenetic analysis of fungal PKSs indicate that the acquisition of an ME domain occurred only once such that all ME-containing fungal PKSs have evolved from an ancient R-PKS progenitor (Kroken *et al.*, 2003). Some R-PKSs also have a non-ribosomal peptide synthase (NRPS) module that catalyse condensation of the polyketide and an amino acid and release the resulting hybrid polyketide-amine molecule. The NRPS module spans ~1400 AA and consists of four functional domains that include a condensation domain (Con), an adenylation domain (A), an ACP domain and a TE domain. The nitrogen-containing moiety of fusarins (Fig. 8.1) is the result of condensation of a polyketide and an amino acid, possibly homoserine, catalysed by a NRPS module in the fusarin PKS (Song *et al.*, 2004).

In contrast to FASs and R-PKSs, fungal NR-PKSs lack KR, DH and ER domains (Fig. 8.2) and cannot catalyse β-keto reductions. As a result, the carbonyl functions on alternating carbons (hypothetical poly-β-lactone; see Fig. 8.6) synthesized by NR-PKSs, facilitate the formation of polycyclic aromatic compounds that are mediated by the product template domain (PT). PT domains are unique to NR-PKSs and are located adjacent to the ACP domain (Fig. 8.2A) (Crawford *et al.*, 2008; Li *et al.*, 2010). Aurofusarin, bikaverin and fusarubin (Fig. 8.1) are examples of *Fusarium* multi aromatic SMs formed from NR-PKS-derived polyketides. Because of the absence of the KR, DH and ER domains, NR-PKSs are usually smaller than R-PKSs, ranging in size from 2000 to 2300 amino acids. NR-PKSs also have two AT domains that flank the KS domain. The first AT domain, referred to as the starter unit acyl-carrier protein transacylase (SAT) domain, transfers the starter unit, a two-, four- or six-carbon acyl compound, to the ACP domain (Crawford *et al.*, 2006, 2008). The second AT domain, referred to as the malonyl acyltransferase (MAT) domain, transfers the extender unit (malonyl) to the ACP domain after transfer of the starter unit to the KS domain. NR-PKS KS domains then catalyse condensation of the starter and extender units in the same manner as FASs and R-PKSs. Similar to R-PKSs, individual NR-PKSs catalyse a fixed number of condensations until the final poly-β-lactone chain is formed.

Once formed, the polyketide products of fungal R- and NR-PKSs typically undergo structural modifications catalysed by other enzymes. Such enzymes are typically encoded by genes that are located adjacent to or near one another as well as the PKS gene. These groupings of genes required for synthesis of the same SM are collectively referred to as SM biosynthetic gene clusters and can also include (1) one or more genes that encode proteins involved in transport of SMs or their biosynthetic precursors across intracellular membranes or the cell membrane; (2) genes that encode transcription factors that regulate expression of all genes in the same cluster; and (3) genes that encode proteins involved in self protection of fungi from toxic SMs (Keller and Hohn, 1997). A comparative genomic analysis of four *Fusarium* genomes revealed that 67% of PKS genes were located within putative gene clusters with an average of eight genes per cluster (Ma *et al.*, 2010).

Fusarium polyketides synthase genes

Early identification

The first *Fusarium* PKS genes were identified as part of research aimed at elucidating biosynthetic pathways for several polyketide-derived mycotoxins and pigments and at a time when there were no genome sequence data available for the fungi. Identification strategies employed knowledge of the previously characterized SMs to identify PKS genes, one gene at a time. The first *Fusarium* PKS gene (*FUM1*) that was identified and functionally characterized encodes an

R-PKS required for synthesis of fumonisin in *F. verticillioides* (Proctor *et al.*, 1999). Initially, a fragment of *FUM1* was amplified by PCR using (1) degenerate primers designed based on conserved sequences within the KS domain of other PKSs and (2) a cDNA template generated from RNA extracted from a culture of *F. verticillioides* that was producing fumonisins (Proctor *et al.*, 1999). The discovery and characterization of *FUM1* quickly led to the identification of a cluster of 16 co-regulated genes flanking *FUM1*. Most genes in the cluster have since been shown to play a role in fumonisin synthesis (Alexander *et al.*, 2009; Proctor *et al.*, 2003). A differential display/PCR approach with *F. fujikuroi* was used to identify the second *Fusarium* PKS gene (*pks4/BIK1*), which encodes an NR-PKS required for synthesis of the red pigment bikaverin (Linnemannstons *et al.*, 2002). Subsequent work identified a cluster of five genes flanking *BIK1* that are also involved in bikaverin synthesis (Wiemann *et al.*, 2009). The third *Fusarium* PKS gene (*pksN/PKSN*) that was identified encodes an NR-PKS involved in synthesis of an uncharacterized red pigment present in perithecial walls of *F. solani* (Graziani *et al.*, 2004). *PKSN* was identified by complementation analysis of a mutant that produced unpigmented perithecia with a cosmid library. Analysis of sequence flanking *PKSN* identified four genes with predicted functions consistent with SM synthesis; one of these genes is predicted to encode a cytochrome P450 monooxygenase and is required for production of pigmented perithecia (Vasnier *et al.*, 2005). The fourth *Fusarium* PKS gene (*fusA/FUS1*) identified encodes an R-PKS in *F. moniliformis* and *F. venenatum* involved in synthesis of the toxin fusarin C (Song *et al.*, 2004). Degenerate PCR primers designed based on conserved amino acids in the ME domain were used to identify *FUS1* (Song *et al.*, 2004). Analysis of sequence flanking *FUS1* identified a number of putative genes with predicted activities that are consistent with SM synthesis.

Identification in the era of whole-genome sequences

Currently, whole-genome sequence data are publicly available for four *Fusarium* species: the *F. solani* sequence at the Joint Genome Institute (http://genome.jgi-psf.org/) and the *F. graminearum*, *F. oxysporum* and *F. verticillioides* sequences at the Broad Institute (http://www.broadinstitute.org/annotation/genome/*Fusarium*_group/MultiHome.html). The *F. graminearum* genome sequence was the first of these to be released (in 2005) and quickly led to the identification of 15 PKS genes based on homology to previously characterized PKS genes (Cuomo *et al.*, 2007; Gaffoor *et al.*, 2005). Subsequently, each gene was inactivated by integrative gene disruption to determine its function (Gaffoor *et al.*, 2005). This led to elucidation of the role of five of the genes in polyketide/secondary metabolite biosynthesis: *ZEA1* and *ZEA2* (two adjacent PKS genes) are required for zearalenone production; *FUS1* is required for fusarin production; *AUR1* is required for aurofusarin (also rubrofusarin) production; and *PGL1* for production of the violet/black perithecial pigment (Gaffoor *et al.*, 2005). Disruption of each of the other 10 PKS genes did not result in any obvious phenotype, and analysis of extracts did not lead to the identification of the polyketide product. Although the disruption analysis did not demonstrate the functions of these PKS genes, Northern analysis revealed that 9 of the 10 were expressed under a variety of conditions, supporting the hypothesis that they are functional (Gaffoor *et al.*, 2005). A number of labs have subsequently verified the role of *ZEA1*, *ZEA2* and *AUR1* and identified flanking genes involved in the synthesis of the respective SMs. Work by two groups independently verified the role of *ZEA1* and *ZEA2* in zearalenone biosynthesis (Kim *et al.*, 2005b; Lysoe *et al.*, 2006). Kim *et al.* (2005b) also found that two genes flanking *ZEA1* and *ZEA2* are required for zearalenone production: one of the flanking genes is predicted to encode an alcohol oxidase and the other a transcriptional regulator. Using a restriction-enzyme-mediated-integration/mutagenesis approach, Kim *et al.* (2005a) and a third group also demonstrated that *AUR1* as well as several genes flanking it are required for aurofusarin production (Kim *et al.*, 2005a; Malz *et al.*, 2005).

As with *F. graminearum*, release of *F. verticillioides* genomic sequence in 2007 spurred a burst in research on PKSs in this second species. The genomic data, coupled with transcriptional data and gene deletion studies in *F. verticillioides*, quickly led to the characterization of the PKS genes involved in the synthesis of the violet perithecial pigment (Proctor *et al.*, 2007), bikaverin (Brown *et al.*, 2008), and fusarin C (Brown *et al.*, 2012b) in this fungus. That the functions in secondary metabolism of only a single additional *F. graminearum* PKS gene, involved in fusarielin synthesis (Sørensen *et al.*, 2011) and one other *F. verticillioides* PKS gene, involved in fusaric acid synthesis (Brown *et al.*, 2012b) have been determined during the intervening years reflects the inherent challenges in determining the functions of PKS genes in polyketide biosynthesis. Although transcriptional evidence (via Northern and microarray data) suggest that most PKS genes are expressed and, therefore, may be functional (Brown *et al.*, 2012b; Gaffoor *et al.*, 2005), linking a PKS gene to synthesis of a polyketide (i.e. demonstrating that a PKS gene is responsible for synthesis of a specific polyketide or polyketide-derived SM) has proven exceedingly difficult due to a variety of reasons, including low levels of production, failure to extract, or inadequate analytical methods.

Analyses of the *F. verticillioides*, *F. graminearum*, *F. oxysporum* (strain FOL 4287), and *F. solani* genomes, identified 58 distinct PKS gene homologues (Brown *et al.*, 2012a; Ma *et al.*, 2010). The genes were identified by BLAST (Altschul *et al.*, 1990) using the predicted KS domain of ~420 amino acids (AA) from an archetypal fungal NR-PKS, (i.e. the *Aspergillus nidulans* WA PKS) and an R-PKS, (i.e. the *F. verticillioides* Fum1 PKS). In November 2011, genome sequences of 10 additional strains of *F. oxysporum* were released by the Broad Institute. In addition, a genome sequence for a 12th *F. oxysporum* strain (Fo5176), also sequenced by the Broad Institute, is now available in the National Centre for Biotechnology Information (NCBI) database. BLAST analysis of these eleven *F. oxysporum* genomes and comparison to the PKSs identified in them to those identified in the original *F. oxysporum* (strain FOL 4287) genome as well as the genomes of the other *Fusarium* species led to the identification of four novel PKS genes, for a total of 18 *F. oxysporum* PKS genes and 62 *Fusarium* PKS genes (Tables 8.1 and 8.2) (Brown, unpublished).

A goal of comparing all PKS genes within a single genome to those in other genomes is to identify PKSs that are unique to a single species as well as PKSs with homologues shared by distantly related species. Conservation of a PKS gene in the four publicly available *Fusarium* genomes (*F. graminearum, F. oxysporum, F. solani* and *F. verticillioides*) suggests that the corresponding polyketide metabolite provides an ecological advantage to a wide range of fusaria. In contrast, conservation of a PKS gene in only two of the *Fusarium* species suggests that the corresponding polyketide metabolite may provide an advantage to a more limited range of fusaria (e.g. pathogenicity on certain host plants). As noted above, only three PKSs, are conserved in *F. graminearum, F. oxysporum, F. solani* and *F. verticillioides* (Brown *et al.*, 2012a). These PKSs are the NR-PKS responsible for fusarubin/violet perithecium pigment, an R-PKS in Clade I and another R-PKS in Clade III. Southern blot analysis, using *F. verticillioides*

Table 8.1 Variation in number of different kinds of *Fusarium* PKSs

Species	Total PKSs	NR-PKSs	R-PKSs	R-PKS-NRPS	R-PKS-CAT
F. verticillioides	16	2	11	3	-
F. graminearum	15	5	8	1	1
F. solani	13	2	8	1	2
FOSC[1]	17[2]	2	10	4	1
	61[2]				

[1]*F. oxysporum* species complex. See Table 2
[2]*F. oxysporum* strain Fo4287 contains two PKS genes that share over 97% nucleotide sequence identity and, although considered homologues, both were included in the phylogenetic analysis.

Table 8.2 Distribution of *Fusarium oxysporum* PKSs in 12 species

F. oxysporum strains	Host	15296[1]	05816	15248	03051	03945	11954	14587	02741	10805	04757	02884	01189	14850/ 15886	SC12[3]	SC4[3]	03069[3]	SC261[3]	Total (17)
4287	*Lycopersicum*	x	x	x	x	x	x	x	x	x	x	x	x	xx					14
MN25	*Lycopersicum*	x	x	x	x	x	x	x	x	x	x		x		x				11
CL57	*Lycopersicum*	x	x	x	x	x	x	x	x	x	x					x			11
FOSC 3-a	Homo	x	x	x	x	x	x	x	x	x	x				x	x			13
Fo47	Soil	x	x	x	x	x	x	x	x	x	x		x		x				12
II5	*Musa*	x	x	x	x	x	x	x	x	x	x	x			x	x			13
HDV247	*Pisum*	x	x	x	x	x	x	x	x	x	x					x	x	x	13
Cotton	*Gossypium*	x	x	x	x	x	x	x	x	x	x	x							11
Melonis	*Cucurbita*	x	x	x	x	x	x	x	x	x	x		x	x	x	x			14
PHW808	*Brassica/ Arabidopsis*	x	x	x	x	x	x	x	x	x		x		x		x	x		13
PHW815	*Brassica/ Arabidopsis*	x	x	x	x	x	x	x	x		x					x		x	11
5176	*Arabidopsis*	x	x	x	x	x	x	x	x	x		x		x		x	x		13
Summary		All	All	All	All	All	All	All	All	11	10	5	4	4	5	8	3	2	
Average																			12.3
Fvert homologues[2]	Maize	12610[2]	03695	12523	01914	11932	10535				03379	01736	10497	13715					

[1]Numbers are gene model numbers predicted for strain Fo4287.
[2]*Fvert* homologues refers to likely homologues in *F. verticillioides* strain M-3125. The numbers refers to specific gene models (FVEG_#####) for each PKS.
3SC12 refers to a putative PKS located on supercontig 12 in the genomic sequence of strain FO47. SC4 refers to a putative PKS located on supercontig 4 in strain FOSC 3-a. 03069 refers to a specific gene model (FOXB_03069) in strain 5187. SC261 refers to a putative PKS on supercontig 261 in strain PHW815.

homologues (FVEG_03695, FVEG_01914 and FVEG_10535) of the PKS genes as hybridization probes, revealed that these three genes are also present in 39 other *Fusarium* species (Proctor, unpublished data). The functions of the two common R-PKSs are still not known, because inactivation of the corresponding genes in *F. graminearum* and *F. verticillioides* did not cause a noticeable change in production of known SMs nor any gross morphological change in either species (Gaffoor *et al.*, 2005; Proctor and Busman, unpublished results).

Phylogenetic relationships of *Fusarium* PKSs

Fig. 8.3 shows the inferred phylogenetic relationships of the 62 PKS homologues identified in publicly available genome sequences of four *Fusarium* species. The phylogenetic tree was based on an alignment of deduced AA sequences of the KS and AT/MAT domains using previously described methodology (Brown *et al.*, 2012a). In addition to the *Fusarium* PKSs derived from genome sequence databases, the alignment included AA sequences for *eqiS*, the equisetin PKS from *Fusarium heterosporum* (Sims *et al.*, 2005), 28 other ascomycetous PKSs, each linked to synthesis of a specific SM, and two other ascomycetous PKSs and five bacterial PKSs that have not yet been linked to synthesis of specific SMs (Table 8.3). Consistent with previous analyses (Brown *et al.*, 2012a; Ma *et al.*, 2010), all but one of the *Fusarium* PKSs were resolved into two clades comprised of NR-PKSs (11) and R-PKSs (50) (Fig. 8.3). The NR-PKSs form a single glade with 100% bootstrap support and include 10 PKSs of known function from a wide variety of very distantly related ascomycetes, including a Dothideomycete and a Eurotiomycete in addition to numerous Sordariomycetes (including *Fusarium*). The R-PKSs also form a single clade (with 84% bootstrap support) that includes three well-supported sister clades with 65%, 99% and 100% bootstrap support. This R-PKS clade also includes 17 PKSs of known functions from distantly related ascomycetes. The high bootstrap values for the NR-PKS clade and R-PKS clade support the hypothesis that each clade is monophyletic (i.e. evolved from a single common ancestor), which is also consistent with previous phylogenetic analysis of PKSs from multiple fungal genera (Baker *et al.*, 2006; Kroken *et al.*, 2003).

The PKS gene homologue, Fo47SC12, occurs in 5 of the 12 *F. oxysporum* genome sequences and is unique among *Fusarium* PKS genes because of its domain organization (KS-AT-DH-ER-KR-ACP) and because in phylogenetic analysis it was not resolved within the R-PKS or NR-PKS clades (Fig. 8.3). The domain organization of Fo47SC12 is superficially similar to numerous R-PKSs present in the three R-PKS sister clades. However, a critical difference is the complete absence of an ME domain in Fo47SC12. All other *Fusarium* R-PKSs that have been described have a region of AA sequence corresponding to a non-functional (ΨME) or functional ME domain located between the DH and ER domains (Brown *et al.*, 2012a; Kroken *et al.*, 2003). One PKS in *Trichophyton tonsurans*, another in *Arthroderma gypseum* and five bacterial PKSs have the same domain content and order as Fo47SC12 and were resolved in the same moderately supported clade (78% bootstrap value) as Fo47SC12 (Fig. 8.3). The presence of ME/ΨME domains in the R-PKS clade (+ME R-PKS lineage) consisting of Clades I, II and III (Fig. 8.3) and the complete absence of ME domains in the R-PKS clade (–ME R-PKS lineage) that includes Fo47SC12 suggests two alternative hypotheses for the origin of the two R-PKS lineages. According to the first hypothesis, the common ancestor of the two lineages had an ME domain, and the –ME R-PKS lineage lost the domain and +ME R-PKS lineage retained the domain as each diverged from the ancestor. According to the second hypothesis, the common ancestor lacked an ME domain and the +ME R-PKS lineage acquired the domain as it diverged from the common ancestor. In addition, because multidomain PKSs are rare in bacteria, the presence of sequences from multidomain bacterial PKSs in the –ME R-PKS lineage indicates that the bacteria could have obtained the PKSs from a fungus via horizontal gene transfer.

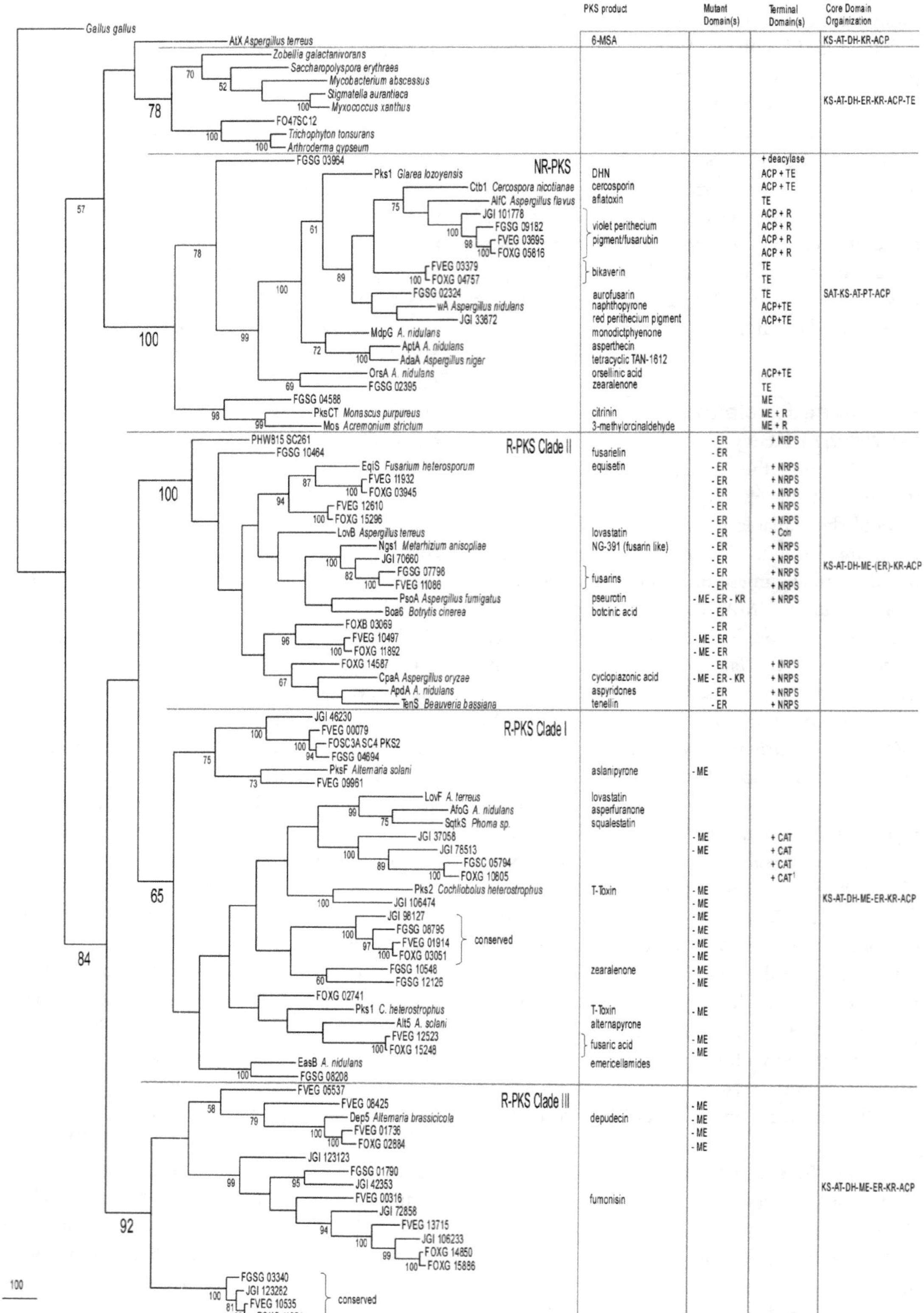
PKS product
Mutant Domain(s)
Terminal Domain(s)
Core Domain Orgainization
Gallus gallus
AtX Aspergillus terreus
6-MSA
KS-AT-DH-KR-ACP
Zobellia galactanivorans
Saccharopolyspora erythraea
Mycobacterium abscessus
Stigmatella aurantiaca
Myxococcus xanthus
FO47SC12
Trichophyton tonsurans
Arthroderma gypseum
KS-AT-DH-ER-KR-ACP-TE
NR-PKS
FGSG 03964
Pks1 Glarea lozoyensis
Ctb1 Cercospora nicotianae
AflC Aspergillus flavus
JGI 101778
FGSG 09182
FVEG 03695
FOXG 05816
FVEG 03379
FOXG 04757
FGSG 02324
wA Aspergillus nidulans
JGI 33672
MdpG A. nidulans
AptA A. nidulans
AdaA Aspergillus niger
OrsA A. nidulans
FGSG 02395
FGSG 04588
PksCT Monascus purpureus
Mos Acremonium strictum
DHN
cercosporin
aflatoxin
violet perithecium pigment/fusarubin
bikaverin
aurofusarin
naphthopyrone
red perithecium pigment
monodictphyenone
asperthecin
tetracyclic TAN-1612
orsellinic acid
zearalenone
citrinin
3-methylorcinaldehyde
+ deacylase
ACP + TE
ACP + TE
TE
ACP + R
ACP + R
ACP + R
ACP + R
TE
TE
TE
ACP+TE
ACP+TE
ACP+TE
TE
ME
ME + R
ME + R
SAT-KS-AT-PT-ACP
R-PKS Clade II
PHW815 SC261
FGSG 10464
EqiS Fusarium heterosporum
FVEG 11932
FOXG 03945
FVEG 12610
FOXG 15296
LovB Aspergillus terreus
Ngs1 Metarhizium anisopliae
JGI 70660
FGSG 07798
FVEG 11086
PsoA Aspergillus fumigatus
Boa6 Botrytis cinerea
FOXB 03069
FVEG 10497
FOXG 11892
FOXG 14587
CpaA Aspergillus oryzae
ApdA A. nidulans
TenS Beauveria bassiana
fusarielin
equisetin
lovastatin
NG-391 (fusarin like)
fusarins
pseurotin
botcinic acid
cyclopiazonic acid
aspyridones
tenellin
- ER
- ME - ER - KR
- ME - ER
+ NRPS
+ Con
KS-AT-DH-ME-(ER)-KR-ACP
R-PKS Clade I
JGI 46230
FVEG 00079
FOSC3A SC4 PKS2
FGSG 04694
PksF Alternaria solani
FVEG 09961
LovF A. terreus
AfoG A. nidulans
SqtkS Phoma sp.
JGI 37058
JGI 78513
FGSC 05794
FOXG 10605
Pks2 Cochliobolus heterostrophus
JGI 106474
JGI 98127
FGSG 08795
FVEG 01914
FOXG 03051
conserved
FGSG 10548
FGSG 12126
FOXG 02741
Pks1 C. heterostrophus
Alt5 A. solani
FVEG 12523
FOXG 15248
EasB A. nidulans
FGSG 08208
aslanipyrone
lovastatin
asperfuranone
squalestatin
T-Toxin
zearalenone
T-Toxin
alternapyrone
fusaric acid
emericellamides
- ME
+ CAT
+ CAT[1]
KS-AT-DH-ME-ER-KR-ACP
R-PKS Clade III
FVEG 05537
FVEG 08425
Dep5 Alternaria brassicicola
FVEG 01736
FOXG 02884
JGI 123123
FGSG 01790
JGI 42353
FVEG 00316
JGI 72858
FVEG 13715
JGI 106233
FOXG 14850
FOXG 15886
FGSG 03340
JGI 123262
FVEG 10535
FOXG 11954
conserved
depudecin
fumonisin
KS-AT-DH-ME-ER-KR-ACP
100

Variation in domain organization in *Fusarium* PKSs

NR-PKS domains

The organization of the SAT, KS, MAT, PT and ACP domains is perfectly conserved in fungal NR-PKSs (Fig. 8.3 and Fig. 8.4A). The most significant differences in domain structure between NR-PKSs occur at or near the carboxyl terminus (C-terminus) of the proteins (Brown *et al.*, 2012a). For example, nine NR-PKSs contain two ACP domains, a feature first observed in the *wA* PKS, which is required for synthesis of spore pigments in *A. nidulans* (Fujii *et al.*, 2001). Site directed mutagenesis of either ACP domain did not affect pigment production, indicating that each ACP is functional and can operate independently of the other (Fujii *et al.*, 2001). The relative widespread occurrence of two ACP domains per protein across the NR-PKS clade indicate that either the ancestral NR-PKS had only one ACP domain and multiple lineages within the clade independently acquired a second ACP domain or the ancestral NR-PKS had two ACP domains and one domain was lost independently by multiple descendent lineages. It is possible that this variation in number of ACP domains reflects the gain and loss of the TE or reductase (R) domains, which are adjacent to the ACP domains and are critical for release of the nascent polyketide from the enzyme. Some NR-PKSs have a TE and others have an R domain at the C-terminus, and both occur in multiple lineages within the NR-PKS clade. It is possible that during their evolution, NR-PKSs exchanged TE and R domains and that an ACP domain was sometimes exchanged along the TE or R domain such that the recipient NR-PKS acquired a second ACP domain as well. The apparent redundancy of the two ACP domains in the *wA* PKS is consistent with this possibility.

As noted above, NR-PKSs can also vary at the C-terminus by whether they have a TE or R domain. In most functionally characterized NR-PKSs, the C-terminal domain is involved in determining chain length, final polyketide cyclization, and release of the polyketide from the PKS (Bailey *et al.*, 2007; Du and Lou, 2010; Fujii *et al.*, 2001; Watanabe and Ebizuka, 2004). Of the 18 known *Fusarium* NR-PKSs, ten have a TE domain, and six have an R domain at the C-terminus. TE domains are the most common C-terminal domain in fungal NR-PKSs, and the mechanism by which they release polyketides from the PKS has been extensively studied (Du and Lou, 2010). TE domains belong to the α/β-hydrolase superfamily of proteins and include a conserved Ser-His-Asp motif within their catalytic region. TE domains can catalyse release of a linear polyketide product with a terminal carboxylic acid or a cyclized product (Du and Lou, 2010) depending on whether water or a hydroxyl group along the acyl chain acts as the nucleophile during release. In contrast, R domains most commonly catalyse the release of linear polyketides, via reduction of the acyl thioester and formation of a terminal alcohol or aldehyde. However, some R domains facilitate formation of an aromatic ring via a non-redox condensation reaction (Du and Lou, 2010). Thus, the presence of either a TE or R terminal domain can provide clues about the structure of the polyketide synthesized by an NR-PKS.

The *Fusarium* NR-PKSs FGSG_04588 and FGSG_03964 are unusual in that the former has a methyltransferase and the latter a deacylase at the C-terminus of the protein (Fig. 8.3). To our knowledge, no *Fusarium* NR-PKS with either of these C-terminal domains has been described in

Figure 8.3 Genealogy of the *Fusarium* PKSs inferred from maximum parsimony analysis of concatenated AA sequences of KS and AT domains. Major clades are delineated with horizontal lines. The protein/gene designations FGSG_##### (*F. graminearum*), FVEG_##### (*F. verticillioides*) and FOXG_##### (*F. oxysporum*) correspond to gene model designations in the *Fusarium* Comparative Database at the Broad Institute. The protein/gene designations JGI_##### correspond to gene model designations in the *F. solani* (*Nectria haematococca*) genome sequence database at the Joint Genome Institute. Statistical support for branches within the phylogenetic trees were generated by bootstrap analysis with 1000 pseudoreplicates. Bootstrap values are indicated below branches. [1]A majority of the CAT domain coding sequence is missing from the FOXG_10805 gene model. See text for additional explanation. The chicken (*Gallus gallus*) FAS served as an outgroup in the phylogenetic analysis.

Table 8.3 PKSs from sources other than *Fusarium* genome sequences

PKS designation and/or organism	Metabolite	Accession	Classification (Kingdom; Phylum/Class)
R-PKS			
AtX *Aspergillus terreus*	6-MSA (6-methylsalicylic acid)	BAA20102	Fungi; Sordariomycetes
Zobellia galactanivorans	Unknown	YP_004734849	Bacteria; Bacteroidetes
Saccharopolyspora erythraea	Unknown	YP_001104811	Bacteria; Actinobacteria
Mycobacterium abscessus	Unknown	EHB99830	Bacteria; Actinobacteria
Stigmatella aurantiaca	Unknown	YP_003950773	Bacteria; Proteobacteria
Myxococcus xanthus	Unknown	YP_632255	Bacteria; Proteobacteria
Trichophyton tonsurans	Unknown	EGD98024	Fungi; Eurotiomycetes
Arthroderma gypseum	Unknown	XP_003175627	Fungi; Eurotiomycetes
NR-PKS			
Pks1 *Glarea lozoyensis*	DHN; dihydroxynaphthalene	AAN59953	Fungi; Sordariomycetes
Ctb1 *Cercospora nicotianae*	Cercosporin	AAT69682	Fungi; Dothideomycete
AflC *Aspergillus flavus*	Aflatoxin	EQ963478	Fungi; Sordariomycetes
wA *Aspergillus nidulans*	YWA1 (yellow spore pigment)	Q03149	Fungi; Sordariomycetes
MdpG *Aspergillus nidulans*	Monodictphyenone	XM_652662	Fungi; Sordariomycetes
AptA *Aspergillus nidulans*	Asperthecin	XM_658512	Fungi; Sordariomycetes
AdaA *Aspergillus niger*	Linear tetracyclic TAN-1612	AEN83889	Fungi; Sordariomycetes
OrsA *Aspergillus nidulans*	Orsellinic acid	XM_676086	Fungi; Sordariomycetes
PksCT *Monascus purpureus*	Citrinin	AAY33862	Fungi; Eurotiomycetes
Mos *Acremonium strictum*	3-Methylorcinaldehyde	CAN87161	Fungi; Sordariomycetes
R-PKS			
EqiS *Fusarium heterosporum*	Equisetin	AAV66106	Fungi; Sordariomycetes
LovB *Aspergillus terreus*	Lovastatin	AAD34559	Fungi; Sordariomycetes
Ngs1 *Metarhizium anisopliae*	NG-391 (fusarin like)	EFY96026	Fungi; Sordariomycetes
PsoA *Aspergillus fumigatus*	Pseurotin	ABS87601	Fungi; Sordariomycetes
Boa6 *Botrytis cinerea*	Botcinic acid	CAP58786	Fungi; Sordariomycetes
CpaA *Aspergillus oryzae*	Cyclopiazonic acid	BAG82673	Fungi; Sordariomycetes
ApdA *Aspergillus nidulans*	Aspyridones	XM_676589	Fungi; Sordariomycetes
TenS *Beauveria bassiana*	Tenellin	CAL69597	Fungi; Sordariomycetes
PksF *Alternaria solani*	Aslanipyrone	BAE80697	Fungi; Dothideomycete
LovF *Aspergillus terreus*	Lovastatin	AAD34559	Fungi; Sordariomycetes
AfoG *Aspergillus nidulans*	Asperfuranone	XM_653548	Fungi; Sordariomycetes
SqtkS *Phoma* sp. C2932	Squalestatin	AY217789	Fungi; Dothideomycete
Pks2 *Cochliobolus heterostrophus*	T-toxin	ABB76806	Fungi; Dothideomycete
Pks1 *Cochliobolus heterostrophus*	T-toxin	AAB08104	Fungi; Dothideomycete
Alt5 *Alternaria solani*	Alternapyrone	BAD83684	Fungi; Dothideomycete
EasB *Aspergillus nidulans*	Emericellamides	XM_650559	Fungi; Sordariomycetes
Dep5 *Alternaria brassicicola*	Depudecin	ACZ57548	Fungi; Dothideomycete

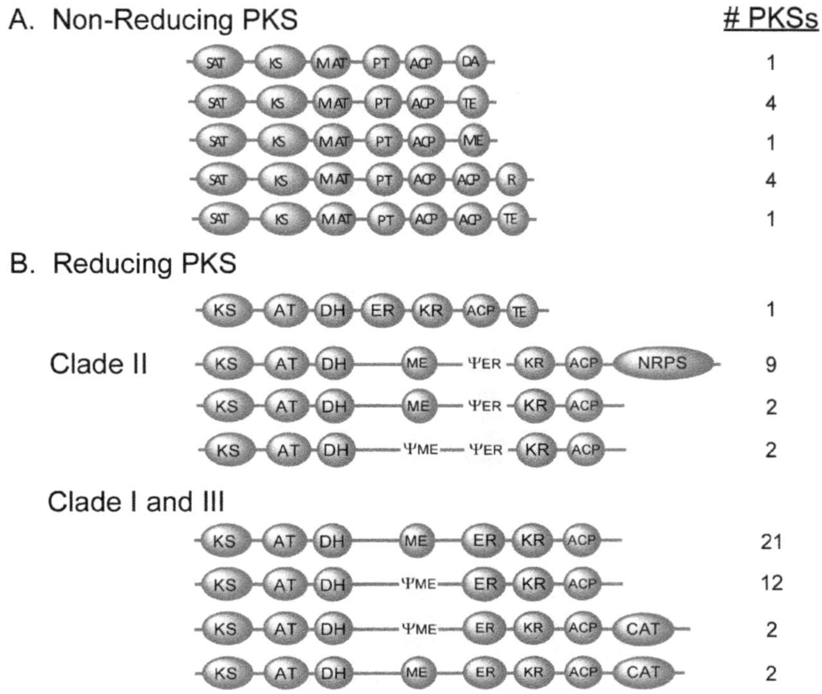

Figure 8.4 Domain organization in *Fusarium* PKSs. Values in the # PKSs column indicate the number of *Fusarium* PKSs that have the domain structure depicted. Clade I, II and III were resolved in the phylogenetic analysis presented in Figure 8.3. Domain abbreviations: SAT=starter unit acyl transferase, KS=keto synthase, MAT=malonyl acyl transferase, PT=product template, ACP=acyl carrier protein, DA=deacylase, TE=thioesterase, ME=methyltransferase, DH=dehydratase, ER=enoyl reductase, KR=keto reductase, NRPS=non-ribosomal peptide synthetase module, and CAT=carnitine acyl transferase. Ψ indicates a non-functional domain. A colour version of this figure is located in the plate section at the back of the book.

the literature. It is possible that enzymes encoded by other genes are responsible for polyketide release from these two PKSs. A similar scenario has been found for three NR-PKSs (MpdG, AptA and AdaA) from *Aspergillus* that have C-terminal ACP domains (Fig. 8.3). The release of the polyketides from MpdG, AptA and AdaA is dependent on a hydrolase encoded by a gene located near the corresponding PKS gene (Li *et al.*, 2011; Sanchez *et al.*, 2011; Szewczyk *et al.*, 2008).

R-PKS domains

In contrast to the NR-PKS clade where all domains appeared to be functional, multiple R-PKSs appear to have non-functional ME, ER and/or KR domains (ΨME, ΨER and ΨKR respectively) (Fig. 8.4B). These domains were noted initially because they appeared as regions with reduced AA sequence homology to corresponding regions in other PKSs but were later determined to be non-functional domains because they were missing AA residues critical for function (Baker *et al.*, 2006; Kroken *et al.*, 2003). For example, alignment of ~120 AA of ΨME and ME domains identified gaps and differences within one, two, three or all four of the AA motifs critical for ME function (Ansari *et al.*, 2008; Brown *et al.*, 2012a). Likewise, alignment of ~230 AA of ΨER and ER domains identified missing AAs required for binding reduced nicotinamide adenine dinucleotide phosphate (NADPH) (Brown *et al.*, 2012a; Song *et al.*, 2004).

The evolutionary histories of ΨME and ΨER domains appear to differ significantly. PKSs containing a ΨME domain were resolved into all three R-PKS clades (i.e. R-PKS Clades I, II and III), and all three clades also include PKSs with functional ME domains. This, and the fact that patterns of AA insertions and deletions (indels) differ in different ΨME domains, suggest that non-functionalization of the ME domain occurred independently in multiple lineages within the R-PKS clade. In contrast, all R-PKSs with ΨER domains were resolved in the same clade, R-PKS Clade II. In fact, all R-PKSs in this clade have a ΨER, suggesting non-functionalization of the ER domain could have occurred only once in a common ancestor of R-PKS Clade II. Distribution of PKSs with ΨME domains further exemplifies the rapid evolution of the PKS gene family overall. For example, the presence of the ΨME-containing R-PKS gene JGI_106474 (R-PKS Clade I) in the *F. solani* genome and its absence in the genomes of the other three *Fusarium* species begs the question: did *F. solani* acquire this gene by horizontal gene transfer or was the gene present in the ancestral *Fusarium* and subsequently lost in the other species after they diverged from *F. solani*? In contrast, a well resolved (100% bootstrap value) clade of four R-PKS orthologues (JGI_98127, FGSG_08795, FVEG_01914 and FOXG_03051 in Clade I) with a ΨME domain is conserved across all four *Fusarium* species. The pattern of AA indels in the orthologues (Brown *et al.*, 2012a) is consistent with the relative phylogenetic relationships, where *F. solani* is most diverged and *F. verticillioides* is most similar to *F. oxysporum* (Ma *et al.*, 2010).

The acquisition and loss of C-terminal domains also appears to have occurred on more than one occasion in the R-PKSs. The presence of the C-terminal NRPS module in R-PKS Clade II and its absence in Clade I and III (Fig. 8.3) suggests that the module was acquired once by a common ancestor of Clade II. However, the complete absence of the NRPS module in five Clade II PKSs and the position of these PKSs within the clade (Fig. 8.3) suggests that loss of the NRPS module has occurred multiple times: once in the branch leading to FGSG_10646, once in the branch leading to the *Botrytis cinerea boa6*, and once in the branch leading to FVEG_10479, FOXG_01189 and FOXB_03069.

The NRPS module is a collection of four domains, spanning 1400 AAs, that are responsible for condensing (forming a covalent bond) the nascent polyketide to an enzyme-bound AA and transfer of the final enzyme-bound product (via a phosphopantetheine prosthetic group) to a domain involved in the release of the polyketide-AA product. In addition to complete loss of the NRPS module, a partial deletion module resulting from the loss of all NRPS-module domains except for the Con appears to have occurred once in the branch leading to *lovB* in *A. terreus* (Fig. 8.3).

The other C-terminal domain present in *Fusarium* PKSs shares significant similarity to carnitine acyltransferases (CAT). In *Fusarium*, there are four PKSs with CAT domains, two in *F. solani* (JGI_78513 and JGI_37058), one in *F. graminearum* (FGSG_05794) (Gaffoor *et al.*, 2005), and one in *F. oxysporum*. In the latter case, putative homologues were identified in 11 of the 12 *F. oxysporum* genome sequences (Table 8.2). Although FOXG_10805 is missing most of the CAT domain, the lack of CAT-encoding sequence is likely an artefact due to a gap in the genome sequence data, represented by 2276 Ns in the Broad database, rather than a deletion of the CAT coding sequence in FOXG_10805. To our knowledge, no polyketide synthesized by a PKS with a CAT domain has been described. However, given the function of the CAT enzyme that are not associated with PKSs, it is easy to rationalize how a CAT domain could function in polyketide biosynthesis. CAT enzymes that are not part of PKSs are involved in transport of acyl compounds (e.g. acetate, fatty acids) across intracellular membranes. The enzymes catalyse covalent attachment of carnitine (L-3-hydroxy-4-aminobutyrobetaine) to the carboxylic acid portion of acyl compounds and thereby generate acyl-carnitine (van Roermund *et al.*, 1999). A specific membrane- transporter then moves the acyl-carnitine across an intracellular membrane. A CAT present on the other side of the membrane, within the cytosol or an organelle, then cleaves the acyl-carnitine to form acyl-CoA and carnitine, which can be recycled (Roze *et al.*, 2011). Based on homology to known fungal CATs, *Fusarium* genomes have two CAT enzyme-encoding genes (Brown *et al.*, 2012a). *CAT1* is likely involved in shuttling acetyl-CoA across intracellular membranes and is required for acetate utilization; while *CAT2* is likely involved in shuttling longer fatty acyl-CoAs and is required for fatty acid utilization (Roze *et al.*, 2011).

Formation of putative hybrid polyketide-carnitine molecules by PKSs with CAT domains could catalyse production of molecules with unique structures and activities that provide an ecological advantage unrelated to the acyl-carnitine transport system. Alternatively, such hybrids could take advantage of the acyl-carnitine transport system to transport polyketides across intracellular or cellular membranes. In one scenario, hybrid polyketide-carnitine molecules would use the acyl-carnitine transport system as a self-protection mechanism to compartmentalize toxic polyketides into organelles or expel them from the cell. In another scenario, the hybrid molecules would exploit the transport system of other organisms to introduce toxic polyketide into host plants or competing microorganisms. The function of the CAT domain is intriguing and work is in progress to over express one of the *Fusarium* PKS-CAT coding sequences in a heterologous host in order to determine the structure of the polyketides synthesized by this class of PKSs as well as to determine the role of the hybrid molecules in the ecology of fusaria that produce them.

Fusarium oxysporum PKSs: insights from multiple genomes of one species

Fusarium oxysporum exhibits a high degree of genetic variability and is often considered a complex of multiple closely related species (i.e. species complex). Collectively, isolates of *F. oxysporum* have a very broad host range, causing disease on over 120 different plant species; however, individual isolates typically have very narrow host ranges (Michielse and Rep, 2009). Thus, according to the Broad Institute database, some of the 12 *F. oxysporum* genome sequences available are from isolates with different host ranges: three strains are pathogens of tomato; three are pathogens of *Arabidopsis*; and one strain each is a pathogen of banana, cotton, cucurbits and pea. In addition, one strain was recovered from a human and another from soil. Variation in host range of *F. oxysporum* isolates has been attributed, at least in part, to the presence of lineage-specific DNA, i.e. chromosomes or chromosomal segments that are unique to *F. oxysporum* or even to certain isolates of the fungus (See Chapter 3) (Ma *et al.*, 2010). Indeed, lineage-specific chromosomes can be transferred between strains and expand the host range of the recipient strain. The tremendous genetic and pathogenic variability of *F. oxysporum* is not reflected in the diversity of PKS genes. Analysis of the 12 *F. oxysporum* genomes indicates that, collectively, they have 17 functionally homologous PKS genes, i.e. homologues that are ≥95% identical and are likely responsible for synthesis of the same polyketide (Brown, unpublished). The minimum number of PKS genes per *F. oxysporum* genome is 11 (in four strains) and the maximum number is 14 (in two strains). In contrast, the genome sequences of single strains of *F. graminearum* and *F. verticillioides*, species that exhibit more genetic and pathogenic uniformity than *F. oxysporum*, have 15 and 16 PKS genes respectively. However, *F. oxysporum* genomes are not uniform with regard to PKS genes: only eight of the 17 PKS genes occur in all 12 of the *F. oxysporum* genomes; five of the PKS genes have homologues in four to 11 genomes; and four of the genes have homologues in two to eight species (Table 8.2).

Analysis of *F. oxysporum* PKSs has provided insight into the dynamic nature of PKS evolution in this species and perhaps *Fusarium* and other fungi in general. The genome of strain Fo4287 contain two PKS genes (FOXG_14850 and FOXG_15886; R-PKS Clade III) that share over 97% nucleotide sequence identity and thus are likely products of gene duplication, i.e. paralogues (Brown *et al.*, 2012a). The high level of identity (~ 98%) of 2.5 kb upstream and 3.2 kb downstream of FOXG_14850 and FOXG_15886 flanking regions indicate that the duplication event likely involved 14.5 kb of genomic DNA. The paralogous regions may represent a three-gene gene cluster, because both the upstream and downstream regions are predicted to encode pfam08242 and pfam13489 domain containing methyltransferases respectively, and such enzymes are often involved in secondary metabolism. The presence of a 767-bp indel within the 5′ flanking region of FOXG_14850 and FOXG_15886 indicates that the two PKSs could be differentially expressed (Brown *et al.*, 2012a). Only three of the other 11 *F. oxysporum*

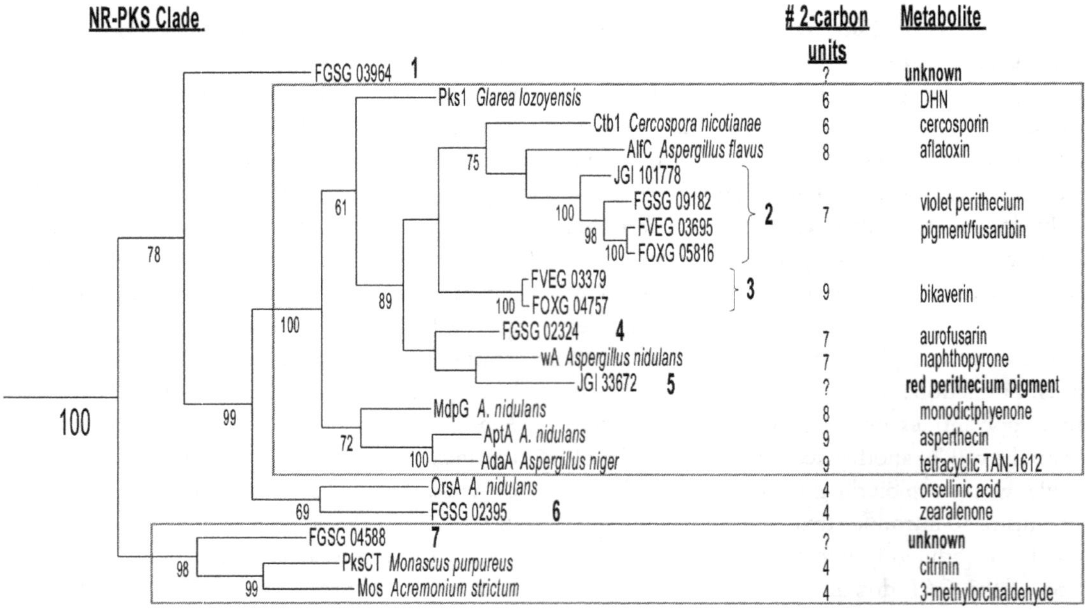

Figure 8.5 The NR-PKS clade (as presented in Figure 8.1) (left), the number of 2-carbon units in the polyketides synthesized by NR-PKSs (middle), and secondary metabolites derived from the polyketides (right). The numbers (1–7) in bold type indicate the seven sets of *Fusarium* NR-PKS homologues. Although PKS JGI_33672 is required for production of a red perithecial pigment, the chemical structures of the pigment and the polyketide from which it is derived have not been determined.

genome sequences have even one homologue of FOXG_14850 and FOXG_15886; none of the other genomes have both paralogues. In FoMelonis and Fo5176, the 5′ flanking region of each homologue is missing the 767-bp sequence and is thus more similar to FOXG_14850. In Fo4287, both paralogues are present in lineage-specific regions of the genome: FOXG_14850 on chromosome 3 and FOXG_15886 on chromosome 2b (http://www.broadinstitute.org). Whether the homologues in other strains are also present in lineage-specific regions and whether their presence in these regions is related to the duplication event remains to be determined. In addition, the failure to find any other strain with two copies of this PKS supports the hypothesis that FOXG_14850 and FOXG_15886 are products of a relatively recent duplication event (Brown *et al.*, 2012a). Gene duplication has been proposed as a mechanism by which gene clusters grow and diverge to increase structural diversity of fungal secondary metabolites (Cary and Ehrlich, 2006; Saikia *et al.*, 2008). Thus, the FOXG_14850 and FOXG_15886 regions may represent an SMB gene cluster that has duplicated and is in the process of diverging and that has the potential to increase the metabolic diversity of *F. oxysporum*. FOXG_14850 and FOXG_15886 are the only obvious example of paralogues among the 61 sets of PKS gene homologues in the four *Fusarium* species considered here. This suggests that duplication of PKS genes can occur in *Fusarium* with some frequency, but that the occurrence of such events is limited and may occur more often in species that have lineage-specific genomic regions. It is interesting to note that the most closely related homologue of FOXG_14850 and FOXG_15886 in *F. solani* (JGI_106233) is also flanked by two methyltransferases that appear homologous to the *F. oxysporum* methyltransferases while the most closely related homologue in *F. verticillioides* (FVEG_13715) is flanked by only one homologous methyltransferase. The lack of a second methyltransferase in *F. verticillioides* suggests that the final SM derived from the *F. verticillioides* PKS is likely structurally different than the SMs derived from *F. oxysporum* and *F. solani* PKSs.

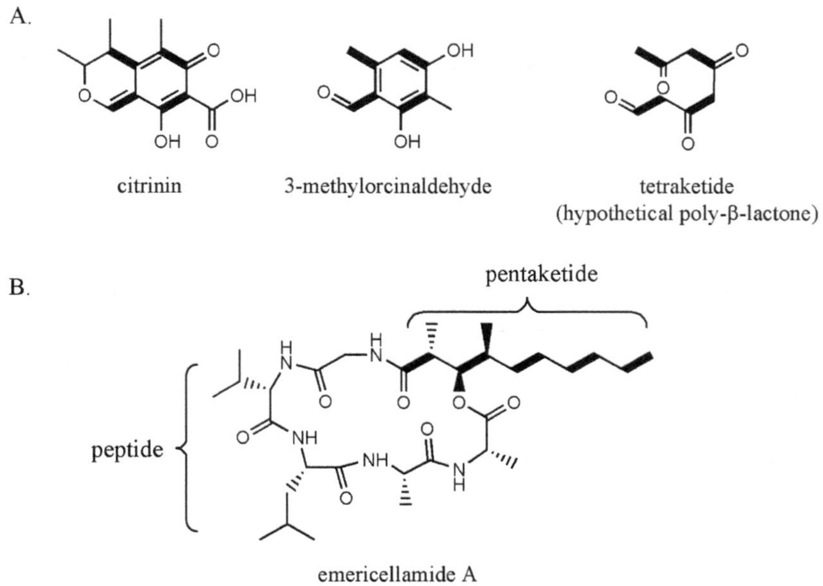

Figure 8.6 (A) The structures of citrinin, a tetraketide from *Penicillium citrinum*, and 3-methylorcinaldehyde, a tetraketide from *Acremonium strictum*. Adjacent carbons arising from each 2-carbon unit (starter or extender unit) are indicated by bold lines. The tetraketide molecule or hypothetical poly-β-lactone, is depicted on the right. (B) The structure of emericellamide A, a hybrid pentaketide-pentapeptide from *Aspergillus nidulans*.

Predicting polyketide structure based on phylogenetic analysis and domain content of PKSs

A sufficient number of PKSs and the polyketides they synthesize are now known such that it is possible to use phylogenetic analysis and comparison of PKS domain content to help predict structures of polyketides synthesized by newly identified PKSs. In the phylogenetic tree presented in Fig. 8.3, the NR-PKS clade includes seven sets of *Fusarium* PKS homologues (numbered 1–7 in Fig. 8.5). The polyketides synthesized by four of the NR-PKSs (#s 2, 3, 4 and 6) have been inferred based on structures of the corresponding polyketide-derived secondary metabolites (Fig. 8.3). Thus, the structures of polyketides synthesized by three *Fusarium* NR-PKSs (JGI_33672, FGSG_04588 and FGSG_03964) remain to be determined. Although the polyketide synthesized by JGI_33672 (#5) is not known, this PKS is required for synthesis of a red perithecial pigment produced by *F. solani* (Graziani *et al.*, 2004). In the phylogenetic analysis, JGI_33672 was resolved on a well-supported branch with 10 other PKSs (large box in Fig. 8.5). These PKSs synthesize polyketides that form structures that consist of multiple aromatic rings derived from between six (hexaketide) and nine (nonaketide) two-carbon units. The presence of JGI_33672 in this clade suggests that it may also synthesize a polyketide that forms multiple aromatic rings derived from between six and nine two-carbon units.

In the phylogenetic analysis, FGSG_04588 was resolved on a well supported (98% bootstrap value) basal branch with PksCT (*Monascus purpureus*) and Mos (*Acremonium strictum*) (small box in Fig. 8.5). PksCT and Mos are required for synthesis of citrinin and 3-methylorcinaldehyde respectively. Both of these compounds consist of an aromatic ring derived from four 2-carbon units (tetrakide) (Fig. 8.6A), even though their KS/AT domains share only 55% AA identity. Given that the FGSC_04588 KS/AT domains share 42% AA identity with those of both PksCT and Mos and that the three PKSs are resolved on the same, well supported branch in the phylogenetic analysis, FGSG_04588 may also synthesize a tetraketide that forms a single aromatic ring. In contrast to JGI_33672 and FGSG_04588, little can be predicted about the structure of the polyketide synthesized by FGSG_03964 because it is relatively distantly related to NR-PKSs with known functions. Nevertheless, the domain content (KS-MAT-PT-ACP-DA) of FGSG_03964 indicates that it likely synthesizes a polyketide in which keto groups are not reduced, and this often gives rise to the formation of compounds with one or more aromatic rings.

Phylogenetic analysis indicate that the three R-PKS clades comprise 28 distinct sets of *Fusarium* PKS homologues and that these PKSs are likely responsible for the synthesis of 28 different polyketides. To date, polyketides synthesized by five of the PKSs have been inferred from the structures of the corresponding secondary metabolites (fusarielin, fusarin, zearalenone, fusaric acid and fumonisin; Fig. 8.3). As with NR-PKSs, the phylogenetic relationships of R-PKSs with known polyketide products and comparison of domain organizations and AA sequences to those with unknown products provide clues to the structure of the polyketides synthesized by the latter. For example, FOXG_02884 and FVEG_01736 exhibit high levels of amino acid identity to (~82% identity across the KS/AT domains) and are resolved on the same branch as the *Alternaria brassicicola* depudecin PKS, Dep5 (Fig. 8.3). This strongly suggests that both FOXG_02884 and FVEG_01736 synthesize the same polyketide precursor that leads to depudecin in *A. brassicicola*. In support of this prediction is the observation that the flanking regions of both FVEG_01736 and FOXG_02884 include closely related homologues of five genes in the *A. brassicicola* depudecin biosynthetic gene cluster (Wight *et al*., 2009). To our knowledge, depudecin production has not been reported in *Fusarium*.

Only two of the remaining *Fusarium* R-PKSs (FGSG_08208 and JGI_106474) were resolved into well-supported clades with PKSs that have known polyketide products. The *F. graminearum* R-PKS FGSG_08208 was resolved into a clade (100% bootstrap) and shares 59% AA identity (KS/AT domains) with the *A. nidulans* EasB, a PKS required for synthesis of emericellamide (Fig. 8.6B) (Chiang *et al*., 2008). Based on this, as well as additional observations described below, it is likely that the polyketide synthesized by the FGSG_08208 PKS is similar in structure to the pentaketide precursor of emericellamide. In emericellamide biosynthesis, the pentaketide produced by EasB undergoes a series of reactions, including condensation with a five amino acid peptide that is catalysed by a five-module NRPS, EasA (Chiang *et al*., 2008). In *A. nidulans*, the genes coding for EasA and EasB as well as two other genes required for polyketide processing are located adjacent to one another in the emericellamide biosynthetic gene cluster. In *F. graminearum*, FGSG_08209, coding for a putative NRPS, is adjacent to FGSG_08208. Although the polyketide products of EasB and FGSG_08208 are likely structurally similar, the secondary metabolite derived from the FGSG_08208 polyketide likely differs markedly from emericellamide, because FGSG_08209 has only three NRPS modules whereas EasA has five modules (Chiang *et al*., 2008).

The *F. solani* R-PKS JGI_106474 was resolved into a clade (100% bootstrap) with *Cochliobolus heterostrophus* Pks2 (Fig. 8.3), and the KS and AT domains of the two PKSs share 45% AA identity. Pks2 along with a second PKS, Pks1, is required for production of a ~40-carbon metabolite known as T-toxin (Baker *et al*., 2006). Analysis of the *F. solani* genome indicates that it does not have a closely related homologue of the Pks1 gene. Thus, it is unlikely that the *F. solani* JGI_106474 PKS synthesizes T-toxin or a secondary metabolite with a similarly long carbon chain. But, JGI_106474 may synthesize a 16 to 18-carbon polyketide with a pattern of functional groups similar to Pks2-derived polyketide of T-toxin.

Future prospects

Fusarium PKSs have proven to be a excellent model system to examine the mechanisms by which fungi increase and/or decrease their secondary metabolite biosynthetic potential. In the future, determination of what polyketides *Fusarium* PKSs synthesize will likely proceed by one of three approaches that are already being used in laboratories around the world. The first approach involves identification of genes encoding transcription factors (TF) within gene clusters that include a PKS gene. The TF gene is then placed under the control of an inducible or constitutive promoter in order to induce high levels of TF expression. This, in turn, should induce high levels of expression of the PKS gene and other genes in the cluster and thereby give rise to high levels of production of the corresponding polyketide-derived secondary metabolite. This strategy has been used to identify the asperfuranone PKS gene in *A. nidulans* (Chiang *et al.*, 2009) and the fusarielin PKS gene in *F. graminearum* (Sørensen *et al.*, 2011). The second approach involves manipulation of global regulators of SM biosynthesis to alter expression of PKS genes and thereby production of polyketide and the secondary metabolites derived from them. Both deletion and over expression of *laeA*, which encodes a histone methyltransferase that affects chromatin structure, led to the identification of a number of PKS genes, including the PKS involved in synthesis of the emodin family of polyketides in *A. nidulans* (Bok *et al.*, 2009). The third approach involves direct expression of PKS genes, as well as flanking genes, in a heterologous host in order to produce polyketide-derived secondary metabolites in organisms that do not normally produce them. This approach was used to identify the tenellin PKS gene in *Beauveria bassiana* by expressing it and three flanking genes in *A. nidulans* (Heneghan *et al.*, 2010).

The next few years will be an exciting time for research on *Fusarium* polyketides and fungal secondary metabolites in general. Whole-genome sequence data have been generated, but are not yet publicly available, for *F. fujikuroi*, *F. circinatum* (Wingfield *et al.*, 2012), *F. mangiferae* and *F. culmorum* and more genomes are likely to be completed within the next few years. A preliminary analysis of the *F. circinatum* and *F. fujikuroi* sequences indicates that these genomes have 15 and 17 PKS genes respectively (Bettina Tudzynski, Westfalische Wilhelms-University and Brenda Wingfield, University of Pretoria; personal communication). Examination of the PKSs corresponding to these genes as well as genes in additional *Fusarium* genomes will likely lead to identification of PKSs required for synthesis of novel polyketide metabolites. Because, fusaria tend to produce a limited number of secondary metabolites under standard laboratory conditions, surveys of genomes for PKS genes with known functions will aid in determining the genetic potential of species to produce known polyketide metabolites, including mycotoxins. Such surveys will also likely lead to an increased understanding of the distribution and diversity of *Fusarium* PKSs and thereby polyketide-derived SMs. Determination of structural differences of polyketides synthesized by PKSs with similar amino acid sequences should facilitate determination of what sequence variation within a given PKS domain can give rise to variation in polyketide structure. This in turn should improve the utility of deduced amino acid sequences of PKSs to predict polyketide structure. An improved understanding of the relationship between sequence variation and polyketide structure should also improve our understanding of the mechanisms by which PKSs catalyse synthesis of polyketides, and this may facilitate the design of novel polyketides with novel pharmaceutical or agricultural properties. Thus, the advent of genome sequences has revolutionized the study of PKSs and will likely continue to improve our understanding of the synthesis and diversity of polyketides produced by *Fusarium* and other fungi of concern to agriculture production and human and animal health.

References

Alexander, N.J., Proctor, R.H., and McCormick, S.P., (2009). Genes, gene clusters, and biosynthesis of trichothecenes and fumonisins in *Fusarium*. Toxin Rev. *24*, 198–215.

Altschul, S.F., Gish, W., Miller, W., Myers, E.W., and Lipman, D.J. (1990). Basic local alignment search tool. J. Mol. Biol. *215*, 403–410.

Ansari, M.Z., Sharma, J., Gokhale, R.S., and Mohanty, D., (2008). *In silico* analysis of methyltransferase domains involved in biosynthesis of secondary metabolites. BMC Bioinformatics 9, 1–21.

Bailey, A.M., Cox, R.J., Harley, K., Lazarus, C.M., Simpson, T.J., and Skellam, E., (2007). Characterisation of 3-methylorcinaldehyde synthase (MOS) in *Acremonium strictum*: first observation of a reductive release mechanism during polyketide biosynthesis. Chem. Commun., 4053–4055.

Baker, S.E., Kroken, S., Inderbitzin, P., Asvarak, T., Li, B.Y., Shi, L., Yoder, O.C., and Turgeon, B.G., (2006). Two polyketide synthase-encoding genes are required for biosynthesis of the polyketide virulence factor, T-toxin, by *Cochliobolus heterostrophus*. Mol. Plant Microbe Interact. *19*, 139–149.

Bok, J.W., Chiang, Y.M., Szewczyk, E., Reyes-Dominguez, Y., Davidson, A.D., Sanchez, J.F., Lo, H.C., Watanabe, K., Strauss, J., Oakley, B.R., *et al.* (2009). Chromatin-level regulation of biosynthetic gene clusters. Nat. Chem. Biol. 5, 462–464.

Bomke, C., and Tudzynski, B., (2009). Diversity, regulation, and evolution of the gibberellin biosynthetic pathway in fungi compared to plants and bacteria. Phytochemistry *70*, 1876–1893.

Brown, D.W., Butchko, R.A., Baker, S.E., and Proctor, R.H. (2012a). Phylogenomic and functional domain analysis of polyketide synthases in *Fusarium*. Fun. Biol. *116*, 318–331.

Brown, D.W., Butchko, R.A., Busman, M., and Proctor, R.H., (2012b). Identification of gene clusters associated with fusaric acid, fusarin, and perithecial pigment production in *Fusarium verticillioides*. Fungal Genet. Biol. *49*, 521–532.

Brown, D.W., Butchko, R.A., and Proctor, R.H. (2008). Genomic analysis of *Fusarium verticillioides*. Food Addit. Contam. *25*, 1158–1165.

Cary, J.W., and Ehrlich, K.C. (2006). Aflatoxigenicity in *Aspergillus*: molecular genetics, phylogenetic relationships and evolutionary implications. Mycopathologia *162*, 167–177.

Chiang, Y.M., Szewczyk, E., Davidson, A.D., Keller, N., Oakley, B.R., and Wang, C.C. (2009). A gene cluster containing two fungal polyketide synthases encodes the biosynthetic pathway for a polyketide, asperfuranone, in *Aspergillus nidulans*. J. Am. Chem. Soc. *131*, 2965–2970.

Chiang, Y.M., Szewczyk, E., Nayak, T., Davidson, A.D., Sanchez, J.F., Lo, H.C., Ho, W.Y., Simityan, H., Kuo, E., Praseuth, A., *et al.* (2008). Molecular genetic mining of the *Aspergillus* secondary metabolome: discovery of the emericellamide biosynthetic pathway. Chem. Biol. *15*, 527–532.

Cole, R.J., Jarvis, B.B., and Schweikert, M.A. (2003). Handbook of secondary fungal metabolites. (Academic Press, San Diego).

Crawford, J.M., Dancy, B.C., Hill, E.A., Udwary, D.W., and Townsend, C.A. (2006). Identification of a starter unit acyl-carrier protein transacylase domain in an iterative type I polyketide synthase. Proc. Natl. Acad. Sci. *103*, 16728–16733.

Crawford, J.M., Thomas, P.M., Scheerer, J.R., Vagstad, A.L., Kelleher, N.L., and Townsend, C.A. (2008). Reconstruction of iterative multidomain polyketide synthase function. Science *320*, 243–246.

Cuomo, C.A., Guldener, U., Xu, J.R., Trail, F., Turgeon, B.G., Di Pietro, A., Walton, J.D., Ma, L.J., Baker, S.E., Rep, M., *et al.* (2007). The *Fusarium graminearum* genome reveals a link between localized polymorphism and pathogen specialization. Science *317*, 1400–1402.

Devereux, J., Haeberli, P., and Smithies, O. (1984). A comprehensive set of sequence analysis programs for the VAX. Nucleic Acids Res. *12*, 387–395.

Du, L., and Lou, L. (2010). PKS and NRPS release mechanisms. Nat. Prod. Rep. *27*, 255–278.

Fink-Gremmels, J., and Malekinejad, H. (2007). Clinical effects and biochemical mechanisms associated with exposure to the mycoestrogen zearalenone. Animal Feed Sci. Tech. *137*, 326–341.

Fujii, I., Watanabe, A., Sankawa, U., and Ebizuka, Y. (2001). Identification of Claisen cyclase domain in fungal polyketide synthase W.A., a naphthopyrone synthase of *Aspergillus nidulans*. Chem. Biol. *8*, 189–197.

Gaffoor, I., Brown, D.W., Plattner, R., Proctor, R.H., Qi, W., and Trail, F. (2005). Functional analysis of the polyketide synthase genes in the filamentous fungus *Gibberella zeae* (anamorph *Fusarium graminearum*). Eukaryot. Cell *4*, 1926–1933.

Glenn, A.E. (2007). Mycotoxigenic *Fusarium* species in animal feed. Animal Feed Sci. Tech. *137*, 213–240.

Graziani, S., Vasnier, C., and Daboussi, M.J. (2004). Novel polyketide synthase from *Nectria haematococca*. Appl. Environ. Microbiol. *70*, 2984–2988.

Heneghan, M.N., Yakasai, A.A., Halo, L.M., Song, Z., Bailey, A.M., Simpson, T.J., Cox, R.J., and Lazarus, C.M. (2010). First heterologous reconstruction of a complete functional fungal biosynthetic multigene cluster. Chembiochem *11*, 1508–1512.

Hopwood, D.A., and Khosla, C. (1992). Genes for polyketide secondary metabolic pathways in microorganisms and plants, in: Chadwick, D.J., and Whelan, J. (Eds.), Secondary Metabolites: Their Function and Evolution. (John Wiley and Sons, Chichester, U.K), pp. 88–112.

Hopwood, D.A., and Sherman, D.H. (1990). Molecular genetics of polyketides and its comparison to fatty acid biosynthesis. Annu. Rev. Genet. *24*, 37–66.

Howard, P.C., Eppley, R.M., Stack, M.E., Warbritton, A., Voss, K.A., Lorentzen, R.J., Kovach, R.M., and Bucci, T.J. (2001). Fumonisin B_1 carcinogenicity in a two-year feeding study using F344 rats and B6C3F1 mice. Environ. Health Perspect. *109*, 277–282.

Keller, N.P., and Hohn, T.M. (1997). Metabolic pathway gene clusters in filamentous fungi. Fungal Genet. Biol. *21*, 17–29.

Keller, N.P., Turner, G., and Bennett, J.W. (2005). Fungal secondary metabolism – from biochemistry to genomics. Nat. Rev. Microbiol. *3*, 937–947.

Kim, J.E., Han, K.H., Jin, J., Kim, H., Kim, J.C., Yun, S.H., and Lee, Y.W. (2005a). Putative polyketide synthase and laccase genes for biosynthesis of aurofusarin in *Gibberella zeae*. Appl. Environ. Microbiol. *71*, 1701–1708.

Kim, Y.T., Lee, Y.R., Jin, J., Han, K.H., Kim, H., Kim, J.C., Lee, T., Yun, S.H., and Lee, Y.W. (2005b). Two different polyketide synthase genes are required for synthesis of zearalenone in *Gibberella zeae*. Mol. Microbiol. *58*, 1102–1113.

Kroken, S., Glass, N.L., Taylor, J.W., Yoder, O.C., and Turgeon, B.G. (2003). Phylogenomic analysis of type I polyketide synthase genes in pathogenic and saprobic ascomycetes. Proc. Natl. Acad. Sci. *100*, 15670–15675.

Lee, K., Pan, J.J., and May, G. (2009). Endophytic *Fusarium verticillioides* reduces disease severity caused by *Ustilago maydis* on maize. FEMS Microbiol. Lett. 31–37.

Li, Y., Chooi, Y.H., Sheng, Y., Valentine, J.S., and Tang, Y. (2011). Comparative characterization of fungal anthracenone and naphthacenedione biosynthetic pathways reveals an alpha-hydroxylation-dependent Claisen-like cyclization catalyzed by a dimanganese thioesterase. J. Am. Chem. Soc. *133*, 15773–15785.

Li, Y., Image, I.I., Xu, W., Image, I., and Tang, Y. (2010). Classification, prediction, and verification of the regioselectivity of fungal polyketide synthase product template domains. J. Biol. Chem. *285*, 22764–22773.

Linnemannstons, P., Schulte, J., del Mar Prado, M., Proctor, R.H., Avalos, J., and Tudzynski, B. (2002). The polyketide synthase gene *pks4* from *Gibberella fujikuroi* encodes a key enzyme in the biosynthesis of the red pigment bikaverin. Fungal Genet. Biol. *37*, 134–148.

Lysoe, E., Klemsdal, S.S., Bone, K.R., Frandsen, R.J., Johansen, T., Thrane, U., and Giese, H. (2006). The PKS4 gene of *Fusarium graminearum* is essential for zearalenone production. Appl. Environ. Microbiol. *72*, 3924–3932.

Ma, L.J., van der Does, H.C., Borkovich, K.A., Coleman, J.J., Daboussi, M.J., Di Pietro, A., Dufresne, M., Freitag, M., Grabherr, M., Henrissat, B., *et al.* (2010). Comparative genomics reveals mobile pathogenicity chromosomes in *Fusarium oxysporum*. Nature *464*, 367–373.

Malz, S., Grell, M.N., Thrane, C., Maier, F.J., Rosager, P., Felk, A., Albertsen, K.S., Salomon, S., Bohn, L., Schafer, W., *et al.* (2005). Identification of a gene cluster responsible for the biosynthesis of aurofusarin in the *Fusarium graminearum* species complex. Fungal Genet. Biol. *42*, 420–433.

Marasas, W.F., Riley, R.T., Hendricks, K.A., Stevens, V.L., Sadler, T.W., Gelineau-Van Waes, J., Missmer, S.A., Cabrera, J., Torres, O., Gelderblom, W.C., *et al.* (2004). Fumonisins disrupt sphingolipid metabolism, folate transport, and neural tube development in embryo culture and *in vivo:* A potential risk factor for human neural tube defects among populations consuming fumonisin-contaminated maize. J. Nutr. *134*, 711–716.

Michielse, C.B., and Rep, M. (2009). Pathogen profile update: *Fusarium oxysporum*. Mol. Plant Pathol. *10*, 311–324.

Morgavi, D.P., and Riley, R.T. (2007). An historical overview of field disease outbreaks known or suspected to be caused by consumption of feeds contaminated with *Fusarium* toxins. Animal Feed Sci. Tech. *137*, 201–212.

Proctor, R.H., Brown, D.W., Plattner, R.D., and Desjardins, A.E. (2003). Co-expression of 15 contiguous genes delineates a fumonisin biosynthetic gene cluster in *Gibberella moniliformis*. Fungal Genet. Biol. *38*, 237–249.

Proctor, R.H., Butchko, R.A., Brown, D.W., and Moretti, A. (2007). Functional characterization, sequence comparisons and distribution of a polyketide synthase gene required for perithecial pigmentation in some *Fusarium* species. Food Addit. Contam. *24*, 1076–1087.

Proctor, R.H., Desjardins, A.E., Plattner, R.D., and Hohn, T.M. (1999). A polyketide synthase gene required for biosynthesis of fumonisin mycotoxins in *Gibberella fujikuroi* mating population A. Fungal Genet. Biol. *27*, 100–112.

Proctor, R.H., Hohn, T.M., and McCormick, S.P. (1995). Reduced virulence of *Gibberella zeae* caused by disruption of a trichothecene toxin biosynthetic gene. Mol. Plant Microbe Interact. *8*, 593–601.

Roze, L.V., Chanda, A., and Linz, J.E. (2011). Compartmentalization and molecular traffic in secondary metabolism: a new understanding of established cellular processes. Fungal Genet. Biol. *48*, 35–48.

Saikia, S., Nicholson, M.J., Young, C., Parker, E.J., and Scott, B. (2008). The genetic basis for indole-diterpene chemical diversity in filamentous fungi. Mycol. Res. *112*, 184–199.

Sanchez, J.F., Entwistle, R., Hung, J.H., Yaegashi, J., Jain, S., Chiang, Y.M., Wang, C.C., and Oakley, B.R. (2011). Genome-based deletion analysis reveals the prenyl xanthone biosynthesis pathway in *Aspergillus nidulans*. J. Amer. Chem. Soc. *133*, 4010–4017.

Seefelder, W., Humpf, H.U., Schwerdt, G., Freudinger, R., and Gekle, M. (2003). Induction of apoptosis in cultured human proximal tubule cells by fumonisins and fumonisin metabolites. Toxicol Appl. Pharmacol. *192*, 146–153.

Sims, J.W., Fillmore, J.P., Warner, D.D., and Schmidt, E.W. (2005). Equisetin biosynthesis in *Fusarium heterosporum*. Chem. Commun. 186–188.

Song, Z., Cox, R.J., Lazarus, C.M., and Simpson, T.J. (2004). Fusarin C biosynthesis in *Fusarium moniliforme* and *Fusarium venenatum*. Chembiochem *5*, 1196–1203.

Sørensen, J.L., Hansen, F.T., Sondergaard, T.E., Wimmer, R., Droce, A., Giese, H., and Frandsen, R.J.N. (2011). Targeted activation of the polyketide synthase 9 gene cluster in *Fusarium graminearum*, 26th Fungal Genetics Conference at Asilomar, p. 55 (Suppl.).

Srobarova, A., Moretti, A., Ferracane, R., Ritieni, A., and Logrieco, A. (2002). Toxigenic *Fusarium* species of *Liseola* section in pre-harvest maize ear rot, and associated mycotoxins in Slovakia. E. J. Plant Path. *108*, 299–306.

Sydenham, E.W., Thiel, P.G., Marasas, W.F.O., Shephard, G.S., Schalkwyk, D.J., and Koch, K.R. (1990). Natural occurrence of some *Fusarium* mycotoxins in corn from low and high esophageal cancer prevalence areas of the Transkei, Southern Africa. J. Agric. Food Chem. *38*, 1900–1903.

Szewczyk, E., Chiang, Y.M., Oakley, C.E., Davidson, A.D., Wang, C.C., and Oakley, B.R. (2008). Identification and characterization of the asperthecin gene cluster of *Aspergillus nidulans*. Appl. Environ. Microbiol. *74*, 7607–7612.

van Roermund, C.W., Hettema, E.H., van den Berg, M., Tabak, H.F., and Wanders, R.J. (1999). Molecular characterization of carnitine-dependent transport of acetyl-CoA from peroxisomes to mitochondria in *Saccharomyces cerevisiae* and identification of a plasma membrane carnitine transporter, Agp2p. EMBO J. *18*, 5843–5852.

Vasnier, C., Graziani, S., Dufresne, M., and Daboussi, M.-J. (2005). Characterization of four clustered genes associated with the biosynthesis of a red perithecial pigment in *Nectria haematococca*. Fun. Genet. News. 555.

Voss, K.A., Smith, G.W., and Haschek, W.M. (2007). Fumonisins: Toxicokinetics, mechanism of action and toxicity. Animal Feed Sci. Tech. *137*, 299–325.

Wakil, S.J. (1989). Fatty acid synthase, a proficient multifunctional enzyme. Biochemistry *28*, 4523–4530.

Watanabe, A., and Ebizuka, Y. (2004). Unprecedented mechanism of chain length determination in fungal aromatic polyketide synthases. Chem. Biol. *11*, 1101–1106.

Wicklow, D.T., Horn, B.W., Shotwell, O.L., Hesseltine, C.W., and Caldwell, R.W. (1988). Fungal interference with *Aspergillus flavus* infection and aflatoxin contamination of maize grown in a controlled environment. Phytopathology *78*, 68–74.

Wicklow, D.T., Jordan, A.M., and Gloer, J.B. (2009). Antifungal metabolites (monorden, monocillins I., I.I., III) from *Colletotrichum graminicola*, a systemic vascular pathogen of maize. Mycol. Res. *113*, 1433–1442.

Wicklow, D.T., and Poling, S.M. (2009). Antimicrobial activity of pyrrocidines from *Acremonium zeae* against endophytes and pathogens of maize. Phytopathology *99*, 109–115.

Wiemann, P., Willmann, A., Straeten, M., Kleigrewe, K., Beyer, M., Humpf, H.U., and Tudzynski, B. (2009). Biosynthesis of the red pigment bikaverin in *Fusarium fujikuroi*: genes, their function and regulation. Mol. Microbiol. *72*, 931–946.

Wight, W.D., Kim, K.H., Lawrence, C.B., and Walton, J.D. (2009). Biosynthesis and role in virulence of the histone deacetylase inhibitor depudecin from *Alternaria brassicicola*. Mol. Plant Microbe Interact. *22*, 1258–1267.

Wingfield, B.D., Steenkamp, E.T., Santana, Q.C., Coetzee, M.P.A., Bam, S., Barnes, I., Beukes, C.W., Chan, W.Y., de Vos, L., Fourie, G., *et al.* (2012). First fungal genome sequence from Africa: A preliminary analysis. S. Afr. J. Sci. *108*, 1–9.

Wu, F. (2007). Measuring the economic impacts of *Fusarium* toxins in animal feeds. Animal Feed Sci. Tech. *137*, 363–374.

Xu, Y., Orozco, R., Wijeratne, E.M., Gunatilaka, A.A., Stock, S.P., and Molnar, I. (2008). Biosynthesis of the cyclooligomer depsipeptide beauvericin, a virulence factor of the entomopathogenic fungus *Beauveria bassiana*. Chem. Biol. *15*, 898–907.

Plant Responses to *Fusarium* Metabolites

9

Takumi Nishiuchi

Abstract

Plant pathogenic species of *Fusarium* produce numerous secondary metabolites during infection of host plants. These metabolites often perturb host defence responses and suppress plant growth. Plant responses to *Fusarium* metabolites can be classified as follows: (1) inhibition of root or shoot growth; (2) inhibition of seed germination; (3) changes in leaf colour such as chlorosis; (4) cell death; and (5) suppression or activation of defence responses. These phytotoxic effects of *Fusarium* metabolites have been reported in various plant species. Two major *Fusarium* metabolites, fumonisins and trichothecenes, induce apoptosis-like programmed cell death and can contribute to virulence of fusaria on some plants. Recently, signalling events have been implicated in plant responses to *Fusarium* metabolites. In contrast, production of the growth-promoting metabolites gibberellins by the rice pathogen *Fusarium fujikuroi* results in the seedling elongation symptom characteristic of bakanae disease of rice. Thus, *Fusarium* secondary metabolites have various effects in host plants. This chapter reviews *Fusarium* secondary metabolites and how plant respond to them.

Introduction

Fusarium species produce a variety of organic compounds that are not essential for growth or reproduction but instead confer selective advantages under certain environmental conditions (Fig. 9.1) (Glenn, 2007). These compounds are referred to as secondary metabolites (SMs) to contrast them from primary metabolites, which are generally required for growth and/or reproduction. Some *Fusarium* SMs are highly toxic to humans and other animals and as a result are referred to as mycotoxins. Infection of cereal crops by some *Fusarium* species often leads to contamination of the resulting grain and grain-derived foods with mycotoxin(s). Consumption of mycotoxin-contaminated grain can cause serious health problems in humans and other animals. The toxic properties of *Fusarium* SMs have been thoroughly studied with the aim of reducing their detrimental effects on health (Gutleb *et al.*, 2002).

Phytopathogenic fusaria can produce SMs during infection of host plants. Some of these metabolites can inhibit plant growth and perturb host plant defence responses. Such SMs are also referred to as phytotoxins (McLean, 1996). Thus, some *Fusarium* mycotoxins are also considered phytotoxins because of their phytotoxicity. Although phytotoxins can contribute to the virulence of fusaria, the phytotoxic modes of action of *Fusarium* SMs are generally poorly understood. However, in order to understand pathogenesis of *Fusarium*, it is important to elucidate the modes of action of *Fusarium* phytotoxins. Recent advances have been made in understanding modes of action of the *Fusarium* toxins fumonisins and trichothecenes by examining the effects of the toxins on plants at the biochemical and molecular level.

Multiple studies have provided insight into the effects of *Fusarium* SMs on animals and other higher eukaryotes. For example, trichothecenes inhibit protein synthesis by binding to and interfering with the function of a ribosomal protein, whereas fumonisins disrupt sphingolipid metabolism by inhibiting the enzyme ceramide synthetase (Gutleb *et al.*, 2002; Nishiuchi *et al.*, 2006). For each of these metabolites, the mode of action is similar in animals and plants. In contrast, the mechanism by which the *Fusarium* SM zearalenone affects animals and plants differs markedly. The primary toxic effects of zearalenone

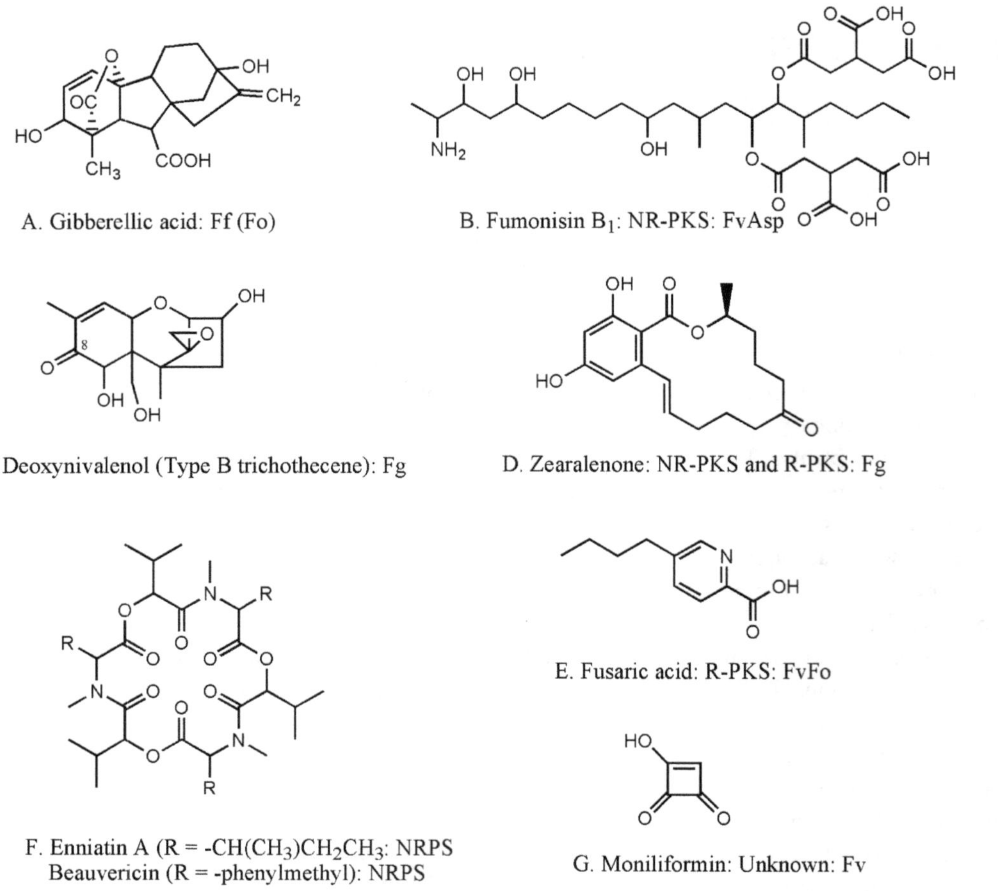

Figure 9.1 Chemical structures of various *Fusarium* metabolites. (A) Gibberellic acid is produced by *F. fujikuroi* and *F. oxysporum*. (B) Biosynthesis of Fumonisin B1 is catalysed by a non-reducing polyketide synthase (NR-PKS) in *F. verticillioides* and *Aspergillus*. (C) Deoxynivalenol is one of type B trichothecenes and is produced by *F. graminearum*. (D) Production of zearalenone is catalysed by a NR-PKS and reducing polyketide synthase (R-PKS) in *F. graminearum*. (E) Fusaric acid biosynthesis is catalysed by R-PKS in *F. verticillioides* and *F. oxysporum*. (F) Biosyntheses of enniatin A and beauvericin are catalysed by a non-ribosomal peptide synthetase (NRPS). (G) Moniliformin is produced by *F. verticillioides*.

in animals is due to its estrogenic activity. However, zearalenone exhibits little phytotoxicity at high concentration and has been reported to promote growth of wheat plants (Filek *et al.*, 2010). In addition to fumonisins, trichothecenes and zearalenone, *Fusarium* produces a variety of other compounds that are referred to as minor metabolites because they are not associated with severe or repeated toxicoses. These minor metabolites can affect growth, cell viability, and defence responses in host plants (McLean, 1996).

The overall phytotoxic effects of *Fusarium* may be due in part to minor metabolites or to combined and/or synergistic effects of multiple metabolites. Metabolic analysis has revealed differences in primary, or basal, metabolism as well as secondary metabolism in different species of *Fusarium* (Lowe *et al.*, 2010). Some differences in metabolomes of *Fusarium* species are consistent with differences in the presence and absence of SM biosynthetic genes and gene clusters that exist in genome sequences of the species (Ma *et al.*, 2010). In the future, advances in metabolomic technologies will likely facilitate identification of novel metabolites that mediate interactions between *Fusarium* and its host plants.

Plant responses to *Fusarium* metabolites can be classified into the following five categories: (1)

phytotoxicity that leads to inhibition of root and/or shoot growth; (2) inhibition of seed germination; (3) chlorosis, necrosis and other changes in leaf health; (4) cell death including programmed cell death (PCD); and (5) suppression or activation of defence responses. This chapter reviews various *Fusarium* metabolites and how plants respond to them, including effects on plant signalling molecules.

Gibberellins

Chemical structure and biosynthesis

Gibberellins (GAs) are tetracyclic diterpenoids with plant growth-promoting activity, especially activity that leads to stem elongation (Fig. 9.1A) (Bömke and Tudzynski, 2009). A characteristic symptom of bakanae disease of rice (*Oryza sativa*), caused by *F. fujikuroi*, is abnormal elongation of seedlings (Wulff *et al.*, 2010). The growth-promoting effect of *F. fujikuroi* has been attributed to its ability to synthesize GAs (Tudzynski, 2005). Although other *Fusarium* species such as *Fusarium verticillioides* have been isolated from rice with bakanae disease, the production of GAs is associated primarily with *F. fujikuroi* (Tudzynski, 2005; Wulff *et al.*, 2010) (See Chapter 7).

Genes responsible for GA biosynthesis have been identified in both *F. fujikuroi* and in the plant *Arabidopsis thaliana* (Tudzynski, 2005). Comparative analysis revealed dramatic differences in the organization of GA biosynthetic pathway genes in the genomes of these two organisms. In *F. fujikuroi*, GA biosynthetic genes are located next to one another in a gene cluster, whereas in *A. thaliana*, known GA biosynthetic genes are scattered, with one or more genes on each of the five chromosomes of the genome (Bomke and Tudzynski, 2009; Tudzynski, 2005). The absence of GA biosynthetic genes in *F. graminearum* and the presence of only one functional GA gene in *F. verticillioides* (Bömke *et al.*, 2008) are consistent with previous observations that these three species differ in their abilities to produce SMs, including GAs (Bömke *et al.*, 2008; Tudzynski, 2005). The difference in the presence and absence of GA genes among these species contrasts the high degree of sequence similarity among primary metabolic genes among the three species. Recent analysis of the genome sequences of 12 isolates of *F. oxysporum*, that varied in host range, identified the entire, apparently functional seven-gene GA cluster in two isolates and partial GA clusters in 11 isolates (D. Brown, personal communication). The ability of *F. oxysporum* isolates with an apparently functional GA cluster to produce GAs and whether production affects growth of the isolates respective hosts are under examination.

Site of action in plants

In plants, the gibberellin insensitive dwarf1 (GID1) protein functions as receptors of GA (Bari and Jones, 2009). GID1 interacts with the protein DELLA in a GA-dependent manner. Research with *A. thaliana* indicates that DELLA promotes resistance to necrotrophic pathogens by activating jasmonic acid (JA)/ethylene (ET)-dependent defence responses. However, the same research also indicates that DELLA allows susceptibility to biotrophic pathogens by repressing salicylic acid (SA)-dependent defence responses (Bari and Jones, 2009). Because GA stimulates the degradation of DELLA through the ubiquitin proteasome pathway, it is possible that GA promotes susceptibility to necrotrophs on one hand and resistance to biotrophs on the other. Thus, necrotrophic fungi like *F. fujikuroi* may produce GAs in part to suppress the defence response in host plants.

Fumonisins

Chemical structure

The chemical structure of fumonisins is similar to that of the sphingolipid biosynthetic intermediate sphinganine. Fumonisins are classified into four groups (A, B, C, and P) according to their chemical structures (Fig. 9.1B) (Gutleb *et al.*, 2002). Among them, the B group, including FB_1, FB_2, FB_3, and FB_4, are the most frequently found contaminants of maize and maize-derived foods. Fumonisin production is discontinuously distributed across lineages of closely related fusaria known as the *Gibberella fujikuroi* species complex. Within the complex production was observed in only *F. verticillioides, F. fujikuroi, F. globosum, F. proliferatum, F. nygamai,* and one strain of *F. oxysporum* (Proctor *et al.*, 2004). The ability to produce fumonisins is conferred by a cluster of 17 co-regulated genes located in a 42-kb region of the fungal genome (Alexander *et al.*, 2009).

Site of action

Fumonisins function mainly by blocking the enzyme ceramide synthase, which catalyses the conversion of the sphingoid bases sphinganine and sphingosine to ceramide. This blockage is facilitated by the structural similarities of fumonisins to sphinganine; fumonisins are 1-deoxy analogues of sphinganine (Desai *et al.*, 2002). The resulting accumulation of free sphingoid bases leads to the disruption of sphingolipid metabolism and this in turn is believed to be the mechanism of fumonisin toxicity. Sphingoid bases can arrest the cell cycle and induce apoptosis (Desai *et al.*, 2002). In animals, for poorly understood reasons, B-series fumonisins are generally more toxic than the other series of fumonisins (Abbas *et al.*, 1998).

Phytotoxic activity

Doehlert *et al.* (1994) first reported phytotoxic effects of FB_1 on germination of maize seeds. Although FB_1 had no effect on the germination rates, shoot and root growth of the seedlings was inhibited in a concentration-dependent manner. Similar phytotoxic effects of FB_1 were noted with *A. thaliana* (Stone *et al.*, 2000). Direct application to intact plants of FB_1 dissolved in water produced several symptoms including chlorosis and the formation of lesions whose severity varied with FB_1 concentration and plant species (Williams *et al.*, 2007). The phytotoxicity of fumonisins suggests that they could contribute to plant pathogenesis of *Fusarium*. It has been possible to assess this contribution with mutant strains of *F. verticillioides* that are unable to produce fumonisins. Although fumonisins were found not to contribute appreciably to the ability of *F. verticillioides* to cause ear rot (Desjardins *et al.*, 2002), they did contribute to the ability to cause maize seedling disease (Glenn *et al.*, 2008).

Cell death-inducing activity in plants

Asai *et al.* (2000) reported that exogenous application of FB_1 induced PCD in *Arabidopsis* protoplasts. Simultaneous application of translation and transcription inhibitors blocked FB_1-induced PCD. In addition, DNA degradation was observed in FB_1-treated protoplasts of leaf cells. These observations suggested that FB_1 causes cell death by activating a controlled program that requires *de novo* transcription and translation. FB_1-induced PCD was less frequent in protoplasts from the JA-insensitive (*jar1*), ET-insensitive (*etr1*), and SA-deficient (*pad4*) mutants as compared to that in wild-type plants (Asai *et al.*, 2000), indicating that pathways leading to PCD and activated by FB_1 are regulated by SA-, JA-, and ET-dependent signalling pathways. Similarly, hypersensitive response (HR)-like lesions were observed in intact *Arabidopsis* leaves infiltrated with 10 μM FB_1. FB_1-induced lesion formation was associated with deposition of callose and generation of reactive oxygen species (Stone *et al.*, 2000). Thus, PCD induction by FB_1 may involve a HR generated by the accumulation of SA. Experiments with maize embryos provided evidence that the frequency of cell death induced by FB_1 was correlated with increases in long-chain sphingoid bases as well as with increases in SA levels and specific endonuclease activity (Torre-Hernandez *et al.*, 2010)

The mechanism of PCD induction by FB_1 may involve caspase-like proteases in *Arabidopsis* leaves. Kuroyanagi *et al.* (2005) reported that FB_1-induced PCD was completely abolished in null mutants of vacuolar processing enzymes (VPE). The four *Arabidopsis VPE* genes (*αVPE, βVPE, γVPE,* and *δVPE*) encode cysteine proteases with caspase-like activity and regulate PCD. In addition, a serine protease (Kunitz trypsin) inhibitor (KTI1) regulates FB_1-induced lesion formation in *Arabidopsis* (Li *et al.*, 2008). Overexpression of *AtKTI1* resulted in reduced lesion formation after infiltration of leaves with FB_1. In contrast, RNAi silencing of *AtKTI1* caused enhanced lesion development in FB_1-infiltrated leaves (Li *et al.*, 2008). Thus, FB_1-induced lesion formation and PCD were positively and negatively regulated by VPEs and AtKTI1, respectively.

Fumonisin-resistant mutants

As mentioned above, FB_1 also causes growth retardation of *Arabidopsis* plants grown at low concentration (0.5 μM) on agar medium. Stone *et al.* (2000) isolated FB_1-resistant (*fbr*) *Arabidopsis* mutants on an FB_1-containing agar medium. The *fbr1* and *fbr2* mutants displayed reduced lesion formation in FB_1-infiltrated leaves compared to wild type and a reduced expression of *PR1* and *PR5* in response to FB_1. They also displayed enhanced resistance to the bacterial pathogen *Pseudomonas syringae* pv. *maculicola*. Thus, FB_1-induced signalling pathways are related to defence signalling pathways including those for disease resistance to bacterial pathogens (e.g. SA-dependent pathways).

Stone *et al.* (2005) reported that the *fbr6* mutant also exhibited an FB_1-resistant phenotype as well as a number of aberrant morphologies. These phenotypes are due to a T-DNA insertion in the SQUAMOSA promoter binding protein (SBP) domain gene, *AtSPL14*. *AtSPL14* encodes a plant-specific transcription factor containing an SBP DNA-binding domain and multiple ankyrin repeats that can function as a transcriptional activator when expressed in yeast. However, the actual role of this transcription factor in FB_1 resistance is still under investigation.

Trichothecenes

Chemical structure

Trichothecenes are sesquiterpenoid compounds characterized by a 12, 13-epoxy-trichothec-9-ene ring system (Fig. 9.1C) (Calvert *et al.*, 2005). More than 200 trichothecenes have been identified and can be classified into four major types, A–D. Type A and B trichothecenes are distinguished by the absence and presence, respectively, of a ketone at the C8 position of the trichothecene skeleton (Mirocha *et al.*, 1994). These two types of trichothecenes are frequent contaminants in cereal grain and food derived from such grain. Type A and B trichothecene production profiles (chemotypes) typically vary among and sometimes within species of *Fusarium*. Relatively few species (e.g. *F. sporotrichioides*) produce the type A trichothecenes T-2 toxin and HT-2 toxin, which have a five-carbon ester moiety (isovaleryl) at C8, whereas a wide range of species (e.g. *F. graminearum*, *F. sambucinum*, and *F. sporotrichioides*) can produce the type A trichothecene diacetoxyscirpenol (DAS), which lacks a functional group at C8 (Alexander *et al.*, 2009). Similarly, a relatively wide range of species (e.g. *F. asiaticum*, *F. graminearum*, *F. culmorum*, *F. equiseti*, and *F. poae*) produce type B trichothecenes, which vary in structure by the presence and absence of an oxygen atom at C4 and of acetyl moieties at C3, C4 and C15. Type B trichothecene production can vary among and within species. For example, in North America and Europe, isolates of *F. graminearum* produce predominantly 3-acetylated and 15-acetylated deoxynivalenol (DON), although individual isolates tend to produce only one or the other of these metabolites. Isolates of *F. culmorum* can produce either nivalenol (NIV) or acetylated derivatives of DON, whereas isolates of *F. asiaticum*, which occurs more frequently in Asia particularly in Japan, produce predominantly NIV.

Biosynthesis

The crop diseases Fusarium head blight (FHB) of wheat and barley, and ear rot of maize can be caused by trichothecene-producing species of *Fusarium* such as *F. asiaticum*, *F. graminearum*, and *F. culmorum* (McLaughlin *et al.*, 1977). Most of the genes required for trichothecene biosynthesis in *F. graminearum* (12 genes) and *F. sporotrichioides* (14 genes) are clustered in a 26-kb region, which is referred to as the core trichothecene biosynthetic gene (*TRI*) cluster (Brown *et al.*, 2004). A second locus encodes a single gene, *TRI101*, responsible for the acetylation of the C3 hydroxyl group (McCormick *et al.*, 1999). A third locus, with two genes in *F. sporotrichioides* involved in T-2 toxin synthesis and a single gene in *F. graminearum* involved in DON synthesis, is responsible for modifications to the C8 carbon (Brown *et al.*, 2003). These *Fusarium* species have strain-specific trichothecene metabolite profiles (Ward *et al.*, 2002), and such chemotype differences may play a role in the phytopathogenicity of individual strains of *Fusarium* (Alexander *et al.*, 2009; Ward *et al.*, 2002).

Site of action

Trichothecenes bind to eukaryotic ribosomes and inhibit protein synthesis (Pestka *et al.*, 2004). In mammals, trichothecenes have been shown to modulate gene expression and induce apoptosis (Moon and Pestka, 2002; Moon and Pestka, 2003). Accordingly, trichothecene contamination in cereal grain-derived products threatens animal and human health (Eriksen and Pettersson, 2004). Trichothecenes may inhibit protein synthesis in plants by binding to ribosomes; they inhibited protein synthesis in *Arabidopsis* suspension-culture cells in a concentration-dependent manner (Nishiuchi *et al.*, 2006). However, the inhibitory effects vary with trichothecene structure. For example, the type A trichothecene T-2 toxin was 10 times more efficient at inhibiting *Arabidopsis* ribosomes than the type B trichothecene DON (Masuda *et al.*, 2007; Nishiuchi *et al.*, 2006).

Trichothecenes induced prolonged activation of MAPKs in animals (JNK and p38 MAPK in mouse) and plants (MPK3 and MPK6 in *Arabidopsis*) (Nagase *et al.*, 2001; Nishiuchi *et al.*, 2006; Shifrin and Anderson, 1999; Yang *et al.*, 2000). Thus, trichothecenes are capable of activating the MAPK cascade leading to cell death in both plants and animals (Nishiuchi *et al.*, 2006). Given that other translational inhibitors such as anisomycin also cause MAPK activation and apoptosis in vertebrates, it is clear that ribotoxic stress can trigger activation of MAPK signalling pathways (Shifrin and Anderson, 1999). However, other translational inhibitors such as anisomycin and cycloheximide did not cause MAPK activation or PCD in plants (Nishiuchi *et al.*, 2006).

Phytotoxic activity

The phytotoxic action of various trichothecenes on seed germination, growth and morphology of *Arabidopsis* grown on Murashige and Skoog (MS) agar medium has previously been reported (Murashige and Skoog, 1962; Masuda *et al.*, 2007). Type A trichothecenes, such as T-2 toxin and DAS, inhibited *Arabidopsis* seed germination in a concentration-dependent manner, with 10 μM DAS decreasing the germination rate approximately 40%. Similar results were reported in tobacco seeds (Muhitch *et al.*, 2000). In contrast, DON did not inhibit germination of *Arabidopsis* seeds even at 10 μM.

All trichothecenes inhibited shoot growth of *Arabidopsis* seedlings in a concentration-dependent manner, and T-2 toxin caused severe inhibition at a low concentration (Masuda *et al.*, 2007). In contrast, 1 μM DON exerted only a minor inhibitory effect. Growth inhibition of DAS-treated shoots was intermediate between that of T-2 toxin and DON (Masuda *et al.*, 2007). A similar pattern of inhibitory effects on root growth was observed for the three trichothecenes. However, 1 μM DAS and 10 μM DON severely inhibited root but not shoot growth, indicating that both DON and DAS preferentially inhibit root elongation. Preferential inhibition of root elongation by DON was also observed in wheat plants (Masuda *et al.*, 2007). Microscopic analysis revealed that DON-treated roots were less organized than control roots. Not only did DON inhibit root elongation, but its phytotoxic effects were more marked for the root than for the shoot elongation and mass in *Zea mays* (McLean, 1996). Among the trichothecenes examined, T-2 toxin most strongly inhibited shoot and root growth in *Arabidopsis* seedlings at low concentrations. Similarly, T-2 toxin inhibited root and shoot growth in wheat seedlings (Wakulinski, 1989). In addition, T-2 toxin-treated *Arabidopsis* plants showed retarded growth and aberrant morphology. Dwarfism and short petioles were observed in 12-day-old T-2 toxin-treated seedlings. The cell size of rosette leaves was reduced in T-2 toxin-treated plants, suggesting that dwarfism in plants is caused by inhibition of cell extension.

Phytotoxic effects of DAS on *Arabidopsis* seedlings are more similar to those of DON and much less than to those of T-2 toxin. Although DAS and T-2 toxin are both type A trichothecenes, they differ in structure by the presence (T-2 toxin) and absence (DAS) of an isovaleryl moiety at the C8 position. T-2 tetraol had only minor effects on growth inhibition at low concentrations. T-2 toxin and T-2 tetraol differ in structure by presence (T-2 toxin) and absence (T-2 tetraol) of acetyl moieties at C-4 and C-15 as well as the presence and absence of the isovaleryl group. Therefore, it is possible that the absence of isovaleryl or one of the acetyls or the absence of two or all three of these functional groups reduces phytotoxicity.

As mentioned above, DAS and T-2 toxin both induce PCD in *Arabidopsis* leaves. Given this and the marked difference in phytotoxicity of these two trichothecenes, it is possible that the mode of action of trichothecene-induced PCD and the mode of action of trichothecene-induced phytotoxicity differ.

The phytotoxicity of trichothecenes suggests that they could contribute to pathogenicity of fusaria on plants. Indeed, analysis of trichothecene-nonproducing mutants of *F. graminearum*, *F. sambucinum*, and *F. sporotrichioides* have demonstrated that trichothecene production can contribute to the ability of *Fusarium* to cause disease on some hosts (Alexander *et al.*, 2009). Accordingly, the trichothecene sensitivity of plants might be expected to be related to their resistance to diseases caused by *Fusarium*. However, Bruins *et al.* (1993) reported that the sensitivity of wheat growth to 40 μM DON was not correlated with levels of resistance to Fusarium head blight (FHB) of the examined genotypes.

Cell death-inducing activity in plants

When *Arabidopsis* leaves were infiltrated with various trichothecene molecules, type A trichothecenes, such as T-2 toxin, induced HR-like lesion formation at concentrations higher than 1 μM (Nishiuchi *et al.*, 2006). Other the type A trichothecenes DAS and HT-2 toxin, induced lesion formation in *Arabidopsis* leaves at 1 μM. Because trypan blue staining revealed many dead cells in these lesions, it was concluded that type A trichothecenes caused cell death (Nishiuchi *et al.*, 2006). The presence of hydrogen peroxide and deposition of callose in the leaves was also consistent with cell death. Application of type A trichothecenes also led to accumulation of SAs, induction of defence genes, and activation of MAPKs in *Arabidopsis* leaves (Nishiuchi *et al.*, 2006). Although type A trichothecenes partially inhibited protein synthesis in *Arabidopsis* cells at 1 μM, they activated transcription of defence-related genes and biosynthesis of defence-related compounds (Nishiuchi *et al.*, 2006). Taken together, these observations indicate that trichothecene-induced cell death is PCD rather than necrosis.

In contrast to type A trichothecenes, activation of defence genes and MAPKs by DON were observed only at 100 μM. DON did not induce cell death in *Arabidopsis* even at a high concentration. However, DON concentrations higher than 5 μM effectively inhibited protein synthesis in *Arabidopsis* cells (Nishiuchi *et al.*, 2006), indicating that the major effect of DON is to block the defence responses of host plants rather than to induce cell death. In contrast, Desmond *et al.* (2008) reported that application of DON induced PCD in wheat leaves. Infusion of DON into the leaves induced generation of hydrogen peroxide, accumulation of defence-related proteins, DNA laddering and cell death (Desmond *et al.*, 2008). These results indicate that DON-induced cell death in wheat is PCD. Thus, the effects of exogenous DON treatment differed markedly between *Arabidopsis* and wheat leaves; high concentrations of DON induce PDC in wheat but do not induce PCD in *Arabidopsis*.

Virulence factor

Trichothecenes are also known to act as virulence factors during *Fusarium* infection of host plants (Bai *et al.*, 2001). In *Fusarium* species, *TRI5* encodes trichodiene synthase, the enzyme that catalyses the first committed reaction in trichothecene biosynthesis. Bai *et al.* (2001) reported the use of a *TRI5* deletion mutant to investigate the role of trichothecene in the virulence of *F. graminearum*. *TRI5* mutant strain of *F. graminearum*, which cannot produce trichothecenes, can infect wheat and cause disease, but the mutant cannot spread to and thereby cause disease in neighbouring florets within the wheat spike (Bai *et al.*, 2002). In addition, exogenous application of 75 μM DON increased the virulence of *F. graminearum* in *Arabidopsis* plants (Chen *et al.*, 2006). Based on these observations, DON appears to be important for disease spread rather than for initial infection of *F. graminearum* in the wheat spike.

The *in vivo* location of DON was examined in *F. culmorum*-infected wheat spikes using an antibody specific to DON (Kang and Buchenauer, 2002). DON was found to be associated with the plasmalemma and in the cytoplasm with ribosomes (Kang and Buchenauer, 2002). The association with the plasmalemma probably results in alterations of membrane permeability, dramatically affecting cell viability. The affinity of DON to ribosomes probably blocks protein synthesis, possibly resulting in severe impairment of host defence mechanisms (Kang and Buchenauer, 2002).

Detoxification in plants

Up-regulation of glutathione S-transferases (GSTs) has been reported in wheat flowers infected with the type B trichothecene-producing species *F. graminearum* (Gardiner *et al.*, 2010). GSTs function to detoxify both endogenous and xenobiotic compounds by catalysing the formation of conjugates of glutathione (GSH) and the compounds (Marrs, 1996). These conjugated compounds can become sequestered in vacuoles and from there released into the extracellular environment. Gardiner *et al.* (2010) found evidence for non-enzymatic formation of DON–GSH conjugates *in vitro* using both liquid chromatography–mass spectrometry (LC-MS) and nuclear magnetic resonance analysis (NMR). Their results suggest that GSTs are involved in detoxification of DON. Plants also decrease the toxicity of trichothecenes by other chemical modifications. Poppenberger *et al.* (2003) reported that *Arabidopsis* UDP-glucosyltransferase can catalyse the glucosylation of trichothecenes such as DON, resulting in a decrease in phytotoxicity.

Chemical modifications of trichothecene play an important role in decreasing its toxicity, not only in plants but also in *Fusarium*. Studies on the self-defence mechanism used by *F. graminearum* against trichothecenes revealed that *Tri101*, which encodes a trichothecene 3-*O*-acetyltransferase, decreases DON toxicity (Kimura *et al.*, 1998). Transgenic plants expressing *Tri101* exhibited increased resistance to *Fusarium* and trichothecenes (Ohsato *et al.*, 2007).

Signalling pathway regulating the response of plants to trichothecenes

Although much is known about the phytotoxicity of trichothecenes, little is known about the signalling pathway(s) that enable plants to respond to trichothecenes. However, treatment of *Arabidopsis* with T-2 toxin resulted not only in reduced growth but also in formation of reactive oxygen species (ROS), accumulation of salicylic acid (SA) and brassinosteroid (BR) inactivation, a group of plant growth regulators involved in multiple physiological processes in plants, including disease resistance (Masuda *et al.*, 2007; Nishiuchi *et al.*, 2006). Thus, ROS, SA and BR inactivation could play a role in the trichothecene signalling pathway in plants. T-2 toxin exposure also induced expression of the *Arabidopsis* gene *AtNFXL1*, a gene that is predicted to encode a transcription factor with similarity to the human transcription repressor NF-X1 (Masuda *et al.*, 2007). The expression of wheat and barley orthologues of AtNFXL1 is induced by DON-producing *Fusarium graminearum* (Gardiner *et al.*, 2010; Jia *et al.*, 2009). In addition, compared to wild-type *Arabidopsis*, *atnfxl1* mutant plants were markedly more sensitive to T-2 toxin, slightly more sensitive to DAS, and not more sensitive to DON (Asano *et al.*, 2000a, 2008b). The *AtNFXL1* repressed T-2 toxin-inducible expression of the SA biosynthesis gene (*ICS1*). The enhanced accumulation of SA in T-2 toxin-treated *atnfxl1* mutant affected its high sensitivity. Thus, *AtNFXL1* is involved in the trichothecene-induced signalling pathway in *Arabidopsis*, but also that the signalling pathway may vary with variation in trichothecene structure.

Zearalenone

Chemical structure

Zearalenone (RAL, F-2 toxin) is a macrocyclic lactone with estrogenic properties (Fig. 9.1D) (Gutleb *et al.*, 2002) that is produced by multiple *Fusarium* species, including *F. graminearum*, *F. equiseti*, *F. culmorum*, *F. tricinctum* and *F. crookwellense* (McLean, 1996). In *F. graminearum*, the zearalenone biosynthetic gene cluster consists of four genes: two adjacent PKS genes, an FAD-binding dehydrogenase gene, and a bZIP transcription factor gene (Kim *et al.*, 2005). Several species, including the FHB pathogen *F. graminearum*, can produce both DON and zearalenone, and this can lead to the frequent co-contamination of cereals with both metabolites.

Site of action

Zearalenone has potent estrogenic activity in animals, owing to its high binding affinity to oestrogen receptors (Kuiper *et al.*, 1997). In contrast, in plants the site of action of zearalenone is essentially unknown.

Effects on plant growth

Zearalenone did not appreciably affect seed germination or root or shoot growth of wheat seedlings at high concentration (Wakulinski, 1989), and its phytotoxic effects were very weak compared with those of other *Fusarium* toxic metabolites (e.g. trichothecenes). When *Arabidopsis* plants were grown on zearalenone-containing agar medium, weak inhibition of plant growth was observed but only at concentrations of more than 100 μM (Nishiuchi, unpublished results). In contrast, zearalenone promoted root and shoot growth of *Z. mays* (McLean, 1996). Thus, although the estrogenic activity of zearalenone is a serious threat to animals, the level of phytotoxicity of zearalenone does not appear to pose a significant threat to plant health.

Fusaric acid

Chemical structure

Fusaric acid was first isolated from *F. heterosporum* as a compound that inhibited growth of rice seedlings (Fig. 9.1E) (Notz *et al.*, 2002; Yabuta *et al.*, 1934). Fusaric acid has also been reported to be produced by *F. crookwellense, F. heterosporum, F. napiforme, F. oxysporum, F. sambucinum, F. solani, F. subglutinans,* and *F. verticillioides* (Bacon et al., 1996; Brown *et al.*, 2012). The chemical structure of fusaric acid (5-butylpicolinic acid) and precursor feeding experiments indicate that fusaric acid is a product of polyketide metabolism. Recently, a polyketide synthase (PKS) gene (*FUB1*) required for fusaric acid production was identified in *F. verticillioides* (Brown *et al.*, 2012). *FUB1* was adjacent to four other genes (*FUB2-FUB5*) that exhibited a similar pattern of expression and were proposed to be a fusaric acid biosynthetic gene cluster.

Toxicity

Fusaric acid exhibits biological activity in both animals and plants. Fusaric acid is only mildly toxic to mammals, and this toxicity is thought to result from inhibition of DNA synthesis (Bacon *et al.*, 1996). Fusaric acid also has an important pharmacological property in that it inhibits dopamine β-hydroxylase, which catalyses the conversion of dopamine to noradrenaline, a neurotransmitter (Nagasaka *et al.*, 1985). Fusaric acid induced inhibition of dopamine β-hydroxylase is associated with lowered blood pressure and subjective improvements in patients with phaeochromocytoma (Nagasaka *et al.*, 1985). In plants, fusaric acid has been implicated in the pathogenesis of tomato wilt symptoms caused by *F. oxysporum* f. sp. *lycopersici* (Bacon *et al.*, 1996). The observation that fusaric acid alone could not induce wilting in maize suggests that the plant wilt symptoms caused by *Fusarium* species may be due to the cooperative action of fusaric acid and other metabolites (Bacon *et al.*, 1996). Fusaric acid and its methyl ester (methyl fusarate) have been proposed as bioherbicides to control witchweed (Capasso *et al.*, 1996). Using a cotton leaf bioassay, Stipanovic *et al.* (2011) recently reported that methyl fusarate was the most toxic naturally occurring fusaric acid derivative. Fusaric acid can also repress production of the antimicrobial compound 2,4-diacetylphloroglucinol in the biological control agent *Pseudomonas fluorescens* strain CHA0 (Notz *et al.*, 2002).

Butenolide

Chemical structure

The water-soluble butenolide 4-acetamido-4-hydroxy-2-butenoic acid γ-lactone belongs to a class of lactones with a four-carbon heterocyclic ring structure. Butenolides may be produced together with DON and 3-ADON by *F. graminearum* (McLean, 1996) and other *Fusarium* species including *F. sporotrichioides, F. equiseti, F. tricinctum, F. culmorum, F. acuminatum, F. lateritium, F. poae, F. semitectum,* and *F. nivale* (McLean, 1996).

Toxicity

In mammals, butenolide can induce dysfunction of mitochondria, and oxidative damage appears to play a crucial role in these deleterious effects. But the mechanism by which the damage occurs is not known (Wang *et al.*, 2009). In the plant *Lemma minor* (duck weed), butenolide caused growth inhibition and decreased chlorophyll content (Vesonder *et al.*, 1992). But butenolide was less toxic to *L. minor* than the *Fusarium* metabolites T-2 toxin and fumonisin B_1 (Vesonder *et al.*, 1992). The butenolide 3-methyl-2H-furo[2,3-c]pyran-2-one, a component of plant-derived smoke, promotes seed germination in some plant species (Flematti *et al.*, 2004). In addition, this butenolide had cytokinin- and auxin-like activity in bioassays of mung bean callus and roots (Jain *et al.*, 2008). However, such activities have not been reported in *Fusarium*-derived butenolide.

Enniatin

Chemical structure

Enniatins are N-methylated cyclic hexadepsipeptides consisting of alternating residues of amino acids linked by peptide and ester bonds (Fig. 9.1F). These metabolites are synthesized by a non-ribosomal peptide synthase (NRPS), known enniatin synthetase. Enniatins appear in nature as mixtures of the four main variants, enniatin A, A_1, B, and B_1 and the minor variants, enniatin C, D, E, and F (Herrmann *et al.*, 1996b). Enniatins are produced by *F. tricinctum*, *F. acuminatum*, *F. avenaceum*, and *F. sambucinum* and may contaminate cereal grain and food derived from it (Logrieco *et al.*, 2002; Tan *et al.*, 2011).

Site of action

Enniatins exhibit both antimicrobial and insecticidal activities. These activities are thought to result from the ability of enniatins to act as ionophores (Ovchinnikov *et al.*, 1974). In a brine shrimp assay, the relative toxicities of enniatins were determined to be, from most to least toxic, enniatin B, B_1, A_1 and A (Tan *et al.*, 2011). Enniatins also interact strongly with the membrane-located ATP-binding cassette transporters in human cells (Dornetshuber *et al.*, 2009).

Effects on plants

Mixtures of enniatins caused necrotic lesion formation in potato tuber tissues in a concentration-dependent manner (Herrmann *et al.*, 1996b). In addition, in the leaf puncture assay, visible lesions formed on leaves of various plants treated with a 1:10 mixture of acetamidobutenolide:enniatin B (Hershenhorn *et al*, 1992). To examine the role of enniatins in pathogenesis of *Fusarium*, enniatin-nonproducing mutants of *F. avenaceum* were generated by targeted gene disruption of the enniatin synthetase gene (Herrmann *et al.*, 1996a). The reduced virulence of the resulting mutants on potato tuber tissue provided evidence that enniatins contribute to virulence of the fungus on potato tubers (Herrmann *et al.*, 1996a).

Beauvericin

Chemical structure

Beauvericin is also a cyclic hexadepsipeptide with alternating methyl-phenylalanyl and hydroxy-isovaleryl residues (Fig. 9.1F) (Logrieco *et al.*, 1998). The metabolite is sometimes found as a contaminant in cereal grain (Logrieco *et al.*, 2002) and is produced by a relatively wide range of *Fusarium* including *F. proliferatum*, *F. subglutinans*, and *F. semitectum* and the relatively closely related fungus *Beauveria bassiana* (McLean, 1996).

Toxicity

Beauvericin has broad antifungal, antibacterial, and insecticidal activities (Logrieco *et al.*, 1998). This metabolite has been shown to exhibit ionophoric properties and is a specific inhibitor of cholesterol acyltransferase (Tomoda *et al.*, 1992). In addition, beauvericin induce PCD similar to apoptosis in animals by elevating intracellular calcium levels (Dombrink-Kurtzman, 2003; Logrieco *et al.*, 1998). Beauvericin application caused a decrease in ascorbate levels in tomato protoplasts, and such imbalances of ascorbate metabolism can lead to cell death (Paciollaa *et al.*, 2004). Beauvericin also caused a decrease in cell viability and DNA fragmentation resulting in cell death in wheat seedlings (Srobarova *et al.*, 2009). The phytotoxic effects of beauvericin suggest that the metabolite may contribute to the virulence of beauvericin-producing species of *Fusarium*. A beauvericin-nonproducing mutant of *B. bassiana* caused less disease on three insect species compared to the wild-type strain from which it was derived, suggesting that beauvericin plays a significant but not an essential role in virulence of the fungus on insects (Xu *et al.*, 2008).

Moniliformin

Chemical structure and toxicity

Moniliformin (1-hydroxycyclobut-1-en-3,4-dione) is a cyclobutane mycotoxin (Fig. 9.1G) that is produced by *F. avenaceum, F. culmorum, F. oxysporum, F. sporotrichioides*, and *F. subglutinans* (Desjardins, 2006; McLean, 1996). Moniliformin can inhibit incorporation of pyruvate into the tricarboxylic acid cycle, resulting in interference in carbohydrate metabolism (Jestoi, 2008). In addition, moniliformin can induce oxidative damage in myoblasts, possibly by inhibiting glutathione peroxidase and glutathione reductase (Jestoi, 2008). In wheat seedlings, leaf growth was significantly decreased by exposure to low concentrations of moniliformin, whereas root growth was significantly decreased only at a high concentration of moniliformin (Wakulinski, 1989). Moniliformin application reduced the biomass of maize calli, although these calli appeared healthy (Van Acsh *et al.*, 1992). In addition, necrosis and chlorosis were also observed in tobacco and maize plants one week after moniliformin treatment. Chlorosis induced by moniliformin was also observed in leaves of Lemna minor (Vesonder *et al.*, 1992).

Future prospects

As stated above, *Fusarium* species produce multiple SMs during infection of host plants. Some of these SMs are highly toxic to plants as well as to animals. *Fusarium* has developed mechanisms, such as chemical modification, to protect itself from these toxic SMs (e.g. trichothecenes) (Kimura *et al.*, 1998). In addition, most *Fusarium* species devote a significant amount of genetic material and energy to production of multiple SMs. Such a significant investment in metabolism and energy suggests that SMs can provide an ecological advantage to the species that produce them. Indeed, there is evidence that enniatins, fumonisins and trichothecene can contribute to the ability of fusaria to colonize plants, thereby providing them with a source of nutrients (Bai *et al.*, 2002; Chen *et al.*, 2006). Genomic and metabolomic approaches have led to the elucidation of the overall biosynthetic pathway of several SMs. However, it is very important to study the production of phytotoxic metabolites in the host plants. Therefore, it is necessary to conduct global expression studies and metabolome analysis of *Fusarium* species in infected plant tissues. These analyses will shed light on the regulatory mechanism of biosynthesis of *Fusarium* metabolites during infection of host plants.

References

Abbas, H.K., Shier, W.T., Seo, J.A., Lee, Y.W., and Musser, S.M. (1998). Phytotoxicity and cytotoxicity of the fumonisin C and P series of mycotoxins from *Fusarium* spp. fungi. Toxicon. *36*, 2033–2037.

Alexander, N.J., Proctor, R.H., and McCormick, S.P. (2009). Genes, gene clusters, and biosynthesis of trichothecenes and fumonisins in *Fusarium*. Toxin Rev. *24*, 198–215.

Asai, T., Stone, J.M., Heard, J.E., Kovtun, Y., Yorgey, P., Sheen, J., and Ausubel, F.M. (2000). Fumonisin B_1–induced cell death in *Arabidopsis* protoplasts requires jasmonate-, ethylene-, and salicylate-dependent signaling pathways. Plant Cell *12*, 1823–1835.

Asano, T., Masuda, D., Yasuda, M., Nakashita, H., Kudo, T., Kimura, M., Yamaguchi, K., and Nishiuchi, T. (2008a). *AtNFXL1*, an *Arabidopsis* homologue of the human transcription factor NF-X1, functions as a negative regulator of the trichothecene phytotoxin-induced defense response. Plant J. *53*, 450–464.

Asano, T., Yasuda, M., Nakashita, H., Kimura, M., Yamaguchi, K., and Nishiuchi, T. (2008b). The *AtNFXL1* gene functions as a signaling component of the type A trichothecene-dependent response. Plant Signal Behav. *3*, 991–992.

Bacon, C.W., Porter, J.K., Norred, W.P., and Leslie, J.F. (1996). Production of fusaric acid by *Fusarium* species. Appl. Environ. Microbiol. *62*, 4039–4043.

Bai, G.H., Desjardins, A.E., and Plattner, R.D. (2002). Deoxynivalenol-nonproducing *Fusarium graminearum* causes initial infection, but does not cause disease spread in wheat spikes. Mycopathologia. *153*, 91–98.

Bai, G.H., Plattner, R., Desjardins, A.E., Kolb, F.L. (2001). Resistance to Fusarium head blight and deoxynivalenol accumulation in wheat. Plant Breeding. *120*, 1–6.

Bari, R., and Jones, J.D. (2009). Role of plant hormones in plant defence responses. Plant Mol Biol. *69*, 473–488.

Bömke, C., Rojas, M.C., Hedden, P., and Tudzynski, B. (2008). Loss of gibberellin production in *Fusarium verticillioides* (*Gibberella fujikuroi* MP-A) is due to a deletion in the gibberellic acid gene cluster. Appl. Environ. Microbiol. *74*, 7790–7801.

Bömke, C., and Tudzynski, B. (2009). Diversity, regulation, and evolution of the gibberellin biosynthetic pathway in fungi compared to plants and bacteria. Phytochem. *70*, 1876–1890.

Brown, D.W., Butchko, E., Busman, R.A., M., and Proctor, R.H. (2012). Identification of gene clusters associated with fusaric acid, fusarin, and perithecial pigment production in *Fusarium verticillioides*. Fun. Genet. Biology *49*, 521–532.

Brown, D.W., Dyer, R.B., McCormick, S.P., Kendra, D.F., and Plattner, R.D. (2004). Functional demarcation of the *Fusarium* core trichothecene gene cluster. Fungal Genet. Biol. *41*, 454–462.

Brown, D.W., Proctor, R.H., Dyer, R.B., and Plattner, R.D. (2003). Characterization of a *Fusarium* 2–gene cluster involved in trichothecene C-8 modification. J. Agric. Food Chem. *51*, 7936–7944.

Bruins, M.B.M., Karsai, I., Schepers, J., and Snijders, C.H.A. (1993). Phytotoxicity of deoxynivalenol to wheat tissue with regard to *in vitro* selection for Fusarium head blight resistance. Plant Science, *94*, 195–206.

Calvert, T.W., Aidoo, K.E., Candlish, A.G., and Fuat, A.R. (2005). Comparison of *in vitro* cytotoxicity of *Fusarium* mycotoxins,deoxynivalenol, T-2 toxin and zearalenone on selected human epithelial cell lines. Mycopathologia. *159*, 413–419.

Capasso, R., Evidente, A., Cutignano, A., Vurro, M., Zonno, C., and Bottalico, A. (1996). Fusaric and 9,10-dehydrofusaric acids and their methyl esters from *Fusarium* nygamai. Phytochem. *41*, 1035–1039.

Chen, X., Steed, A., Harden, C., and Nicholson, P. (2006). Characterization of *Arabidopsis thaliana-Fusarium graminearum* interactions and identification of variation in resistance among ecotypes. Mol. Plant Path. *7*, 391–403.

Desai, K., Sullards, M.C., Allegood, J., Wang, E., Schmeiz, E.M., Hartl, M., Humpf, H.U., Kiotta, D.C., Peng, Q., and Merrill, A.H. (2002). Fumonisins and fumonisin analogs as inhibitors of ceramide synthase and inducers of apoptosis. Biochim. Biophys. Acta. *1585*, 188–192.

Desjardins, A.E., Munkvold, G.P., Plattner, R.D., and Proctor, R.H. (2002). *FUM1*--a gene required for fumonisin biosynthesis but not for maize ear rot and ear infection by *Gibberella moniliformis* in field tests. Mol. Plant Microbe. Interact. *15*, 1157–1164.

Desjardins, A.E., Maragos, C.M., and Proctor, R.H. (2006). Maize ear rot and moniliformin contamination by cryptic species of *Fusarium subglutinans*. J. Agric. Food Chem. *54*, 7383–7390.

Desmond, O.J., Manners, J.M., Stephens, A.E., Maclean, D.J., Schenk, P.M., Gardiner, D.M., Munn, A.L., and Kazan, K. (2008). The *Fusarium* mycotoxin deoxynivalenol elicits hydrogen peroxide production, programmed cell death and defence responses in wheat. Mol. Plant Path. *9*, 435–445.

Doehlert, D.C., Knutson, C.A., and Vesonder, R.F. (1994). Phytotoxic effects of fumonisin B_1 on maize seeling growth. Mycopathologia. *127*, 117–121.

Dombrink-Kurtzman, M.A. (2003). Fumonisin and beauvericin induce apoptosis in turkey peripheral blood lymphocytes. Mycopathologia. *156*, 357–364.

Dornetshuber, R., Heffeter, R., Sulyok, M., Schumacher, R., Chiba, R., Kopp, S., Koellensperger, G., Micksche, M., Lemmens-Gruber, R., and Berger, W. (2009). Interactions between ABC-transport proteins and the secondary *Fusarium* metabolites enniatin and beauvericin Mol. Nutr. Food Res. *53*, 904–920.

Eriksen, G.S., and Pettersson, H. (2004). Toxicological evaluation of trichothecenes in animal feed Animal Feed Sci Tech. *114*, 205–239.

Filek, M., Biesaga-Kościelniak, J., Marcińska, I., Cvikrová, M., Macháčková, I., and Krekule, J. (2010). Contents of polyamines during vernalization in wheat and the effect of zearalenone. Biologia Plantarum. *54*, 483–487.

Flematti, G.R., Ghisalberti, E.L., Dixon, K.W., and Trengove, R.D. (2004). A compound from smoke that promotes seed germination. Science. *305*, 977.

Gaffoor, I., and Trail, F. (2006). Characterization of two polyketide synthase genes involved in zearalenone biosynthesis in *Gibberella zeae*. Appl. Environ. Microbiol. *72*, 1793–1799.

Gardiner, S.A., Boddu, J., Berthiller, F., Hametner, C., Stupar, R.M., Adam, G., and Muehlbauer, G.J. (2010). Transcriptome analysis of the barley–deoxynivalenol interaction: evidence for a role of glutathione in deoxynivalenol detoxification. Mol. Plant Microbe Interact. *23*, 962–976.

Glenn, A.E. (2007). Mycotoxigenic *Fusarium* species in animal feed. Animal Feed Sci. Tech. *137*, 213–240.

Glenn, A.E., Zitomer, N.C., Zimeri, A.M., Williams, L.D., Riley, R.T., and Proctor, R.H. (2008). Transformation-mediated complementation of a *FUM* gene cluster deletion in *Fusarium verticillioides* restores both fumonisin production and pathogenicity on maize seedlings. Mol. Plant Microbe Interact. *21*, 87–97.

Gutleb, A.C., Morrison, E., and Murk, A.J. (2002). Cytotoxicity assays for mycotoxins produced by *Fusarium* strains: a review. Environ. Toxicol. Pharmacol. *11*, 309–320.

Herrmann, M., Zocher, R., and Haese, A. (1996a). Effect of disruption of the enniatin synthetase gene on the virulence of *Fusarium avenaceum*. Mol. Plant Microbe Interact. *9*, 226–232.

Herrmann, M., Zocher, R., and Haese, A. (1996b). Enniatin production by *Fusarium* strains and its effect on potato tuber tissue. Appl. Environ. Microbiol. *62*, 393–398.

Hershenhorn, J., Park, S.H., Stierle, A., and Strobel, G.A. (1992). *Fusarium avenaceum* as a novel pathogen of spotted knapweed and its phytotoxins, acetamido-butenolide and enniatin B. Plant Sci. *86*, 155–160.

Jain, N., Ascough, G.D., and Van Staden, J. (2008). A smoke-derived butenolide alleviates $HgCl_2$ and $ZnCl_2$ inhibition of water uptake during germination and subsequent growth of tomato-possible involvement of aquaporins. J. Plant Phys. *165*, 1422–1427.

Jestoi, M. (2008). Emerging *Fusarium*-mycotoxins fusaproliferin, beauvericin, enniatins, and moniliformin: a review. Crit. Rev. Food Sci. Nutr. *48*, 21–49.

Jia, H.Y., Cho, S.H., and Muehlbauer, G.J. (2009). Transcriptome analysis of a wheat near-isogenic line pair carrying Fusarium Head Blight-resistant and -susceptible alleles. Mol. Plant Microbe Interact. *22*, 1366–1378.

Kang, Z., and Buchenauer, H. (2002). Studies on the infection process of *Fusarium culmorum* in wheat spikes: Degradation of host cell wall components and localization of trichothecene toxins in infected tissue. Eur. J. Plant Path. *108*, 653–660.

Kimura, M., Kaneko, I., Komiyama, M., Takatsuki, A., Koshino, H., Yoneyama, K., and Yamaguchi, I. (1998). Trichothecene 3-O-acetyltransferase protects both the producing organism and transformed yeast from related mycotoxins. Cloning and characterization of *Tri101*. J. Biol. Chem. *273*, 1654–1661.

Kuiper, G.G., Carlsson, B., Grandien, K., Enmark, E., Haggblad, J., Nilsson, S., and Gustafsson, J.A. (1997). Comparison of the ligand binding specificity and transcript tissue distribution of estrogen receptors alpha and beta. Endocrinology. *138*, 863–870.

Kuroyanagi, M., Yamada, K., Hatsugai, N., Kondo, M., Nishimura, M., and Hara-Nishimura, I. (2005). Vacuolar processing enzyme is essential for mycotoxin-induced cell death in *Arabidopsis thaliana*. J. Biol. Chem. *280*, 32914–32920.

Li, J., Brader, G., and Palva, E.T. (2008). Kunitz trypsin inhibitor: an antagonist of cell death triggered by phytopathogens and fumonisin B_1 in *Arabidopsis*. Molecular Plant. *1*, 482–495.

Logrieco, A., Moretti, A., Castella, G., Kostecki, M., Golinski, P., Ritieni, A., and Chelkowski, J. (1998). Beauvericin production by *Fusarium* species. Appl. Environ. Microbiol. *64*, 3084–3088.

Logrieco, A., Rizzo, A., Ferracane, R., and Ritieni, A. (2002). Occurrence of beauvericin and enniatins in wheat affected by *Fusarium avenaceum* head blight. Appl. Environ. Microbiol. *68*, 82–85.

Lowe, R.G., Allwood, J.W., Galster, A.M., Urban, M., Daudi, A., Canning, G., Ward, J.L., Beale, M.H., and Hammond-Kosack, K.E. (2010). A combined ^{1}H nuclear magnetic resonance and electrospray ionization-mass spectrometry analysis to understand the basal metabolism of plant-pathogenic *Fusarium* spp. Mol. Plant Microbe Interact. *23*, 1605–1618.

Ma, L.J., van der Does, H.C., Borkovich, K.A., Coleman, J.J., Daboussi, M.J., Di Pietro, A., Dufresne, M., Freitag, M., Grabherr, M., Henrissat, B., *et al.* (2010). Comparative genomics reveals mobile pathogenicity chromosomes in *Fusarium oxysporum*. Nature. *464*, 367–373.

Marrs, K.A. (1996). The functions and regulation of glutathione S-transferases in plants. Ann. Rev. Plant Phys Plant Mol. Biol. *47*, 127–158.

Masuda, D., Ishida, M., Yamaguchi, K., Yamaguchi, I., Kimura, M., and Nishiuchi, T. (2007). Phytotoxic effects of trichothecenes on the growth and morphology of *Arabidopsis thaliana*. J. Exp. Bot. *58*, 1617–1626.

McCormick, S.P., Alexander, N.J., Trapp, S.E., and Hohn, T.M. (1999). Disruption of *TRI101*, the gene encoding trichothecene 3-O- acetyltransferase, from *Fusarium sporotrichioides*. Appl. Environ. Microbiol. *65*, 5252–5256.

McLaughlin, C.S., Vaughan, M.H., Cambell, I.M., Wei, C.M., Stafford, M.E., and Hansen, B.S. (1977). Inhibition of protein synthesis by trichothecenes. Pages 263–273 in: Mycotoxin in Human Animal Health. Rodericks, J.V., Hesseltine, C.W., and Mehlman, M.A., eds. Pathox Publishers, Park Forest South, I.L., U.S.A.

McLean, M. (1996). he phytotoxicity of *Fusarium* metabolites: An update since (1989). Mycopathologia. *133*, 163–179.

Mirocha, C.J., Xie, W., Xu, Y., Wilcoxson, R.D., Woodward, R.P., Etebarian, R.H., and Behele, G. (1994). Production of trichothecene mycotoxins by *Fusarium graminearum* and *Fusarium culmorum* on barley and wheat. Mycopathologia. *128*, 19–23.

Moon, Y., and Pestka, J.J. (2002). Vomitoxin-induced cyclooxygenase-2 gene expression in macrophages mediated by activation of ERK and p38 but not JNK mitogen-activated protein kinases. Toxicol. Sci. *69*, 373–382.

Moon, Y., and Pestka, J.J. (2003). Deoxynivalenol-induced mitogen-activated protein kinase phosphorylation IL-6 expression in mice suppressed by fish oil. Nutr. Biochem. *14*, J., 717–726.

Muhitch, M.J., McCormick, S.P., Alexander, N.J., and Hohn, T.M. (2000). Transgenic expression of the *TRI101* or *PDR5* gene increases resistance of tobacco to the phytotoxic effects of the trichothecene 4,15-diacetoxyscirpenol. Plant Sci. *157*, 201–207.

Murashige, T., and Skoog, F. (1962). A revised medium for rapid growth and bioassays with tobacco tissue cultures. Physiol. Plant. *15*, 473–497.

Nagasaka, A., Hara, I., Imai, Y., Uchikawa, T., Yamauchi, K., Suzuki, S., Takatsuki, K., Tomita, A., Niinomi, M., Nakagawa, H., *et al.*, (1985). Effect of fusaric acid (a dopamine beta-hydroxylase inhibitor) on phaeochromocytoma. Clin. Endocrinol. *22*, 437–444.

Nagase, M., Alam, M.M., Tsushima, A., Yoshizawa, T., and Sakato, N. (2001). Apoptosis induction by T-2 toxin: activation of caspase-9, caspase-3, and DFF-40/CAD through cytosolic release of cytochrome c in HL-60 cells. Biosci. Biotechnol. Biochem. *65*, 1741–1747.

Nishiuchi, T., Masuda, D., Nakashita, H., Ichimura, K., Shinozaki, K., Yoshida, S., Kimura, M., Yamaguchi, I., and Yamaguchi, K. (2006). *Fusarium* phytotoxin trichothecenes have an elicitor-like activity in *Arabidopsis thaliana*, but the activity differed significantly among their molecular species. Mol. Plant Microbe Interact. *19*, 512–520.

Notz, R., Maurhofer, M., Dubach, H., Haas, D., and Defago, G. (2002). Fusaric acid-producing strains of *Fusarium oxysporum* alter 2,4-diacetylphloroglucinol biosynthetic gene expression in *Pseudomonas fluorescens* CHA0 *in vitro* and in the rhizosphere of wheat. Appl. Environ. Microbiol. *68*, 2229–2235.

Ohsato, S., Ochiai-Fukuda, T., Nishiuchi, T., Takahashi-Ando, N., Koizumi, S., Hamamoto, H., Kudo, T., Shinozaki, K., Yamaguchi, I., and Kimura, M. (2007). Transgenic inactivation of the *Fusarium* mycotoxin deoxynivalenol in rice plants as a model of cereal crops. Plant Cell Report. *26*, 531–538.

Ovchinnikov, Y.A., Ivanov, V.T., Evstratov, A.V., Mikhaleva, I.I., Bystrov, V.F., Portnova, S.L., Balashova, T.A., Meshcheryakova, E.N., and Tul'chinskii, V.M. (1974). Enniatin ionophores. Conformation and ion binding properties. Int. J. Pept. Protein. Res. *6*, 465–498.

Paciollaa, C., Dipierroa, N., Mule`b, G., Logriecob, A., and Dipierroa, S. (2004). The mycotoxins beauvericin T-2 induce cell death and alteration to the ascorbate metabolism in tomato protoplasts. Phys. Mol. Plant Path. *65*, 49–56.

Pestka, J.J., Zhou, H.R., Moon, Y., and Chung, Y.J. (2004). Cellular and molecular mechanisms for immune modulation by deoxynivalenol and other trichothecenes: unraveling a paradox. Toxicol. Lett. *153*, 61–73.

Poppenberger, B., Berthiller, F., Lucyshyn, D., Sieberer, T., Schuhmacher, R., Krska, R., Kuchler, K., Glossl, J., Luschnig, C., and Adam, G. (2003). Detoxification of the *Fusarium* mycotoxin deoxynivalenol by a UDP-glucosyltransferase from *Arabidopsis thaliana*. J. Biol. Chem. *278*, 47905–47914.

Proctor, R.H., Plattner, R.D., Brown, D.W., Seo, J.A., and Lee, Y.W. (2004). Discontinuous distribution of fumonisin biosynthetic genes in the *Gibberella fujikuroi* species complex. Mycol. Res. *108*, 815–822.

Shifrin, V.I., and Anderson, P. (1999). Trichothecene mycotoxins trigger a ribotoxic stress response that activates c-Jun N-terminal kinase and p38 mitogen-activated protein kinase and induces apoptosis. J. Biol. Chem. *274*, 13985–13992.

Srobarova, A., da Silva, J.A., Kogan, G., Ritieni, A., and Santini, A. (2009). Beauvericin decreases cell viability of wheat. Chem. Biodivers. *6*, 1208–1215.

Stipanovic, R.D., Puckhaber, L.S., Liu, J., and Bell, A.A. (2011). Phytotoxicity of fusaric acid and analogs to cotton. Toxicon. *57*, 176–178.

Stone, J.M., Heard, J.E., Asai, T., and Ausubel, F.M. (2000). Simulation of fungal-mediated cell death by fumonisin B_1 and selection of fumonisin B_1-resistant (fbr) *Arabidopsis mutants*. Plant Cell. *12*, 1811–1822.

Stone, J.M., Liang, X., Nekl, E.R., and Stiers, J.J. (2005). Arabidopsis AtSPL*14*, a plant-specific SBP-domain transcription factor, participates in plant development and sensitivity to fumonisin B_1. Plant, J. *41*, 744–754.

Tan, D.C., Flematti, G.R., Ghisalberti, E.L., Sivasithamparam, K., and Barbetti, M.J. (2011). Toxigenicity of enniatins from western Australian *Fusarium* species to brine shrimp (*Artemia franciscana*). Toxicon. *57*, 817–825.

Tomoda, H., Huang, X.H., Cao, J., Nishida, H., Nagao, R., Okuda, S., Tanaka, H., Omura, S., Arai, H., and Inoue, K. (1992). Inhibition of acyl-CoA: cholesterol acyltransferase activity by cyclodepsipeptide antibiotics. J. Antibiot. *45*, 1626–1632.

Torre-Hernandez, M.E., Vicente, M.R., Greaves-Fernandez, N., Cruz-Ortega, R., and Plasencia, J. (2010). Fumonisin B_1 induces nuclease activation and salicylic acid accumulation through long-chain sphingoid base build-up in germinating maize. Phys. Mol. Plant Path. *74*, 337–345.

Tudzynski, B. (2005). Gibberellin biosynthesis in fungi: genes, enzymes, evolution, and impact on biotechnology. Appl. Microbiol. Biotechnol. *66*, 597–611.

van Asch, M.A.J., Rijkenberg, F.H.J., and Coutinho, T.A. (1992). Phytotoxicity of fumonisin B_1, moniliformin, and T-2 toxin to corn callus cultures. Phytopathology *82*, 1330–1332.

Vesonder, R.F., Labeda, D.P., and Peterson, R.E. (1992). Phytotoxic activity of selected water-soluble metabolites of *Fusarium* against *Lemna minor*, L. (Duckweed). Mycopathologia *118*, 185–189.

Wakulinski, W. (1989). Phytotoxicity of the secondary metabolites of fungi causing wheat head fusariosis (head blight). Acta Physiologia Plantarum. *11*, 301–306.

Wang, H.J., Wang, Y.M., and Peng, S.Q. (2009). Repeated administration of a *Fusarium* mycotoxin butenolide to rats induces hepatic lipid peroxidation and antioxidant defense impairment. Food Chem. Toxicol. *47*, 633–637.

Ward, T.J., Bielawski, J.P., Kistler, H.C., Sullivan, E., and O'Donnell, K. (2002). Ancestral polymorphism and adaptive evolution in the trichothecene mycotoxin gene cluster of phytopathogenic *Fusarium*. Proc. Natl. Acad. Sci. *99*, 9278–9283.

Williams, L.D., Glenn, A.E., Zimeri, A.M., Bacon, C.W., Smith, M.A., and Riley, R.T. (2007). Fumonisin disruption of ceramide biosynthesis in maize roots and the effects on plant development and *Fusarium verticillioides*-induced seedling disease. J. Agric. Food Chem. *55*, 2937–2946.

Wulff, E.G., Sorensen, J.L., Lubeck, M., Nielsen, K.F., Thrane, U., and Torp, J. (2010). *Fusarium* spp. associated with rice Bakanae: ecology, genetic diversity, pathogenicity and toxigenicity. Environ. Microbiol. *12*, 649–657.

Xu, Y., Orozco, R., Wijeratne, E.M.K., Gunatilaka, A.A.L., Stock, S.P., and Molnár, I. (2008). Biosynthesis of the cyclooligomer depsipeptide beauvericin, a virulence factor of the entomopathogenic fungus *Beauveria bassiana*. Chem. Biol. *15*, 898–907.

Yabuta, T., Kambe, K., and Hayashi, T. (1934). Biochemistry of the bakanae fungus. I. Fusarinic acid, a new product of the bakanae-fungus. J. Agric. Chem. Soc. Jpn. *10*, 1059–1068.

Yang, G.H., Jarvis, B.B., Chung, Y.J., and Pestka, J.J. (2000). Apoptosis induction by the satratoxins and other trichothecene mycotoxins: relationship to ERK, J., p38 MAPK, and SAPK/JNK activation. Toxicol. Appl. Pharmacol. *164*, 149–160.

Index

N

O

P

Q

R

S

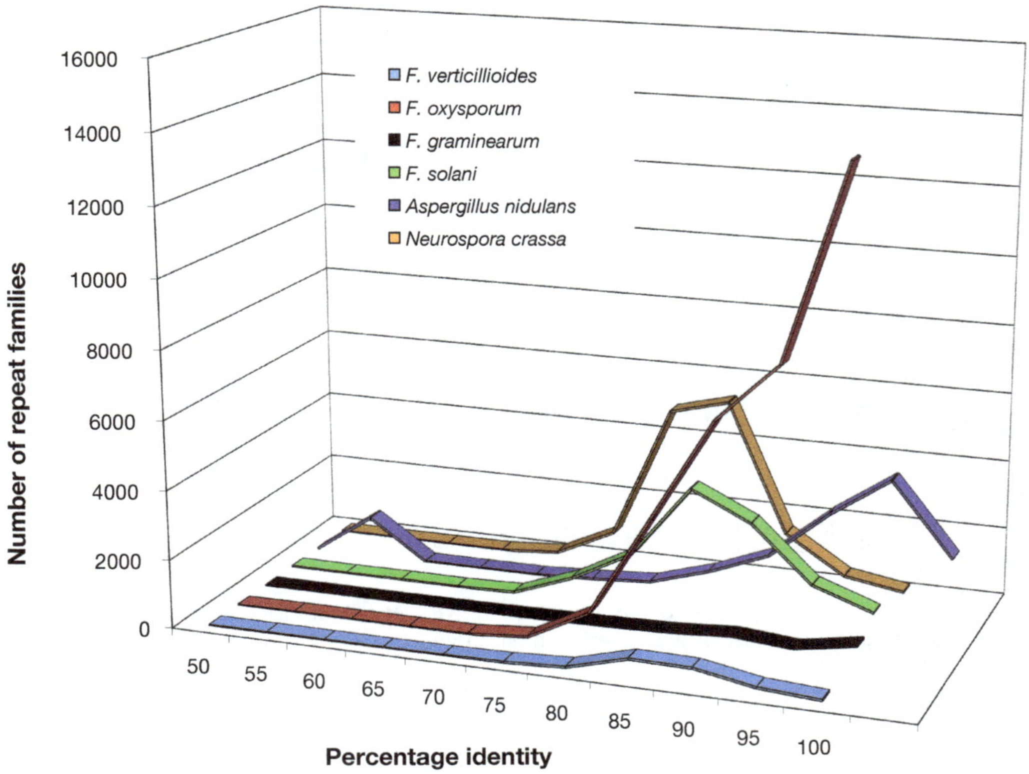

Figure 3.2 Degree of sequence identity for repeated sequences within six fungal species.

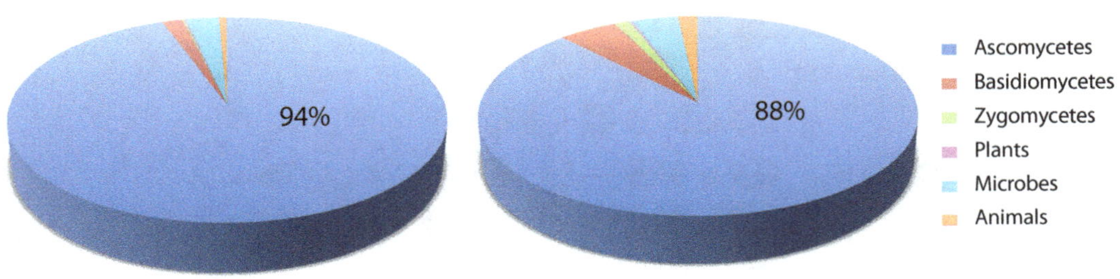

Figure 3.3 Homologous profile of *F. oxysporum* genes encoded in the conserved versus LS regions comparing to proteomes across different kingdoms. The *Fol* proteins were searched using BLASTP (1e-20) against the NCBI metazoan, plant, microbial gene sets available at ftp://ftp.ncbi.nlm.nih.gov/gene/DATA/GENE_INFO (21 February 2008 version) and the non-*Fusarium* fungal database including the protein sets from Ascomycete: two fungal genomes from each subphylum Sordariomycetes (*Magnaporthe grisea*, *Neurospora crassa*), subphylum Leotiomycetes (*Botrytis cinerea*, *Sclerotinia sclerotiorum*), and Eurotiomycetes (*Aspergillus fumigatus*, *A. oryzae*); Basidiomycete: *Ustilago maydis*, *Coprinus cinereus*, *Cryptococcus neoformans* serotype A; and a zygomycete fungal *Rhizopus oryzae* protein set.

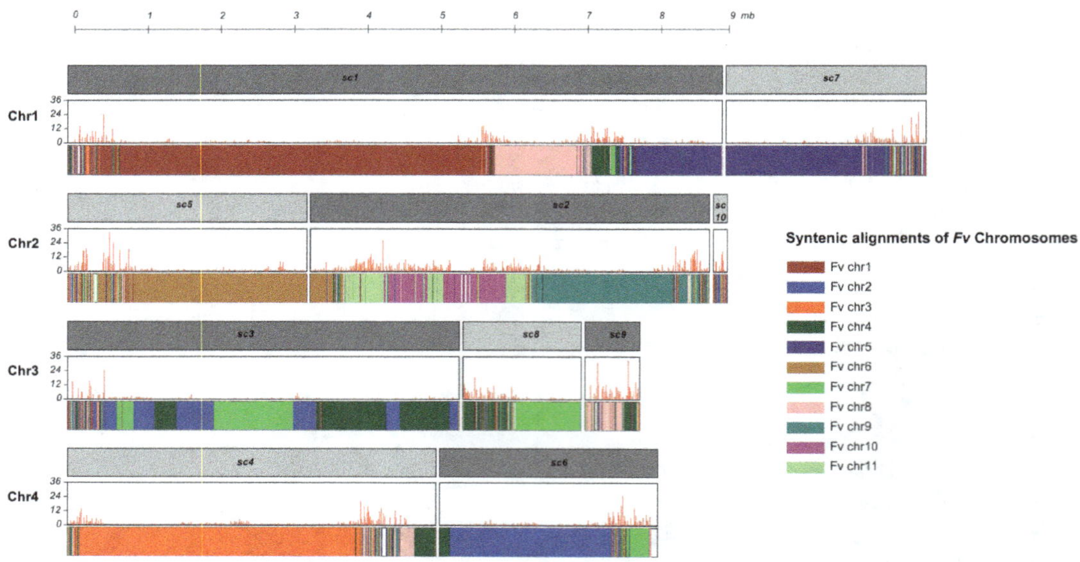

Figure 3.4 Whole genome alignments of *F. graminearum* with *F. verticillioides* illustrating chromosomal fusion events in FG. The alignments display end-to-end synteny in large blocks with the exception of *F. verticillioides* chromosome ends and reveal multiple chromosome fusions in *F. graminearum*. The previously described highly polymorphic and recombinogenic regions of *F. graminearum* (Cuomo *et al.*, 2007) correspond to *F. verticillioides* chromosome ends, including the implied interstitial fusion sites.

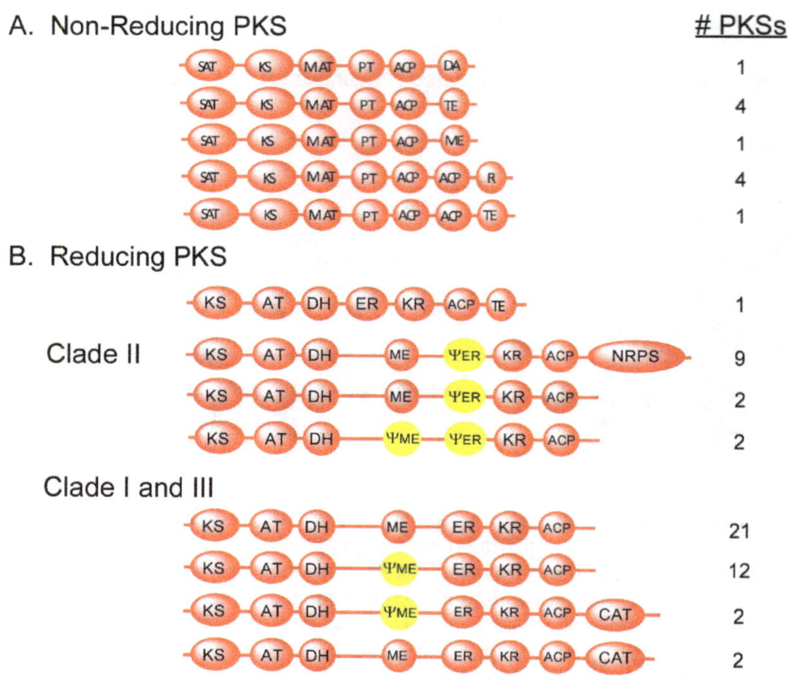

Figure 8.4 Domain organization in *Fusarium* PKSs. The numbers in the # PKSs column indicate the number of *Fusarium* PKSs that have the domain structure depicted. Clade I, II and III were resolved in the phylogenetic analysis presented in Figure 8.3. Domain abbreviations: SAT=starter unit acyl transferase, KS=keto synthase, MAT=malonyl acyl transferase, PT=product template, ACP=acyl carrier protein, DA=deacylase, TE=thioesterase, ME=methyltransferase, DH=dehydratase, ER=enoyl reductase, KR=keto reductase, NRPS=non-ribosomal peptide synthetase module, and CAT=carnitine acyl transferase. Ψ indicates a non-functional domain.

www.ingramcontent.com/pod-product-compliance
Lightning Source LLC
LaVergne TN
LVHW062239200926
848484LV00006B/1572
9781908230256